WITHDRAWN

Cultural Forests of the Amazon

CULTURAL FORESTS OF THE AMAZON

A Historical Ecology of People and Their Landscapes

William Balée

The University of Alabama Press/Tuscaloosa

Copyright © 2013
The University of Alabama Press
Tuscaloosa, Alabama 35487-0380
All rights reserved
Manufactured in the United States of America
Typeface: Garamond and Helvetica

Cover photograph: A gargantuan hardwood tree found in ancient anthropogenic forests. Photo by Osmar Ka'apor; from the collection of Meghan Kirkwood.
Cover design: Todd Lape/Lape Designs
Author photograph: Photo by Paula Burch-Celentano

∞

The paper on which this book is printed meets the minimum requirements of American National Standard for Information Sciences—Permanence of Paper for Printed Library Materials,
ANSI Z39.48-1984.

Library of Congress Cataloging-in-Publication Data

Balée, William L., 1954-
　Cultural forests of the Amazon : a historical ecology of people and their landscapes / William Balée.
　　　p. cm.
　Includes bibliographical references and index.
ISBN 978-0-8173-1786-7 (trade cloth : alk. paper) — ISBN 978-0-8173-8655-9 (ebook)
　1. Urubu Kaapor Indians—Ethnobotany. 2. Urubu Kaapor Indians—Philosophy. 3. Urubu Kaapor Indians—Social conditions. 4. Indigenous peoples—Ecology—Amazon River Region. 5. Traditional ecological knowledge—Amazon River Region. 6. Cultural landscapes—Amazon River Region. 7. Rain forest ecology—Amazon River Region. 8. Amazon River Region—Social conditions. 9. Amazon River Region—Environmental conditions. I. Title.
　F2520.1.U7B33　2013
　581.6'309811—dc23

Permissions on page 248

CONTENTS

Preface — vii
Acknowledgments — xvii

PART I
LANDSCAPE TRANSFORMATIONS: Overview — 1
1. Villages of Vines and Trees — 7
2. An Estimate of Anthropogenesis — 32
3. Comparison of High and Fallow Forests — 53

PART II
CONTACT AND ATTRITION: Overview — 71
4. People of the Fallow Forest — 75
5. Vanishing Plant Names — 89
6. Conquest and Migration — 103

PART III
INDIGENOUS SAVOIR FAIRE: Overview — 119
7. From Their Point of View — 123
8. Retention of Traditional Knowledge — 132
9. Confection, Inflection — 144

PART IV
DIMENSIONS OF DIVERSITY: Overview — 159
10. Discernment of Environmental Variation — 161
11. Rethinking the Landscape — 174

Appendix I. Guajá Generic Plant Names — 185
Appendix II. Trees of the Anthropogenic Forest — 203
Notes — 207
Works Cited — 213
Permisions — 248
Index — 249
Illustrations follow page — 111

PREFACE

The point of this book is to share certain insights I have had in researching and thinking about Amazonian forests during the past quarter century. I use the term *Amazon* as the English name of the river that has the greatest water volume in the world, as well as to label the entire land surface it drains and the adjoining hinterlands. I hope this volume can contribute not only to understanding the past of these forests, and their associated cultures and peoples, but also to current management and policy concerning the same, which are today among the most threatened landscapes and social systems left on the planet.

This book is directed to readers with an interest in understanding better the long-term human engagement with the Amazon. Although the Amazon Basin is often regarded as a "rich realm of nature," as in part of the famous title by historian David Sweet (1974), my focus for a long time has been on how people who have lived there ended up changing it, both now in the observable, ethnographic present and in the remote, prehistoric past. Specifically, I have been asking about how culture has made inroads into nature, perhaps making it even richer than it was to begin with. Although I consider myself a materialist, not an idealist, in the way I conceive of the history of landscapes and societies, I am also convinced that Plato was essentially right about something that is as equally ideational as it is physical: namely, diversity—both of people and species. It is a good in the Platonic sense, in other words, in and of itself. I believe that traditional Amazonian societies fulfill that philosophical sense of the good empirically, because of how they influenced cultures and landscapes and the distribution of languages and biota. I put forth arguments and data on the crucial question of human influences on what we often call nature in what follows.

Forest people of the Amazon deployed technologies that many scholars in diverse fields, including biology, archaeology, and geography, until recently, thought were not sufficiently complex or sophisticated to have altered, in any fundamental way, the layout of the land. Many also have not realized the extent to which people of the past modified the distributions of plants and animals across multiple, distant landscapes.

This book sets out an explanation of what I believe really happened in the forests of the Amazonian past and what makes them so biologically and culturally complex today. I do not propose to capture every detail of this complexity, and I certainly don't have all the answers (I am reminded here of a useful subtitle by Steve Beckerman, "Hold the answers, what are the questions?") I do think I have a handle on the questions intrinsically related to diversity. A full explanation of it is still elusive, however, because science is trending behind the reality. The microbes of the fertile Amazonian soil called *terra preta,* which is a human artifact, are substantially different from those of the surrounding, natural soil. Yet the species of these microbes number in the millions and have not been identified except at the most general phyletic levels, namely, as bacteria, archaea, and fungi—all the major divisions of life except for all eukaryotes other than fungi. The complexities have yet to be fully worked out, for systematics of these microbes is tracking far behind diversity of gargantuan scope (Tsai et al. 2009).

Luisa Maffi's (2001) insight that language, culture, and environment are profoundly entangled empirically is convincing. I see my contribution to that ensemble as concerning specifically how people impacted and transformed landscapes before the entire Amazon got completely integrated into the global economy, which it certainly is now. I describe a time before the advent and expansion in the region of logging interests, placer mining, soybean fields, and other examples of capital-intensive, industrial, and mechanized agribusiness, together with the usual cattle ranching on gigantic estates (*latifundia*). These transformations in the aggregate are really a product of the past thirty years or so only, so the time frame of the essays in this volume is contemporary with those events. The Amazon of then, not really so far back, particularly if one thinks in terms of the *longue durée,* is a time and place to which I have borne witness, through study in the field and in collaboration with and reading the works of other specialists on the region, from whom I have learned greatly. I have also learned about the Amazon, its landscapes and biota, together with its people, cultures, and languages, from many indigenous persons who have helped educate me on the subject matter at hand. To a large extent I mean to authenticate an aim, namely, to validate the inimitable, rich relationship between the forests and the peoples I know, before they undergo completely the profundity of change one can expect based on the reach of globalization, the world system of our time. Robert F. Murphy used to say some anthropologists are interested in why things stay the same; others are interested in why they change. This book is really about both a time having its status quo, or what seemed to be one more or less that, and dynamic, driving landscape transformations of the most severe sort, a hallmark of globalization.

This book in the final analysis is about where the forests we see today came from and where they might be going. Many of the landscapes I discuss in this book, located inside indigenous lands, are very palpably threatened with corporate logging, ranching, and industrial and postindustrial interests. They are rapidly disappearing, in spite of years of efforts by many concerned individuals, both indigenous and not, to stem the tide of demand for nonrenewable forest products, such as the timber taken from species-rich forests. There are also rising waves of interest across the Amazonian hinterlands in international projects aimed at stemming the flow of greenhouse gases, as through the REDD program. The people who live in those forests are clearly changing the way they perceive and interact with the forests that remain in their environs. In some cases, they are valuing those forests more, as part of their cultural heritage and as a future investment, because of the possibility of building landscape and carbon-sequestration capital; in others, they are letting the forests be depleted, in the short term, in order to generate income to pay for various expenses.

Perhaps now is a good time to make available again some of my documentation of Amazonian forests and forest peoples' practices of managing these forests, for I am convinced these are not going to be easily available for eyewitness study, simply because they have been altered by the increasing demands for tropical products, including wood from ancient trees, of our global economy. I am taking some inspiration for this task from an Amazonianist colleague of mine, who once remarked, "Bill, your papers are scattered in quite different venues, and a number of them are hard to find: why don't you publish them together in a single volume, so people can have easier access to them?" I was very honored when The University of Alabama Press agreed with my colleague's suggestion.

I am here conjoining several chapters that were, in fact, already published in slightly different form, either as articles in journals or as pieces that appeared in volumes of contributed chapters during the years 1989–2010. Those chapters are bookended by new ones written expressly for this volume. The chapters are related by a theme: the forests made by human hands and how people of the present perceive, categorize, cogitate on, utilize, and manage them.

My field research in Amazonia has consciously been focused on numerous groups affiliated with a single language family: Tupí-Guaraní. That focus is partly pragmatic. In following the language learner's cliché that if you learn one Romance language, others should follow without too much duress, I kept a focus on one language family. Specifically, that was for the purpose of grasping essential commonalities in the history and ecology of a language family spread across diverse environments in the vastness of a continental, tropical region. It is possible some bias in my perspective has been introduced thereby; if so, at least it might complement and contrast with the remarkable body of evidence on prehistoric expansions of the Arawakan language family across the Amazon Basin into the Caribbean, south to the edge of Patagonia, and west to the bottom of the Andean escarpment. That expansion included the diffusion of artifactual, agricultural, engineering, and political elements of Arawakan origin (Eriksen 2011; Heckenberger 2005; Heckenberger et al. 2008; Hornborg 2005; Lathrap 1970; Neves and Petersen 2006).

In this book, I have been less focused on these issues, except where they reflect, or where they can be illuminated and understood from, living things in the environment and contemporary Tupí-Guaraní peoples and cultures, which in some cases may have come after Arawakan transformations of forests. South of the Amazon River itself, the Tupí-Guaraní family is as impressively distributed as is Arawakan, if not more so, because they occupied every possible environment, including Amazonian forests, *cerrado* country, Atlantic Coastal Forest, Chaco, and even lower slopes of the Andes.

I am not, of course, the only aficionado of things Tupian. Essentially, there are two phases of Tupinology. The first phase was a literary and nationalist one, whereby Brazilian identity was defined in terms of its indigenous past, with the indigene seen as speaking a Tupí-Guaraní language, more or less the same one encountered by the first Portuguese and Africans in the Americas during the sixteenth century (Wagley 1971, 251–253). That was the "good talk" (*Nhe'éngatu*). The second phase was basically a scholarly and historical quest to contextualize ethnographic materials on Tupí-Guaraní groups, essentially taking shape in the twentieth century, with the work of social anthropologists such as Curt Nimuendaju, Herbert Baldus, Florestan Fernandes, Eduardo Galvão, Darcy Ribeiro, Egon Schaden, and Charles Wagley, in the early years, and Eduardo Viveiros de Castro, Mércio Gomes, Françoise and Pierre Grenand, Aryon D. Rodrigues, and Carlos Fausto, among others, in the latter part of the twentieth century and down to the present day. There is no one school of Tupinology, as this mélange of names suggests. Its practitioners are divided by theoretical leanings, institutional training, political comprehensions, and national traditions of scholarship. My work in this vein has fallen into a niche concerned with relationships between the Tupí-Guaraní family of languages and the forest biota—mostly but not entirely botanical—that comprise their numerous, diverse landscapes, and how they engage with and influence the distribution of these.

The chapters herein are arranged basically, but not entirely, by date of original publication. I have made many cosmetic revisions throughout. I changed all the original chapter and essay titles for the purpose of coherence. I also consolidated the references into a single, consistently formatted bibliography, *mutatis mutandis*.

I consolidated the usage of scientific names for plants. I decided to try to keep these names up to date, rather than use some of the now obsolete synonyms that were in vogue at the time of the original publications. Generally speaking, I have updated names of angiosperm (flowering plant) families in the tables, text, and appendixes, in trying to follow as closely as possible the Angiosperm Phylogeny Group and their latest assessments of similarities and differences in the angiosperm genome (Stevens 2001 onward). For example, the families Bombacaceae, Sterculiaceae, and Tiliaceae of a few years ago are now all subsumed in the cotton family, Malvaceae. The generic and species names have not in principle changed so much, but there are exceptions. *Tabebuia* spp., which denoted a tremendously important genus of neotropical trees, has now become *Handroanthus*. Angiosperms, of course, are the most common vascular plants in the Amazon. For the others, including gymnosperms, ferns, mosses, and allies, I have tried to rely on the latest terms available to me. I have used guides, both online and not, for this purpose. For example, for the palm family, I have used Henderson (1995), because this work is most sensitive to the Amazon region itself. In numerous cases, I have benefited from recent identifications and reidentifications of my voucher (herbarium) specimens by botanists such as Andrew Henderson himself on palms, and so I have sought to update names for older materials as much as possible, based on the identifications I have either received or retrieved directly from the online herbarium at the New York Botanical Garden. I cannot be certain that all the scientific names employed herein are still correct for referencing taxa I collected with voucher specimens or otherwise recorded as present in a particular milieu of research at a past time. Systematic botanists have been changing plant names rather rapidly in the past twenty years, and many changes have occurred with respect to materials I collected, referenced in these pages. At any rate, synonymy should help the interested reader in looking more deeply into almost any taxon listed in these pages if there is a question about it. Almost fifteen hundred of my voucher specimens from the Amazon are described online in the searchable C. V. Starr Virtual Herbarium of the New York Botanical Garden (http://sciweb.nybg.org/science2/VirtualHerbarium.asp), and a good number of these are accompanied by interactive imagery of the specimen. It is a most useful resource.

On indigenous terminology, the book contains many minor revisions. I have systematized the orthography of Tupí-Guaraní terms in the many languages of that family with which I came into contact over the past thirty years, or with which I otherwise have obtained some reasonable familiarity based on substantial literature and known similarities to Ka'apor, such as Wayãpi of French Guiana.

I have used the International Phonetic Alphabet fairly systematically, but not entirely so. I allowed for a couple of kinds of exceptions. The first kind of exception concerns spelling of indigenous societal names. I have used what is most recognizable to readers, not being consistent as to the group's own term for themselves or outsiders' terms. For example, the Guajá (and numerous other Tupí-Guaraní groups) call themselves *awá* ("people"), but I have let the term *Guajá* designate the group. Other scholars use *Awá-Guajá,* but I don't think hyphenated, hybrid terms really capture indigenous reality any better for the reader. That is also true of Araweté: their name for themselves is *büde*. I have let *Araweté* as the group term stand, for that is how they are known in the literature, not by their own, unique term for themselves. I have indicated stress in indigenous group names where in the indigenous language it would be morphologically unlikely otherwise. This is the case in Sirionó, the accepted name in Bolivia for that indigenous society, who would be linguistically more likely to call themselves Sirióno. In fact, they call themselves *mbía*, which is not used in the literature to designate them, though it is cognate with the names of certain indigenous societies in Paraguay, Argentina, and Brazil, and is clearly a legitimate Tupí-Guaraní ethnonym for this group, whereas Sirionó is an outsiders' term. I am spelling the indigenous name Asurini with two *s*'s, that is, *Assurini* or *Assurini do Xingu,* even though in much of the literature, even in Portuguese, it is spelled Asurini in both cases (e.g., Müller 1985), because that is how Lusophone readers will best perceive how to pronounce it: an *s* between two vowels in Portuguese sounds like a *z* (that is, it is voiced). A double-*s* in English between two vowels is likely to sound like an *s* also (as in "posse" or "missive") and that is how the *s* in Asurini or Assurini is pronounced in Portuguese. If anyone, including and especially indigenous readers now and in the future, is offended by my liberty in ethnonyms, I offer this explanation of my procedure, and my sincere apologies in advance now.

The second kind of exception is I have not indicated stress in indigenous words for things where it is regular (that is, predictable). In Assurini, stress is canonically on the penultimate syllable of the word, as in Portuguese. In Ka'apor, Guajá, Tembé, and the language of Aurê and Aurá (Chapter 1), stress falls on the final syllable. In Araweté, Sirionó, and Wayãpi, stress is irregular. I have only indicated stress in the languages in which it is irregular when I knew the location of the stress, based on what I had recorded phonetically or what I have read. I have also tended only to indicate primary stress, when known, not secondary stress.

Ka'apor is the only language in this group in which I have reasonably good functional ability today. There are nevertheless snippets of a fast conversation among a group of Ka'apor that suddenly lose me. Such losses can then cost me the entire rest of the discussion, as I struggle to decipher the meaning of the word or phrase that tripped me up initially, while the conversation itself continues, several sentences ahead. On the other hand, I can usually understand my interlocutors on a one-on-one, face-to-face basis, and I can hold my own end of a phone conversation in the language, with an occasional request to speak more slowly or to repeat something.

I have been speaking Ka'apor over about a thirty-year period, and although I am rusty right now, upon returning to the villages and forests of the Ka'apor people, which is the only place in the world where one can practice their language in a more or less natural setting, I could be back up and speaking it passably well again in a few weeks. Many Ka'apor people seem to define one's indigeneity, incidentally, in terms of language, and they are on the whole very proud of theirs. I have been truly honored by several Ka'apor in recent years who spontaneously told me, "*Mil, nde Ka'apor!*" (Bill, you are one of us).

The orthography of most of the languages in this book is more or less phonemic; in several cases, such as Guajá, Araweté, and the language of Aurê and Aurá (Chapter 1), phonological descriptions were not yet available at the time I did research, or are still not available, so I could not adopt an orthography apart from what I considered to be a systematic use of symbols in recording speech and, frankly, educated guesswork about what were phonemes and what were their allophones. I knew enough Guajá at the time I collected Guajá plants and plant names in 1987, 1989, and 1991 that I could gloss many of these (Appendix I). I had a firmer grasp of both phonological and morphological matters obviously with Ka'apor; as to phonology, I had some security with Assurini, Sirionó, and Wayãpi, which have relatively (for Amazonia) substantial literatures and published phonemicizations available. That includes the culturally sensitive and linguistically insightful dictionary by Françoise Grenand (1989) and the comprehensive ethnobiological and ethnohistorical treatises of Pierre Grenand (1980, 1982) on the Wayãpi, as well as substantial lexical and textual materials available on the Sirionó (as in Priest and Priest 1980 and Schermair 1958, 1962).

On a more mundane yet necessary level, I have taken the opportunity permitted by this publication to correct certain typos, misspellings, punctuation errors, and the like that afflicted the original work. In addition, I took some things out, such as the original acknowledgments printed in each individual publication, as well as most of the original photographs. In fact, I have added several images never before published. I have also cross-referenced the chapters herein where it seemed expedient.

Some areas of the volume I have left in more or less pristine condition. I did not make reference to pertinent, exogenous publications more recent than those cited in the original work inside the newly reprinted content. Instead, I have tried to keep things fresh by citing the most recent, relevant work in this preface, in the overviews to the four parts, and in the original, new chapters (Chapters 1 and 11). I also did not consolidate original citations of different editions of the same work employed in different publications herein at different times. Hence, in a few cases, different editions of the same work have worked their way into the bibliography.

This laissez-faire approach leads to some repetition of situational information and phraseology, along the lines of "the Ka'apor people live in Maranhão, Brazil" or "the Ka'apor speak a Tupí-Guaraní language." The reason I didn't eliminate such redundancies is that the chapters would have lost their original context. Doing so would also have interrupted the sequential flow of the chapters. There is also a redundancy in the first twenty species names listed in Tables 3.3 and 10.1 (with Table 3.3 being longer), which I could not avoid without, it seemed to me, losing sense, logic, and narrative flow in the two respective chapters (3 and 10). Those are my explanations and apologies. Let me proceed to the substance.

The chapters that follow address specifically different aspects of the question concerning human influence on Amazonian nature and the substantive and cultural effects of that influence, as these have played out over a long period of time. These different aspects of the problem are reflected in the layout of the book, which is divided into four parts: (1) landscape transformations, (2) contact and attrition, (3) indigenous savoir faire, and (4) dimensions of diversity. Apart from situational repetition (as of geographic locatives, other identifiers, and linguistic affiliations), I think the book can be read as a fairly succinct compendium of these topics.

The issues raised herein are relevant for beginning to understand ways in which we can conceive of people living in the threatened biomes of today's tropical forests worldwide, starting with the biggest one of them, the Amazon. It does not seem practical any longer to follow the wishes of many—though certainly not all—in the conservation biology community to exclude traditional, indigenous people from all these biomes. In many cases, the biomes are being overrun anyway by commercial logging, farmer-colonists, cattle ranchers, public works projects (such as dams and mining operations), and industrial agriculture, not the native people who have already lived there, who are rather helpless as they face the expansion of the populations of nation-states that have nowhere else, it seems, to go, nor any other kinds of less destructive, more sustainable livelihoods to engage in. Globalization needs to fix these problems. I believe it can, but it will take time, perhaps exceeding the longevity of the character of the forests and peoples I describe herein.

Because native people have occupied these ancient forests since prehistory, and even though they have changed in their social and political organization, in their cultures, and in their languages, lessons can be learned from the technologies and societies of the past by studying their effects in the present. Ancient technologies allowed humanity to live in highly fragile, species-rich tropical regions without denaturing them. This book contains evidence that people occupied Amazonian forests for a long time without degrading, destroying, and converting them to species-poor environments, like the extensive grasslands we see in many parts of the region, especially south of the Amazon River proper, replete with alien wildlife such as cows, pigs, goats, sheep, and chickens living on burnt grasslands, with charred tree stumps that testify in their starkness to the mighty evergreen hardwood forests that once stood there. In some ways, this book can be used to provide an alternative pathway to such destructive land-use regimes, by appreciating technologies of the past that allowed for both human occupation and development as well as coexistence with a wealth of species of plants and animals.

The book is also designed to be read by scholars, scientists, and students who would like to deepen their knowledge of the Amazon region and of the origins, development, and sometimes major modification of habitats, often called by archaeologists "constructed landscapes" or "cultural landscapes," once deemed to have been due to entirely natural causes, not anthropogenic ones. Those constructed environments actually support a surprising diversity of forms of life. Proving that proposition is one of the tasks I have endeavored to accomplish in this book.

This book is both a retrospective and current summary of my research, for it reflects the long-term study I have undertaken, including several cumulative years of fieldwork in the forests that are discussed herein, while living among six different indigenous forest peoples, in both Brazil and Bolivia. I should mention here that I have also traveled extensively throughout Amazonia, up and down most of the length of the Amazon River itself, from its mouth on the Canal Norte at Macapá to the floating neighborhood of Belén, at Iquitos, Peru. I have traveled in its interiors especially from east to west, south of the Amazon River. And I have made numerous visits and occasional research forays to the Atlantic Coastal Forest (Brazil and Argentina) in the south and French Guiana in the north. I have sojourned in most of the countries in the greater Amazon region also, except for Venezuela, the two Caribbean rim Guianas (Guiana and Suriname), and Ecuador.

The chapters herein focus on the concept and reality of what have been variously called "anthropogenic forests," "domesticated landscapes," and the term I use in the title to this collection, "cultural forests," which I first employed in an article published in *Garden* magazine (Balée 1987c). People changed the composition of forests around them, including the living things in these forests, such that Amazonian forests that were originally classified by scientists as pristine and untouched actually are indexical of footprints of the past, that is, of historical-ecological markings etched onto living nature. Some living things, in other words, are also artifacts.

The other important and completely unexpected finding is that these forests are not more impoverished in species numbers than the natural forests that show no or very little human modification, now or in the past. The data that support this assertion are historically encouraging: people do not have to destroy forests and species diversity, even if their populations are large and their societies are complex. Finally, *Cultural Forests of the Amazon* attempts an exposé of what people today know about the history of the forests that surround them. In these pages, I am seeking to answer questions like "What exactly do they know?" "How do they know what they know?" and "When did they know it?" This attempt to understand other (non-Western, non-Eurocentric) cognitive approaches to environmental reality is called the investigation of traditional knowledge (TK) (Chapter 10), also referred to as traditional ecological knowledge (TEK) or even traditional ethnobiological knowledge (as the term appears in Chapter 8). I believe the understanding and translation of TK to be a key to unlocking a new strategy of conserving the species-rich, decidedly fragile, and immensely valuable forested landscapes of Amazonia and other forested regions straddling the equator—as a world legacy of nature and culture. The region is profoundly threatened with a primary landscape transformative process that will lead it to much lower species and cultural diversity if that process is not halted or modified. I think the case to be made for protection of this fragile domain of culture and nature is of a very high order, and it cannot be ignored without global consequences.

The book further examines how past landscape changes have affected the way modern forest dwellers classify the forests, the biota these contain, and even how the changes accumulated in the forest, in terms of transformations of soil, plants, and animals, have had an effect on the very languages people speak in the region and the cultural behaviors they exhibit.

Speaking of languages, I am proposing some new terms for analysis of the relations between people and environments. Ultimately, these suggestions are derived from my theoretical and methodological approach to the Amazon region, its landscapes, and the people who engage those landscapes, which is called "historical ecology" (Balée 1995, 1998, 2006; Crumley 1994, 1998, 2006; Erickson 2003, 2008). It is a research program based on assumptions about how people behave in the environment and how best to describe, measure, interpret, and analyze this behavior and its effects, both on the landscape and on culture, through time. Historical ecology as I understood it from these sources holds essentially that people virtually always have an impact on the environment and biota that surround them, and in turn, the changes they instigate in nature, over time, have implications for the very ways later generations perceive, think about, and classify that nature. It is a way of looking at the juxtaposition of culture and nature, and it is appropriate to a postmodern twenty-first century. I believe cultural forests, as a legacy of indigenous forest dwellers' livelihoods and actions, in both the recent and remote past, can be best viewed, interpreted, and understood through that particular lens afforded by historical ecology.

ACKNOWLEDGMENTS

This book came together in part because of a colleague's advice some years back. She suggested my papers were scattered and hard to find and that it would be useful for those who might be interested in reading them if they were reprinted in a single volume. She said this, in part, because I have indeed published articles and chapters, as some of us do from time to time, in a variety of venues, including rather hard-to-find and out-of-the-way ones. All the essays herein, including two new ones, were arranged with the purpose of showing certain threads of my thinking in historical ecology and Greater Amazonia as these have unfolded over the past thirty years, and the central aspects of these ideas that remain in academic discourse.

Of course, I have all the individuals, institutions, groups, and societies to acknowledge for the original papers that appeared over the years and are here reissued in modified form to fit the layout and concept undergirding this book. They have been thanked as individual persons in the original publications, so I won't repeat their specific identifiers here. I would merely state that I could not have done the work obviously without the acceptance and to a large extent expert tutelage in forest intricacies by the indigenous communities who are the focus of the book, especially the Ka'apor, Sirionó, Araweté, Assurini, Guajá, and Tembé. I am also indebted to longstanding support by the New York Botanical Garden, the Museu Paraense Emílio Goeldi in Belém, Brazil, and the Stone Center for Latin American Studies at Tulane University. I am also grateful to the collegial ambience consistently provided by my colleagues and students in the Department of Anthropology at Tulane University.

In putting the volume together, I have tried to take note of current findings and trends in related work, and several scholars have helped me stay, as much as possible, au courant. The people I must thank the most for this are the two anonymous reviewers whom The University of Alabama Press chose for the volume. They supplied valuable insights and helpful suggestions, by far most of which I incorporated herein, with the objective of improving the work. As far as it goes for the identifiable people who got closest to this project, I am indeed grateful for various specific kinds of assistance. First, I would like to thank photographer and art historian Meghan Kirkwood for permission to reproduce two of her photographs. I would like to acknowledge Kathy Cummins, who copyedited the manuscript in a thorough and timely fashion. Tulane University research assistants Nicole Katin and Dustin Reuther deserve my gratitude for their insights regarding several images as well as in the occasionally vexing, though usually merely boring, task of rekeying of several chapters. Finally, I thank my editor, Joseph Powell of the University of Alabama Press, for his initial interest in and thereafter unflagging support getting this project into print. I am also grateful to him for a number of suggestions regarding the images included herein.

Cultural Forests of the Amazon

PART I

LANDSCAPE TRANSFORMATIONS
Overview

The following three chapters contain background data on the emergence of anthropogenic forests in the Amazon region, before these and natural (or high) forests were subjected to the ravages of modern industrial agriculture, commercial logging, and conversion to bovine pasturelands. The origins of cultural influences on Amazonian forests have a deep past, far beyond the memory of the people who live there today. That does not mean all indigenous groups recognize the differences between forests that have a historical legacy and those that do not, or seem not to. Whether they distinguish anthropogenic impacts from other sources of forest diversity, however, they know much more about these living landscapes than anyone else, including scientists and scholars of Amazonia. That knowledge they have can be demonstrated.

Chapter 1 presents my own path of discovery into these landscapes and the indigenous societies that inhabit them, starting with the first indigenous people I came to know in the Amazon region, the Ka'apor, a Tupí-Guaraní–speaking group of the extreme east of the region. Although for my doctoral dissertation (Balée 1984) I had studied their ethnohistory, social organization, and ritual life, and had lived with them in the field for more than one year, I had not systematically investigated the forest types that surround their villages, gardens, streams, rivers, and swamps. I did undertake such work after graduate school, when I held a research fellowship at the Institute of Economic Botany of the New York Botanical Garden, from 1984 to 1988. That was a fruitful time for fieldwork. It was then that I undertook a careful examination of the species contents and distributions of different kinds of forests in the Ka'apor habitat and the influences, if any, that the Ka'apor and their ancestors had had on those arrangements.

In 1984, I met Darrell Posey (1947–2001), who was then working on ethnoecology of the Kayapó Indians in central Brazil. In his relatively short life, Posey became a pioneer in demonstrating the complexity, depth, and empiricism of indigenous knowledge of the environment and biota. He was finding evidence of Kayapó modification of well-drained savanna country (called *cerrado* in Brazil) landscapes into patches of woodland called forest islands, or *apêtê,* in Kayapó (Posey 1984a, 1984b, 1985a, 2002; cf. Parker 1992). He took to the Kayapó colleagues from diverse disciplines, including geography (Susanna Hecht), ethnopharmacology (Elaine Elizabethsky), botany (Anthony Anderson), plant genetics (Warwick Kerr), entomology (João Maria Franco de Camargo, William Overal), and many others. Posey was a charismatic individual and sought protection of indigenous intellectual property rights. He was also a pioneering scholar of ethnobiology, particularly ethnoentomology: one of his major findings is that the Kayapó discriminate more species of bees and wasps than Western zoology, and the Kayapó were right. Posey's contributions are neatly summarized in a fine selection of his papers, edited by Kristina Plenderleith, and published posthumously (Posey 2002).

In the mid-1980s, Posey's and his many colleagues' influential findings—if later a bit controversial (especially Parker 1992)—regarding the Kayapó had an effect on my own thinking about where the tropical forests of the Ka'apor Indians, as well as other Amazonian groups not living in savannas, had come from, and also where they were going. I found certain Ka'apor forests to be similar in species composition to the Kayapó's forest islands (Chapter 7), but different in the time frame by which they unfolded and in the exact mechanisms involved in their formation. I now believe the forest islands to have been the results of *primary landscape transformation,* and Ka'apor cultural forests derive from *secondary landscape transformation*. Primary landscape transformation is the complete or near-complete change in species composition in an environment effected by certain kinds of human activities; secondary landscape transformation is a partial change in the same. We will return to these concepts in the final two chapters of this volume.

Chapter 1 renders the background of my initial foray into the question of cultural forests, which the Ka'apor, in their language, call *taper*. The fieldwork described in that chapter was part of an intellectual, peripatetic journey into forests that harbored inscriptions, stories, and memories in the living vegetation itself. It is an account of field research on questions that today are widely in the public eye but that were then rather heretical in certain venues, particularly those of conservation biology and systems ecology. As I mentioned before, historian David Sweet (1974) once referred to Amazonia as a "rich realm of nature." I gradually became convinced it is rich in nature, yet also rich in culture. Moreover, the idea that ancient societies of the region could have contributed enhancements to the diversity of the forests—that they could have managed resources, rather than merely adapted to given constraints of nature—was unorthodox then, yet that is exactly where my data were leading.

I put this into perspective of time and place by describing the essential points of one big journey, consisting of multiple passages, into certain tropical timberlands. This field journey is being revisited in these pages for the purpose of clarifying the beginnings and early development of my approach to Amazonian realities, which is called "historical ecology."

Historical ecology is a research program that situates our species in a material and diachronically sensitive cultural dialogue with the environment, which results in new and changing landscapes, some of which are richer, some poorer, in species diversity than the natural landscapes that had to have preceded them. Historical ecology is the approach that undergirds the remaining chapters of the book.

Chapter 2 is a slightly revised version of a 1989 paper on the existence of anthropogenic forests in Amazonia. I proposed that *at least* 11.8 percent of the well-drained forests of Amazonia are anthropogenic, or cultural, in origin. According to science journalist Charles Mann, in his highly regarded book *1491*, the article that appears here as Chapter 2 had been "widely cited" by the beginning of the twenty-first century (Mann 2005, 305), even though it has never been easy to find. It is not online nor is it available in other easily accessible media. In his book, Mann (2005, 305) also pointed out that other scholars now think my original estimate to have been "conservative." A belief that cultural forests are more common than I suggested is also clear from a widely viewed documentary film called *The Search for El Dorado* (Sington 2002). The problem is no one has put forth a believable alternative. That is because the exact parameters of cultural forests are hard to pinpoint: it is difficult to say where they might have expanded and where they have shrunk (cf. Bush et al. 2007).

Let us examine the other side of the argument. One can ask whether paleoecologist Mark Bush and colleagues (Bush et al. 2007) were correct concerning a very simple successional process that they proposed might have occurred millennia ago in western Amazonia. They and McMichael et al. (2012) argue for there having been virtually no long-term impact on forests of western Amazonia by ancient agricultural societies that might have existed there. In other words, ancient peoples had nothing to do with current biological diversity in and the physical structure of the forests we see today, in their view (also see Barlow et al. 2011). Bush et al. (2007) maintain that small groups exploiting no more than 140 hectares for hunting, fishing, and agriculture would have left no signature in the species and structural composition of the forests that have survived into our times. After a few hundred years, the effects of their intervention would have, in their words, "worn off" (Bush et al. 2007). That is a good description of simple, secondary succession. The forest would leave no indicators of past disturbance, and in a sense, would become primeval again. It is mistaken, however, to presuppose that these early agriculturalists were the first people to make their home in the scant and hypothetical 140 hectares. Humans have lived in the Amazon for at least eleven thousand years (Roosevelt et al. 1996). Perhaps—and it is even likely—the postulated agriculturalists, who occupied and disturbed an area of 140 hectares, were actually preceded by simple hunter-gatherers or trekking people, that is, people who utilize the landscape and its resources without necessarily domesticating the flora and fauna around them. What Bush et al. (2007) seem to ignore is that even simple hunter-gatherers and trekkers can alter forest composition by occupation of sites and encouraging, if not domesticating, the presence of some species yet not others (Politis 2010; Rival 2002; Zent and Zent 2004). This means the forest encountered by the ancient agrarian societies proposed by Bush et al. (2007) might not have been primeval at all but rather had elements of aboriginal culture—namely, species of plants and animals that are found in association with small-scale incursions into and manipulations of the forest. These include many plant species, even trees, which we will discuss in this section and others, some of which eventually were even domesticated, as with Brazil nut trees (Shepard and Ramirez 2011) and peach palms (Clement et al. 2010). Considering such incursions to have resulted in altered landscapes, the forest cut down in the 140 hectares of the hypothetical ancient society was already a cultural forest, not a primary or primeval one.

The mechanisms of landscape transformation, apart from human and contingent ones, are not, however, easy to identify. Actually, I think it's appropriate in these pages to reinitiate the discussion of just how much of Amazonia has been transformed. That is another reason for reprinting the essay in Chapter 2, even though findings in anthropogenic soils, size and complexity of archaeological sites, known ages of indicator tree species, and degree of domestication of some crop trees and their association with anthropogenic soils and other patches of human occupation in the past have advanced considerably since the time of the paper's original publication (e.g., Clement et al. 2010; Clement and Junqueira 2010; Denevan 2001, 2006; Erickson 2003, 2006; Glaser and Woods 2004; Junqueira et al. 2010; Kawa and Oyuela-Caycedo 2008; Lehmann et al. 2003; Shepard and Ramirez 2011; Woods and McCann 1999; Woods et al. 2009). Those findings, most of which in one way or another we will touch upon or discuss later in this volume, by and large do not contradict Chapter 2; they lend support to its principal suggestions, in the big picture if not in detail. A conservation biologist, Carlos Peres, and his colleagues recently showed their misunderstanding of the analytic relationship of anthropogenic forests to Amazonian Dark Earths when they wrote, "In fact, contrary to several interpretations of Balée's (1989[a]) 'anthropogenic forest' hypothesis, > 85% of Amazonia almost certainly did not sustain permanent settlements practicing perennial agriculture" (Peres et al. 2010, 2315). I never said that, and I don't know of anyone else who did. Although I have often been asked in conferences and lectures whether I would care to revise the original estimate, usually upward, such as to 100 percent, I have tended to decline those invitations, and I still would demur on it. The reason is I believe it is too early to change it.

Only since the publication of my 1989 article mentioned by Mann, for example, have networks of gigantic geometric ditches and walls been revealed to the world, and seen from the air, due unfortunately to deforestation in the western Brazilian Amazonian state of Acre. There are about three hundred of them known already, in a relatively confined region of the eastern part of the state; from the air they look like monumental circles, squares, and rectangles, with perfect or near-perfect geometric form and alignment; at present, these geometric earthworks, comparable to "enclosures" in Neolithic Europe, are being called geoglyphs (or *geoglifos,* in Portuguese) (Pärssinen et al. 2009; Schaan 2011), which is the term originally applied by their principal discoverer, Alceu Ranzi. We still do not know whether the forests that lie atop the remaining geoglyphs, yet to be discovered, will display a cultural origin or not. That is a question being addressed in ongoing research being led by archaeologist Denise Schaan of the Universidade Federal do Pará (in Belém, Brazil) and her collaborators, including geographer/paleoecologist Alceu Ranzi of the Universidade Federal do Acre (Rio Branco, Brazil), archaeologist José Iriarte (University of Exeter), paleoecologist Francis Mahle (University of Edinburgh), geographer William Woods (University of Kansas), and soil scientist Wenceslau Teixeira (EMBRAPA/Rio de Janeiro), several scholars from the University of Helsinki, myself, and others. No definitive answer yet exists. In other words, the data are not yet completely in; the work goes on, it moves forward. Because there may be other such regions of intense prehistoric activity, in which it is still too early to know whether the forests that cover them are the result in part of past human manipulation of the environment, I have concluded it is simply too early to elaborate on or to be much more specific about supporting an upward revision of my earlier estimate of anthropogenesis of Amazonian forests. For that reason, the paper may still be germane to discussion, so I will leave it as it stands for now, as Chapter 2, with a new title and only the slightest tweaking (of the sort described in the preface) in order to be comparable to the other chapters, such as updating of scientific (Latin) names.

For the most part, that paper's other specific findings have not been disproven, either. At least, the most recent evidence on such forests influenced by human action tends to support the findings of Chapter 2. One of the most important findings is the affirmation that Brazil nut trees are associated with anthropogenic forests (Shepard and Ramirez 2011). Adolpho Ducke (1946, 8, cited in Balée 1992a, 51) appears to have originated the idea; Shepard and Ramirez (2011) have made their point empirically very well. Domestication of the Brazil nut tree, therefore, in their line of reasoning, helps explain why this massive tree species is only found in such areas indicative of human-mediated disturbance in the past.

The last chapter in this exposition, Chapter 3, was originally published in a special issue of *L'Homme* in 1994. It treads logically behind Chapter 2, insofar as it identifies mechanisms in the development of cultural forests and how these systematically differ from those of natural or pristine or high forests. The most important finding of this chapter is that the Ka'apor did not diminish but rather augmented locally the abundance of species, by creating a new kind of old growth forest, known as fallow or anthropogenic forest. It is a harbinger of the notion that indigenous occupations of Amazonian forests have not necessarily been destructive of what most people value—diversity, in a Platonic sense—and may even have enhanced it. That issue of missing destructiveness will be revisited in the final chapters of the book, when diversity *contingent upon* human activity is taken up again in an analytical model of landscape transformation.

1
Villages of Vines and Trees

The Ka'apor cultural consultants whom I considered to be the most knowledgeable on the subject of forest types and vegetative associations told me our destination, the old growth forest, looked like a true forest, but that it was in reality an old village, long abandoned by any human occupants. They called it *taper* (pronounced *ta-pair*), and for them it constituted more than the continual triumph of luxuriant vines and spiny palms over the deteriorating thatch and rotting posts and beams of what had once been people's homes (Balée 1987b, 1988b, 1989a). These consultants live in the eastern Amazon region; they have no illusions on the permanence of things, but life-giving forces do not so easily dissipate. Even though they did not at any time bury their dead inside the houses of the living, so common to ancestor worshipers elsewhere, the Ka'apor people seemed to believe in the presence of ancestral ghosts, whose earthly personas had departed time and who haunted the places where their bodies had once lived. One seemed to always have the feeling of being watched, not only in the village of the living but even in the "Village of the Bones" (the Ka'apor term for a cemetery, *kangwer-rendá*), hidden in the shade of a forest. For though they did not bury the dead in the village, they knew that villages had once been fields, and it was in fields recently put to fallow that they interred the deceased. So any old village could also be a cemetery.

Cemeteries are not to the Ka'apor people themselves comforting places to visit. The panegyrics that attend Euro-American funerals—at least some of them—are entirely lacking in Ka'apor burial practices. It is enough to inter the dead, and hyperbolic praise is virtually never accorded any human being, living or dead. Even the gods, with the possible exception of the creator god, *Ma'ir*, are treated to polite, if sincere, respect, but not supplication. Because the Tupí-Guaraní peoples of coastal Brazil in the sixteenth century did not pray to or adore any deities, the early missionaries thought they had no religion (Clastres 1995, 8); the Sirionó people, also a Tupian society, of eastern Bolivia, were likewise thought, mistakenly, to have had no religion (Holmberg 1969; Nordenskiöld 1924).

But Tupí-Guaraní religion is of another sort altogether. The scions of the ancestral deities—some of whom were animals, some of whom were plants, and some of whom looked like people—are, after all, ordinary people living and breathing in ordinary time, year in, year out. The delectation one experiences in visiting friends and relatives in other villages in the dry season (when trails are not flooded) comes from the perceived meeting of the minds over shared memories, sometimes of long-deceased loved ones. The shared relational experience constitutes the fulcrum by which Ka'apor people define themselves—they are first of all *anam*—relatives—

to one another, and second, they are a society that consists of relatives on the one hand and in-laws and potential in-laws on the other.

The second condition of a society divided into relatives and affines (i.e., marriageable persons or in-laws) is a necessary, presupposed one for avoiding incest and for guaranteeing proper heterosexual unions. The first condition is the fictional bliss wherein everyone is related in one larger, more encompassing unit of shared substance, emotion, experience, and language, with the proper Ka'apor accent. No sinecures exist in Ka'apor society and all people, male and female, ideally proceed through the same life cycle, the same sets of rights and duties (as child, adolescent, young adult, old adult) and together bear the essentials of life in the forested mosaic that has for them historically constituted their center of the universe. One of their metaphorical names for themselves, according to Yupará, now in his seventies and son of a renowned headman from the 1940s, is *Ya-ko-Piter*, which literally means "We here, of the Center." In many ways, ancient fallows, cultivated and lived in as villages in the past by the ancestors, symbolize the historical nucleus of this perceived world. Many indigenous societies worldwide, incidentally, consider themselves to be situated at the center of the known world.

The evening before my first excursion to the old growth forest called *taper*, Serō, who is one of the two headmen of the Ka'apor Indian village of Urutawi (that is, village of the long-tailed potoo, a nocturnal Amazonian bird related to owls), and I spoke by the dying embers of his hearth about what one might find there, other than mounded graves that might be in any case hard to distinguish from natural berms and burrowing of an assortment of critters. It had not rained for days, and the wet season now, in June 1985, was tapering off and grading into the time of the dry season (*warahi haku rahã*, "when the sun is hot," as the people say), when the nights have their low dew points and are even cool enough for one to need a blanket or to want to sleep by the fire. The smell of burning wood permeated the air as we spoke, some of it wafting in from other nearby houses that, like this one, had no walls. Serō, a well-built man about five feet, eight inches tall (170 cm), thirty-five years old, wore the traditional Ka'apor tonsure, his hair being cut to shoulder length on the sides and back with the bangs cut straight across the eyebrows. Each cheek had a vertical red line painted down it with urucu (annatto or achiote) dye, the juice called *uruku-tikwer*, and there was a horizontal, red urucu line painted between his lower lip and chin. Both women and men paint a few identical designs on their faces, and they have done so for a long time.

In the perforation of his lower lip, he wore a labret made from a long yellow tail feather of an oropendola (or cacique), called *yapu*, a bird related to orioles that is associated with headmanship and masculinity. One calls the feather *hē rembepipo* (my lower lip feather). Serō wore cotton trousers and a short-sleeved, button-up cotton shirt, otherwise indistinguishable from the outerwear of the peasant men on the nearby frontier. He sat across his hammock, as his wife and daughter dozed in theirs a few feet away. We spoke in his language that I had come to speak and understand reasonably well over the few years of my association with him and his people. He said the *taper* we would visit tomorrow, which lies some four miles (about 6.4 km) north of the village of Urutawi, had been a settlement perhaps sixty or seventy years ago. A massacre of Ka'apor people had occurred there, he claimed, shortly before peace between the Indians and the Brazilians came to the region, which I knew from old newspaper clippings to have been in 1928. The whites had killed men, women, and children, as was their way in those bitter times, and those who survived had fled to a new site. Tomorrow we would see what the old village had become since that time.

Serõ and Lusiã, a youth of about seventeen years of age and the son of the village's other headman, were my guides, and we took to the trail shortly after sunrise. We walked the undulating, hardscrabble trail, with manioc fields on either side interspersed with patches of true forest until we reached a stream in a depression densely covered with açaí palms; tall, slender, elegant, and smooth-barked, their white-gray trunks were crowned with dark green, fernlike leaves and purple bunches of fruit. These bunches of berry-sized fruits grow on the rachis and are much sought after this time of year, their prime growing season. We would not collect them today, but the news would arrive in the village. Some enterprising youths might trudge up here on the morrow to shinny up some of these palms and bring back caches of the fruit. Women would boil and then knead the fruits, which separates the edible oily skin from the large, fibrous seeds. They would then let the grape-colored, viscous product cool for people to later drink with their manioc flour. It is one of those dishes consumed throughout the Lower Amazon that people have been eating since before there were any cities here.

We waded knee-deep across the stream, stopping briefly on the other side for a drink of water. Serõ quickly cut an arrowroot plant from its muddy substrate, fashioned a cup from the long, ovoid leaf, and we all drank in turn from it. We then walked uphill with dense forest on both sides of the trail, and in a few minutes we arrived at an almost imperceptible opening on the left-hand side of the trail. We had arrived at the *taper*, Serõ said, and we stepped into it. Machetes were needed to cut our way through the vines and spines for some one hundred feet. Meanwhile, the lighting grew less and less, since the farther one gets from the main trail, the more the canopy covers and shades the traveler. Yet the forest here was like none I had ever seen, or if I had seen such before, I had been too naive to distinguish it from the abundant greenery of all the rest of the forest. There had been a time when it all seemed pretty much the same to me—everlasting green, green, green—broken only by an occasional black fallen log, white mushroom, red balanophore (a parasitic herb without chlorophyll), or yellow or red fruit. To be sure, well into the dry season—in the month of October—one can see isolated patches of the forest floor carpeted with bright yellow or purple tubular flowers of pau d'arco trees, called *tayi*, but one has to be there in that season for such color, a visual treat. Otherwise, the ingenuousness most temperate travelers feel in the Amazon today is probably of the sort that prompted European pioneers to call Amazon forests a green hell.

Serõ and Lusiã led the way, warning me in their hospitable manner of spines (*yu*) and other obstacles as they appeared. I have heard Amazon researchers sometimes speak with pride of the palm spines embedded in their skin, like the lead some old warriors bear beneath theirs with the knowledge of having fought the good fight. But it is also true that those who know their way around best in the forest rarely get stuck by a spine or bitten by a snake, though an accident can happen at any time. In this milieu, forest knowledge confers at least a partial shield from nature's thorns.

We progressed on through the thicket, until Serō stopped and kicked at a patch of ground. Then with his machete he began to scoop out clumps of blackened dirt and there, at a depth of about half an inch (1.25 cm), lay a nearly complete ceramic griddle, about four feet in diameter. That was what they had cooked their manioc flour on, in order to burn off the free cyanogenic gases inside it and render it the edible staple of the Ka'apor people, otherwise quite poisonous to those who do not know how to prepare it. Not far from the griddle, a luminescent object caught my eye—it was a mirror the size and shape of a silver dollar, with light blue backing on it. It was one of those manufactured baubles that the people used to seek afar regardless of the hardships, which sometimes involved epidemic disease or violent death.

Contact had those twin results: material wealth on the one hand, at least temporarily, and premature death on the other. The classic statement on how contact with peoples of Old World origin had a devastating effect on Amazonian populations is Darcy Ribeiro's (1956), with specific reference to a disastrous measles epidemic among the Ka'apor that he witnessed in 1949, when he was doing pioneering fieldwork among them. The susceptibility of indigenous groups to viral syndromes such as smallpox, measles, and colds has been abundantly documented (e.g., Black 1992; Cook and Lovell 1992).

Someone here a long time ago had looked into that mirror and seen more light than there was today. The griddle and the mirror were reminders of a time when people laughed, shouted, cooked, ate, groomed, slept, talked, exchanged glances, gestures, and hugs, and breathed in the open air of a village fit for people to live in. Some had been born here, some had died here, and their names were no longer matchable to faces and voices, their bones long since dissolved in an old field, somewhere—maybe in an old field like this one. But all was not death, decay, and mute, manmade artifacts. The *taper* itself—that is, the forest—gave living, green testimony to those earlier human lives. Long since gone were the houses, the pets, and the dooryard gardens. But the trees that stood all around us were an index of past events in human society.

The *taper* consists of many trees not encountered in the dense forest. On that day, we saw and collected the *taper* fruit tree (called *taperebá* in Portuguese and meaning fruit tree of *taper* forest), the tucumã palm, a wild cherimoya, and a sapote tree. The *taper* fruit tree (*taperiwa'i* in Ka'apor) is hog plum in the cashew family; it has tart, juicy, yellowish-orange fruits that are high in vitamin C content. Numerous game animals also like to eat these fruits. The tucumã palm is a large, multistemmed, colonial plant from whose trunks grow long, flat, black spines that can easily stick into the skin of an unwary, naive hiker; it has edible fruits displaying significantly high vitamin A content, and its leaves are used for making resilient cordage by other Amazonian Indians. The wild cherimoya tree has sweet, pulpy fruits enjoyed in the wet season, as does the sapote tree. These plants I knew to be absent in the high forests I had already been studying, though in less abundance any one of them might be found along a creek's edge. Here in the *taper* they formed an orchard of fruit trees. It then dawned on me that the *taper* was a forest type in its own right, not just one more vegetational zone easily distinguished from any other. It was a *cultural forest*.

The Ka'apor forest is one of the most verdant, biologically rich, and culturally complex islands of Amazonian forests I have ever seen. Surrounded on all sides, except for part of the Gurupi River, by settlers, loggers, ranchers, and other non-Indian people, it is a giant arboretum in the midst of a massive development, which is causing profound alteration of eastern Amazonian landscapes. Regrettably, environmental destruction is the continuing state of things, and I cannot intellectually hold out much hope for the future of cultural and biological diversity in this area, at least not the immediate future. The demands for timber and land, emanating from outside the reserve, are intense. In 1985, the Luso-Brazilian frontier was far less close to the Tupí-Guaraní–speaking Ka'apor, Guajá, and Tembé peoples who inhabit the reserve and its adjacent reserves than it is today. To reflect upon that isolated arboretum known as the Terra Indígena Alto Turiaçu, as it was in 1985, replete with forests of diverse ages and anthropogenic influences, seems like a voyage into another world.

At that time, I realized that if there were cultural forests here in the extreme east of the Amazon, it was likely one would encounter them also elsewhere in the vastness of the greatest river basin in the world.

And if there was one cultural forest here, there could be others, in Ka'aporland and beyond—in any place where people had lived and gardened in Amazonia in the time-tested ways of the ancestors. That moment in 1985 changed my way of viewing Amazonian rainforests and their human inhabitants profoundly from then on. I would visit numerous other indigenous groups in Amazonia after that time, and in all cases I was keen to determine the presence or not of cultural forests.

The culture of the Ka'apor long ago had sent a completed message into the living forest of today, not just a signature, down to the time when forests everywhere were beginning to vanish. Knowing that helped stimulate the sense of urgency that I, and many of my colleagues, felt about our research. The late 1980s saw also the beginning of the end of what Brazilian social anthropologist Eduardo Viveiros de Castro (1996) appositely denoted as the "standard model" of Amazonian ethnology. That model, elsewhere called the adaptationist approach, was mostly propounded by cultural anthropologists and archaeologists inspired by and indoctrinated in the techno-environmental determinism of Julian Haynes Steward, the visionary and systematic editor of the monumental *Handbook of South American*

Indians (1946–1950). Steward (1902–1972) cast a shadow over anthropological and ecological understandings of human adaptation in South America for half a century; unquestionably a great deal of work in ethnography, ethnohistory, archaeology, and biological anthropology was carried out beneath the penumbra of the comparative and analytical entries in the *Handbook*. The standard model presupposed severe limits on indigenous population growth, population density, and political complexity because of the widespread occurrence of soils low in fertility (as in the classic account by Betty Meggers [1971, 1996]) or, in a later version of the model, scarcity of game animals (Gross 1975).

The standard model assumed these conditions, and erroneously, the forest itself, to be primeval relics of the Pleistocene or some earlier epoch over which native people had had no managerial impact in the past. The standard model is collapsing in light of a vast array of recent evidence that increasingly implicates humans and their cultures in environmental contexts thought previously to be pristine. These contexts include Amazonian Dark Earths and telltale (though often subtle) earmarks of human manipulation over time seen in unexpected species distributions, species diversity, and altered waterways (Balée 1994; Denevan 1992, 2001; Eriksen 2011; Erickson 1995, 2000a, 2006, 2008; Heckenberger et al. 2003, 2008; McEwan et al. 2001; Posey 2002; Raffles 2002; Schaan 2011; Stahl 2002; Viveiros de Castro 1996). The rise of these new interpretations of diachronic relationships between indigenous Amazonians and their local environments has even acquired cachet in popular print and film media (Mann 2005; Sington 2002; Toniolo 2011).

A nonindustrial message, inscribed in the living concatenation of trees and vines, to be sure, was unmistakable at the *taper* on that memorable day, to me at least, in 1985—people had diversified the landscape at one time without deforesting and denaturing it. No inherent mechanism in the genome of the species had led humankind to perceive the world's richest complex of organisms as an obstruction to progress that needed to be waylaid and razed. On the contrary, they had enriched it even more. And my research would never be the same, for I began to be able to read that message in other Amazonian forests where it had been written, still remote from the arriving peasants, colonists, settlers, loggers, and ranchers, but utilized by people—Indian people in this case.

And I would find out that not only had the people of the past written, as it were, a message into the living forest as if it were a palimpsest (Erickson 2006), but also that forest had worked its way into their languages and cultures of today. The past was present, and however nameless and faceless they might be now, the ancestors had not really gone away totally. I had begun to realize that my research could take place only where people and forests in Amazonia still coexisted. One without the other seemed rather boring.

How the Lost Survived

A little more than two years later, in October 1987, I found myself in the front passenger seat of a Toyota four-wheel-drive pickup truck, riding up a dusty, single-lane road out of Marabá, on the Tocantins River, heading due west, toward the Xingu River. It was a logging road, not yet completed but recently opened up to large flatbeds that were now hauling hundreds of tons of mahogany out of that forest every day to sell in Marabá, Belém, São Paulo, and other cities, and some of it would eventually get to the United States, Japan, and Europe. The driver and the Indian agent in the back seat, fortyish Genésio Carvalho, worked for the National Indian Foundation of Brazil, a government agency like the U.S. Bureau of Indian Affairs, better known to all as FUNAI (pronounced *foon-eye*).

I had been invited by the FUNAI administrator in Belém, Mr. Salomão Santos, to accompany Genésio to a recently contacted "wandering" tribe of Indians (*índios arredios*) of the Tapirapé River. That body of water was, in fact, not a river, but a narrow stream that drained part of the vast middle area between the Xingu and Tocantins Rivers, ultimately pouring into the Itacaiunas River, a major tributary of the Tocantins. The Tocantins itself is a tributary of the Amazon River and one of the largest rivers in the world: its mouth can be seen as coterminous with the lower mouth of the Amazon itself. Some years before this it had been dammed for an electric plant that now supplied energy to millions of people in northern Brazil. I had seen the huge transmission lines from the air on my flight from Belém to Marabá.

The hinterland to the west had been partly opened in the 1960s by the Trans-Amazonian Highway, along which most of the forest had vanished by the 1980s. We were traveling on a newer road far to the south and roughly parallel to the Trans-Amazonian Highway. Small frontier towns had arisen already in the two years the road had been cut. The road came to an abrupt halt, because the government interdicted further traffic in the area due to *índios arredios* ("wandering" Indians), in the forest some 160 miles (256 km) west of Marabá—it had taken us seven hours to reach that point, and there had been several hairpin curves with oncoming, mahogany-laden flatbeds that we had somehow negotiated along the way. We got out of the truck at 3:20 P.M. and hiked through the forest until reaching a small adobe-and-thatch house set amid a field of about three acres at 4:30 P.M. Waiting for us on the earthen porch were another FUNAI worker and two near-naked men, who laughed uproariously as we climbed the little hill to the house. They were the Wanderers of the Tapirapé River, the only ones known to date.

FUNAI concerned itself not only with Indians in more or less continual contact with Brazilian society; it also had a branch that dealt exclusively with *índios arredios* who had or were thought to have had little or no contact with the outside, Luso-Brazilian world. The job of men like Genésio was to approach such people, "attract" them, and offer them peaceful accommodations with FUNAI. Once good relations could be established, new FUNAI agents would come onto the scene, and the wandering folk would then be more settled in a certain place that could be indicated with exactitude on maps and charts, and question marks around their people's identity then could be removed. The present problem was, no one could figure out who these two men were.

A month before my trip, the linguist Dr. Denny Moore of the Goeldi Museum and a student of his had determined for the FUNAI agents that the two men spoke a language of the Tupí-Guaraní family of languages. Like the Romance languages, whose ancestor was Latin, the ancestral tongue of Tupí-Guaraní languages was spoken beginning around twenty-five hundred years ago. The Tupí-Guaraní family today is spread across much of Amazonia and lowland South America, just as Romance is found over much of Europe. Moore had also found that the language of the two men was a new Tupí-Guaraní tongue, never even alluded to before, much less described by linguistic science. New languages and new cultures are not often found in the world today—but Amazonia is one place where such discoveries occur from time to time. Because I spoke Ka'apor, and at the time I was somewhat conversant with Araweté, Assurini, and Tembé, all of which are Tupí-Guaraní languages, it was thought I might be able to help communicate with the two men to find out whether there were any more of their kind, hiding in the forest.

Genésio, short, mustachioed, and rugged, was a fraternal nephew of another famous Indian-attractor (or *sertanista,* in Brazil), João Carvalho, who for his part had been the government agent of the Ka'apor during the 1950s and 1960s. I met him briefly, at his home, in 1987; he was thanked profusely in books by Darcy Ribeiro related to the Ka'apor, for logistical and other help rendered as far back as 1950. Genésio enjoyed a free and easy relationship with the two wandering men, though they could not understand each other's speech. The two men had come to be called Aurê and Aurá. Their names have since changed, though none of their names as yet known appear to have been original to them. In any case, they did not correct their interlocutors with any hint of their real names. They seemed fairly free and easygoing. Aurê was about thirty to thirty-two; he walked with a wooden staff, because one of his legs was shriveled, perhaps from infantile paralysis (i.e., polio) or a childhood accident, snakebite or something along those lines. He was thin, with long, rangy hair, a sparse moustache and goatee, and a ready smile. Aurá was about twenty-eight or thirty, sturdily built with close-cropped hair, darting eyes, a quiet smile, and overall nervous and jumpy demeanor. Some of the FUNAI personnel thought they might be brothers; I was never sure of this. They resembled each other, but not overly so. They spoke often to each other in an agitated, rapid-fire way, and I understood almost nothing of their conversation together. When I interviewed Aurê, the lame man, that night, though, it became clear that something terrible had happened to them both, long ago. Genésio had suspected that they were the only wandering people around here. I began to think he was right.

The story had begun in June 1987, when the house we were now occupying had been that of a family of Brazilians, who had followed the logging road to make their claim to land and livelihood here, only a few months before that. They had already cleared the surrounding forest and had a small field of manioc going, along with some rice, sweet potato, West Indian gherkin, chile pepper, cacao, gourd, cotton, and other crops. These were growing when they built a house in the middle of the field. By June, the men were harvesting rice and preparing some of the surrounding ground for more planting, while the women remained home attending to other chores. One afternoon a woman saw two naked strangers at the edge of the woods, some two hundred feet from the house. She uttered a cry as the two men began to advance rapidly toward the house; the women took refuge inside the house; they used a piece of rope hanging from the doorjamb to tie the palm-woven door shut. The two men reached the house and had little trouble pulling down a corner of the thatched door, and as they peered inside, they were shouting *kuyũ! kuyũ!* ("women! women!"). The women's subsequent shrieks alerted their menfolk at the opposite edge of the woods, who soon arrived, axes, machetes, and knives in hand. The two wanderers stood still while the Brazilian armed men and they eyeballed one another.

Some side must have given, or blinked, because no one was killed or hurt in the encounter. The two men at some point returned to the forest; the owner of the house made a humanitarian decision. He hitched a ride to Marabá and alerted the FUNAI administration there of "wild Indians" (*índios bravos*) near his farm. He did so to avoid the possibility of needless interethnic violence.

Soon the attraction agents arrived, met the two wanderers at the edge of the woods, and began what they evidently thought would be a routine process of pacification of a newly discovered tribe, as had been the case so many times previously. That process involved evacuating the farmers, who could be in danger and who might also frighten the other Indians believed to be hiding in the nearby woods. But as the days and weeks went by, no one else appeared, only the two wanderers. Other bilingual Indians—Gê and Tupí-Guaraní speakers—from all around were brought in, and none it seemed could communicate with the wanderers, the objective always being to determine who they were and where the rest of their kinfolk were. The others were assumed to be still hiding somewhere in the forest.

At that point, Denny Moore had been contacted. By the time he arrived, the two men were spending the night with the FUNAI personnel in the farmer's erstwhile house. Before that, they appeared every day and received gifts, such as metal tools, but had always returned to the forest in the afternoon, signaling upon departure for no one to follow.

I communicated better with Aurê, the lame wanderer, who had a more relaxed manner than the obviously more high-strung Aurá. By pointing at different objects, I elicited his words for them, and began to put together a rudimentary dictionary. Many of the nouns I got were clearly cognate with other Tupí-Guaraní languages. Cognate terms are words in diverse, related languages that refer to the same or similar things and that are derived from a common mother language. In the language of Aurê and Aurá, these include words for some of the farm's cultigens I showed them, such as sweet potato (*yiči*), squash (*i'úv* for the leaves), gourd (*ïru* or *tuni*), and field corn (*awači*). The field corn word is also the same as the word for sugarcane—both, of course, are in the grass family; the word for rice, also in the grass family, was modeled on the word for field corn, as *awači-rũ* ("maize-inauthentic" or "plant that looks like maize"). They also had an obviously cognate word for bitter manioc (*mani'a*).

That meant to me that the two men were not hunter-gatherers but came from a village with domesticated plant foods, not a camp. They would have been from a sedentary society, in fact, and not one of true wanderers—or hunter-gatherers. The next day, I and one of the FUNAI workers who had remained while Genésio picked me up in Marabá, walked into the forest from which Aurê and Aurá had first stepped out back in June. We gesticulated for them to follow us into the forest, but their smiles dimmed as Aurê and Aurá looked askance at these futile signs, indicating their rejection of the offer. I found out later there was a religious reason for their refusal to return to the forest.

The FUNAI worker and I conducted ourselves along the narrow, barely visible footpath that had been used by Aurê and Aurá to make their first contact with Brazilians. We advanced through a forest of lianas where Aurê and Aurá had lived since who knows when, until we reached a campsite; it was where they had lived before the initial contact in June. It was a little more than two miles (3.2 km) southeast of the house of the farmers (which had by the time I arrived become the headquarters of the Attraction Post for the Wandering Indians of the Tapirapé River).

The campsite's central structural feature was a rectangular lean-to of about two hundred square feet; its thatched roof of babaçu palm leaves was caving in. In the language of Aurê and Aurê, the babaçu palm (*Attalea speciosa* Mart. ex Spreng.) is called *waí*, which is very close to the Guajá word (*wa'ɨ'ɨ*) for the same plant (Chapter 5 and Appendix I). Inside the temporary shelter were forty-five old arrows and a bow of about six feet in length. The arrows had points made from deer bone and bamboo. The men had slept on babaçu palm thatch, and their former beds were now becoming sodden and rank, with the browning decay of moist palm leaves.

In the garbage dump outside the house were many tortoise shells and Brazil nut shells—remains of food that can be gathered by hand easily. Numerous old carrying baskets strewn about suggested the site to be no more than about two years old. After the initial contact, it is surmised, they had moved a little closer to the farmers' house. We reached that location by means of another trail. Here was another rectangular lean-to of a couple hundred square feet surrounded by a fence made from unsecured saplings. An extremely worn steel axehead was found inside. The house was a lean-to—not a freestanding structure—and like the other, its makeshift roof was tied by vines to nearby, growing trees. Here, too, we found tortoise shells, as well as armadillo skulls, again, game animals that are not too difficult to capture and kill, when available. A wooden fire drill was also present on the floor of the camp. We returned to headquarters near dusk, with the bow and a few arrows. Aurê and Aurá indicated they no longer wanted these artifacts; I later donated them to the Goeldi Museum (Museu Goeldi, in Portuguese, which is the Museu Paraense Emílio Goeldi, located in Belém, Pará, Brazil). These artifacts have been stored in the ethnographic collections of the museum, and I had the opportunity to see them again, well curated, in 2009.

The bow of Aurê and Aurá was made from a liana. It was curved and wanting in tensile strength. The arrows were all top-heavy. Bow and arrow making is a skill involving a learning curve not mastered by boys until late adolescence in most parts of Amazonia. I began to suspect what Aurá would confirm later that night—the wanderers were feral children, orphans, in fact, who had not grown to psychological and cultural adulthood. The two men seemed childlike in some ways; when not eating or talking, they would swat continually at flies, using self-carved slats of bamboo. It seemed like a nervous habit, formed during long, lonely hours inside their lean-to.

In my halting wanderer speech, on our second night at the farmhouse, I asked Aurê about his father, as follows: "Your father?" (*Nde ru*?), not knowing the interrogative construction equivalent to "Who is . . . ?" For that question, the verb is understood. It's an empty set verb. Aurê immediately answered, "*Manū*" ("dead," an apparent cognate term to Ka'apor *manõ*, "s/he's dead" or "s/he's out of it"; in Ka'apor, a lunar eclipse is called *yahɨ-manõ*, "moon-death," and an epileptic seizure is dubbed *manõ-manõ*, "dying, dying" or "out of it, out of it") (There are several epileptic persons among the Ka'apor, and there are a significant number also of deaf persons, such that they have a sign language of their own for the deaf; the two conditions appear to be caused by tropical yaws). I then asked (having learned the word for "what" through trying out conversation earlier in the day), *Ma'a yukā* ("what killed him?" or "what did he die from?"). Aurê's reply was instantaneous and emphatic, "*Awa-yu yukā*" ("the Yellow People killed him"). His father had been murdered; but who were the Yellow People? Not the white man, it seemed, for if whites, even jaundiced whites, killed Indians in their villages, they would have done so with guns, a point Denny Moore had made to me when I told him this story. And Aurê and Aurá, in their innocence, had no fear of FUNAI's guns when these were fired at a guan or some other game bird that might have unluckily found itself near that house. In fact, the two men laughed at the sound of gunfire. They didn't have any idea how deadly gunfire can be.

I began to think that they and their kinfolk, who were certainly not in the area, if any were left at all, had been attacked by another indigenous group. Antônio Carlos Magalhães, an anthropologist at the Goeldi Museum, also came to this conclusion (Magalhães 1991) and even specified whom he thought they were, a lost and never really known tribe of the 1600s. It is a fascinating hypothesis, if hard to prove. After Aurê and Aurá were evacuated from the logging trail's end and taken to Marabá, agents of FUNAI tried to resettle them among a nearby group of Parakanã, also Tupí-Guaraní speakers, who once had roamed between the Xingu and Tocantins Rivers. But Aurê and Aurá refused to leave the car that had driven them into the middle of the village, shouting "*Awa-yu! Awa-yu*!" the name for their hated enemy, the Yellow People. Magalhães (1991), who saw Aurê and Aurá at the Parakanã village in November 1987, believed, based on what he thought was their fear of the Parakanã, that the *Awa-yu* had likely been the Parakanã. That makes as much sense as anything else in this atomistic fragment of forest history.

That same day, Aurê and Aurá were returned to Marabá. Later they would be welcomed among the Assurini on the Xingu River, and then expelled from there when in a rage, Aurá chopped off an Assurini man's arm with a machete. The man bled to death on the boat ride downstream to Altamira. Upon being returned to Marabá another time, they were refused entry to the Tembé people. Finally, they ended up in a house upstream from the Guajá Post Awá on the Caru River. And they had five FUNAI attendants. I saw them there again in January 1991. They were still killing flies and deerflies with small slats of bamboo. The FUNAI personnel confirmed they could not fletch arrows or make a bow suitable for hunting. Nor were they trying to, either. The FUNAI attendants cooked their rations every day; perhaps the Wanderers of the Tapirapé River had finally gained some respite from the struggle to survive on limited knowledge in a forest that to be deciphered required long, patient study and exposure to the knowledge of the elders.

More Survivors

I had known the Assurini do Xingu people since March 1986, when I stayed among them briefly while riding up to the Araweté for the purpose of conducting a forest inventory. They lived in a village on the right bank of the Xingu River, about six miles (close to 10 km) downstream from the mouth of the Ipiaçaba River, which was the tributary of the Xingu far into whose headwaters they had lived before FUNAI had persuaded them, some six months before my visit, to live on the Xingu, a little closer to civilization. There were at that time only fifty-seven Assurini, thirty-eight females and nineteen males. They are much more numerous than that now. At that time, like so many other Indian villages in the Xingu basin, theirs was situated atop what had been another village, long ago. The ground in and around the village was pregnant with potsherds—the people say after a heavy downpour, the potsherds are especially abundant on the surface. They call these material vestiges of a past culture *Maï me'e* ("spirit things"), since they claim to have no knowledge that any humans made them, not even their own ancestors. And the Assurini are famous in the Brazilian Amazon for their beautiful pottery, which is lacquered and painted in black with elaborate bird and animal motifs. I picked up one nearly complete bowl with three double upside-down black *V*'s painted into the inside of it; my interlocutor said that wasn't their own people who had made that pot, since they have since the beginning of time only painted their earthenware on the outside. At that point, I decided to return later in the year and do a forest inventory to see if the remains of a culture had left its mark on the vegetation in that area.

I returned to the Assurini in June 1986 and carried out that forest inventory. The forest contained markers of past human occupation such as babaçu palms, tucumã palms, Brazil nut trees, and others. It was another grove of fruit trees that to the untrained eye seemed like pristine tropical rainforest, imposing and forbidding, perhaps untouched by human hands. The inventory was six-tenths of a mile long and quite narrow, only some thirty-two feet wide, yielding a hectare (i.e., about 2.2 acres). So many vines grew in this forest that the transect we cut looked more like a tunnel than a trail. One tended to be surrounded by vines, and we walked on a mat of them and supported ourselves by leaning from time to time on walls made of them. It is not surprising to see a profusion of vines and bramble in areas long ago cleared for cultivation. It even happens in temperate zones of the world. The bloody, week-long Battle of the Wilderness in Virginia, 1864, between Grant and Lee during the Civil War, was fought in a vast bramble patch and vine forest in which it was hard to see the enemy at anything except close range. That "Wilderness," as the area was known by locals, was actually land that had originally been put to the plow, former farmland. True wilderness usually affords better views than a vine forest, when one is inside it. In the Amazon, palms, such as babaçu, are among the trees that tend to grow well in a vine forest. That is because they lack branches and usually have smooth barks. Whereas a heavy load of vines can bring a magnificent Brazil nut tree crashing to the ground in a matter of seconds, the vines are usually unable to sustain themselves structurally on straight-growing palms such as babaçu. There is another reason babaçu can do well and become common in such milieus, which has to do with how the seed germinates. By far most forest tree seeds germinate near the surface and then grow upward toward the life-giving source of forest energy, sunlight. But babaçu palm seeds germinate underground and then the young stem at first grows downward deeper into the soil. Anthony Anderson, a researcher at the Goeldi Museum in 1986, had shown that babaçu palm seedlings could live underground for up to seven years after germination (Anderson 1983; Anderson and Anderson 1983; Anderson et al. 1991). Babaçu palms are often common in forests once cleared for slash-and-burn agriculture. That is because when the felled vegetation is fired by farmers, the seedlings of most species are killed; underground babaçu seedlings will turn and grow upwards, breaking the surface long after the ground has cooled and the farmers have moved on. In such a way of germination and early growth, babaçu has been called a pyrophyte (a fire-loving plant). Actually, it was especially well-adapted to humans and their agriculture, and it must have flourished more after agriculture began in the Amazon, perhaps six thousand years ago, than it did before that time. One might think of it as a weed, despite its massive size, great longevity, and multitude of uses (as plant food, roofing thatch, and home of tasty, edible beetle larvae).

In the course of collecting and studying the trees and vines of the Assurini forest, I took five soil samples along the tunnel transect through the vines. Two of these samples were what the Assurini call *iwi-una* ("black earth"). Indeed, the soils were charcoal black, here in the middle of what has often been considered to be a wasteland of poor soils. These charcoal black soils, called *terra preta do índio* ("Indian black earth"), are manmade, the result of many years of forest burning for agriculture, and they long preceded the arrival of the Luso-Brazilians (Woods and McCann 1999). I had once again found myself in a forest where nature and culture could not be separated into analytic categories. The presence of the forest would not be explained without including people from the past, and the people of the present who lived there would not be there without that forest. Human history and Amazonian trees were tangled up in the Assurini land and culture of today; one could not seem to find refuge in the Amazon wilderness without being in a cultural forest. And whereas every cultural forest has its own history, its own native people of the past (who were often linguistically and ethnically distinct from the native people of the present who used that cultural forest), and its own suite of plants and animals, all share one broad feature: agriculture, or some form of significant disturbance in the forest originating in human activity, had been once practiced in each of them.

The Assurini grow sweet manioc, yams, sweet potatoes, squashes, beans, corn, and numerous fruit trees. They have a particular fondness for a variety of manioc that has a large tuber and contains a fairly sweet-tasting aqueous exudate (which nevertheless needs to be boiled before consuming). They call it *maniakáwa* (*mandiocaba* in Portuguese). The Ka'apor have the same manioc variety, which they call *maniaká,* but only eat it occasionally. The Assurini cook up porridge from it almost every day. They also grow tobacco (*petim*), smoked by shamans, and cotton (*aminiyu*), used by women for making thread for hammocks.

These are all native Amazonian crops. At the time I visited, they were also experimenting with pigeon pea (or Congo pea) (*Cajanus cajan*), a plant originally from Africa that had been introduced by Tapirapé Indians (not to be confused with the Wanderers of the Tapirapé River—these people are from a river of the same name, but it exists about two hundred and fifty miles [close to 400 km] to the south and drains into the Araguaia River). These Tapirapé Indians were working for FUNAI and have been in continuous, basically peaceful contact with Brazilian society since the 1930s (Wagley 1977). Their job was to build a canoe for the Assurini, now that the Assurini lived on a great river and could use one. The Assurini named this pigeon pea *iwira-kumaná* ("tree-bean"), that is, more idiomatically in translation, "bean tree," even though like any other bean, it grows on a vine, not a tree. I would later learn that other groups apply life form words (words such as "tree," "vine," and "herb") to domesticated plants that have been introduced, but never or almost never do they use those terms as part of the names for local domesticated plants. Calling something a "tree" or a "vine" or an "herb," which are all forest things, is a way of marking an introduced domesticate as foreign and new. One can use names like that in reverse, too, for the purpose of identifying things that had been introduced.

It was also at this time among the Assurini that I began to realize in earnest that agriculture and its thousands of years in Amazonia had actually affected the way people speak about plants grown by people versus plants that thrive on their own in the forest or in the swamp. I noted that the Assurini had three varieties of corn: *awači-pinimu* ("striped corn"), *awači-ete* ("true corn"), and *awači-yuwu* ("yellowish corn"). All these names begin, in the Assurini language, with the generic word for corn—*awači* (Balée 1989b; Balée and Moore 1991, 1994).

In Ka'apor, the corn word is *awaši;* in Tembé, a Tupí-Guaraní language that neighbors Ka'apor, it is *awači;* in Guajá, also neighbors of the Ka'apor, *wači;* in Araweté, *awaci*. These are cognates, not words borrowed among these languages. These words are all descended from the common mother tongue of these languages, Proto-

Tupí-Guaraní. The corn word in that language is reconstructed by the methods of historical linguistics as *aßati* (where *beta*—the *ß*—sounds like a *v* except the lips close together lightly [Rodrigues 1986]).

The corn word means that the cultivated maize plant existed at the time the mother tongue was spoken, now believed to have been about twenty-five hundred years ago. The modern languages—Assurini and its sister tongues—are related in the way the Romance languages Spanish, Portuguese, and French are related. While Latin is the mother language of the Romance languages, at the same time it was being spoken in Rome twenty-five hundred to two thousand years ago, Proto-Tupí-Guaraní was being spoken somewhere in Amazonia, likely in the southwest. And it was starting to spread outward, toward the coast, and south, and north toward the Amazon River itself. Assurini and its sister languages (such as Ka'apor, Tembé, Guajá, and Araweté) were not yet born, so to speak, at that time. What glued my interest to linguistics at this time, in the Assurini village, were the cognate (sister) forms of words for Amazonian crops, or at least for crops that had been in Amazonia for thousands of years, whatever their ultimate source. That is where I first started noticing and thinking about patterns with the Ka'apor and other indigenous groups of Tupí-Guaraní speakers in terms of plant nomenclature. Common, cognate forms are to be found also for cotton, gourd, squash, manioc, sweet potato, yam, and many other crops in Tupí-Guaraní languages (Chapter 6).

Agriculture and the alteration of Amazonian landscapes by human hands were not only ancient. Landscape management, together with domestication and semidomestication of a staggering array of plant species, was reflected in people's everyday speech. The concept of the garden and foods, textiles, and other useful products that came from it were embedded in subconscious linguistic processes that resulted in an extraordinarily stable vocabulary for native crops over time. This vocabulary sometimes survived temporary setbacks among individual groups, when for one reason or another they could not maintain agriculture in practice, or even in thought.

Food for Survival

The Assurini in 1986 had only a few years before that been wanderers themselves, for a temporary and sad time. And their village of today was not of their choosing. It was the buggiest place I have ever been. The rainy season was well under way at the time of my visit, and their river edge location was conducive to swarms of blackflies (that came in three sizes there) by day and mosquitoes with preposterously long, needlelike probosci at night. Those mosquitoes quickly targeted any spot not drenched in DEETS, whereas the blackflies seemed not to care—they pursued, unapologetically, an *r*-strategy of reproduction and fitness, one might say. The Assurini were experiencing some discomfort with these bloodsucking dipterids, and indeed it was not their idea to have moved to this location in the first place. I don't know of any indigenous group who would have done that.

In this regard, I must linger on bugs and villages for a moment. The film *The*

Emerald Forest (John Boorman, director, 1985), which was otherwise a fine and sensitive portrait of the destructiveness of Amazonian development at that time, contributed to a widespread misconception that forest dwellers in the Amazon (the "Invisible People" in that film) had a penchant for living in houses built on stilts and gangplanks over stagnant water. I actually saw the muddy site of the village in the film, on the grounds of the Brazilian agricultural agency, EMBRAPA, in Belém, in 1985. In fact, native peoples in Amazonia, like folks most everywhere, assiduously avoid pest-ridden sites for villages, especially when other sites may be available. The Tikuna of the Upper Amazon are famous for living along blackwater tributaries where these discharge into the whitewater Amazon River proper, since although they utilize the whitewater swamps and floodplains for various activities, they also know that mosquitoes, which transmit malaria, yellow fever, and filariasis (which can result in a kind of elephantiasis), breed much more readily in that nutrient-rich milieu than they do in the chemically depauperate and acidic blackwater creeks nearby.

The Assurini were familiar with the Xingu River, and their ancestors had lived in that basin, between it and the Tocantins to the east, since before recorded memory (Lukesch 1976; Müller 1993). And so had their enemies, the Araweté, the Parakanã, and the northern Kayapó, with the irony here being that Parakanã and Araweté are Tupí-Guaraní languages, like Assurini, and Parakanã is reportedly especially close linguistically to Assurini. But native warfare in Amazonia has been maintained for centuries, as strangers are typically classified as enemies until proven otherwise and the slightest difference of speech is enough to cause one to fall under the foreigner rubric. There has been this tendency to Balkanization in the tropical forest, and it no doubt predates the formation of late prehistoric and early colonial chiefdoms and confederacies. Native alliances are intrinsically fragile, because of the etiquette of reciprocity and pervasive fiction of equality. Complex alliances bringing hundreds or thousands of people under their political umbrellas have been the exception in the Amazonian past.

In any event, a swampy creek mouth juxtaposed to the Xingu River is the kind of village location one ends up with when people without the knowledge of the elders are in charge of influencing and even coercing one's own people's decisions. Yet at one time some people had lived here, and perhaps with more clearing for village and garden space, and a once faster flow of water in the Xingu nearby, things might have been different, as concerns the insect pests. Some people had lived before in an earlier village, as well as in a village of the 1970s, where they were massacred by marauding Araweté warriors on, as the Assurini call it, the "Day of the Bloody Leaves." The peaceful village was taken by surprise, surrounded by Araweté, who entered the site rampaging and killing indiscriminately men, women, and children. It was an attempt at indigenous ethnic cleansing, and it was brutal, but there were Assurini survivors who regrouped in the forest. Then they came downstream and sought out help—from Italian missionaries, FUNAI, and Brazilian society itself. Some of the Araweté victors set up a village on the site of the massacre, a place now abandoned that I visited on one of my two trips to the Araweté during the 1980s. After living off forest foods, and having had to abandon their plantations of food and other crops, the Assurini got agriculture back again. People in the Amazon do not choose to give up agricultural lifestyles willingly.

Several Assurini survivors have scars from the arrow wounds they received on the Day of the Bloody Leaves, and one of these was the old woman Jacundá. She and an elderly friend, Takomãi, told me that the other survivors went for about a year wandering in the forest, afraid to make gardens for fear the clearings, fire, and smoke would alert lurking enemies as to their whereabouts. They lived off game meat (especially collared and white-lipped peccaries, agoutis, tortoises, and various pheasants). They did not and never did eat macaws, monkeys, armadillos, alligators, and electric eels, all of which were tabooed by the ancestral Assurini and their mentors in the spirit world at the beginning of time. The most important plant food, they said, during this time of hardship was the babaçu palm, and in particular, its nuts (i.e., seeds or kernels). The babaçu palm seems to be there when all else fails.

The principal plant food of the hunting-and-gathering Guajá, and a backup food for people who are temporarily forced into foraging full-time, as were the Assurini, babaçu as food remains in the store of conscious plant knowledge that becomes active in times of crisis. Whereas agricultural folks, such as the Ka'apor, often look scornfully upon babaçu eaters, such as the Guajá, they also know that if they ever needed to, they could rely on that resource for sustenance through hard times. In the case of the Guajá, they could not for a long time conceive of the good life without babaçu, so important was it in their diet, shelters, and culture. And to some extent, though it is not by any means a "food of the gods," such as was a chocolate confection in Linnaeus' conception (hence the Greek name for the genus *Theobroma,* "god-food"), it is where it grows in large stately numbers in the high ground of the rainforest a gift, albeit an unconscious one. It is a gift from prehistoric agricultural folks, not deities, who may or may not have been ancestors of the current occupants of the land. Little else is owed them except some piety, some symbolic deference to a past peoples' forest transformations.

Babaçu was common in the Xingu, and around the Assurini village in particular. So were Brazil nut trees. Brazil nut trees are in an even patchier distribution than babaçu palms. Babaçu palms are limited to the south of the Amazon River, but they occur widely in the swamps, creek margins, and old fallow forests of that region. Although Brazil nut trees occur both north and south of the Amazon River, they are clumped in discrete zones, known as *castanhais* (Brazil nut groves). Major Brazil nut groves are found in the Guiana region north of the Amazon and in the Tocantins, Xingu, and upper Madeira River basins south of it, extending from Brazil into northern Bolivia. Also, individual plantings of Brazil nut trees are to be seen in villages and towns along the upper course of the Amazon itself. Based on recent genome data and better understanding of its distribution, it now seems Brazil nut trees were a domesticated tree crop of ancient Amazonian people (Shepard and Ramirez 2011).

Very few people had that idea in the 1980s, however. The great field botanist of the Missouri Botanical Garden, Alwyn Gentry (1945–1993), casually told me at a reception in Belém for participants during an international symposium on moist tropical forests held there in October 1984 that he thought the Brazil nut tree might have been partly determined, in its patchy distribution, by humans. The idea had been briefly touched on by Adolfo Ducke, one of the most important botanists and systematists of Amazonian landscapes of the mid-twentieth century.

Ironically, in his many publications, Al Gentry had been one of the leaders of the school of thought that would explain Amazonian plant diversity solely in relation to climate, soils, latitude, and other geophysical factors (what are often called environmental gradients), not attributing any of the current distribution patterns to ancient native peoples (e.g., Gentry 1988). Had he lived, I think he would have been more receptive to a historical-ecological point of view. By the time I had reached the village of the Assurini people's former enemies (now "pacified" and at war with no one per se, but wary themselves of attacks by warrior bands of northern Kayapó), the Araweté, Al Gentry's hunch about Brazil nut trees seemed reasonable to me.

Brazil nut trees grew in the Assurini forest, which like the Araweté forest in the Xingu basin that I was also studying, was a vine forest, though not so dense with vines. The Araweté call the Brazil nut tree *ya'i*. In other Tupí-Guaraní languages spoken far away, the term is cognate, which is to say it is related. In Parintintin (which includes Tupí-Kawahib, Uru-eu-wau-wau, and Yupaú as dialects), a Tupí-Guaraní language of the Madeira River valley to the west, it's called *ña'i;* in Assurini, *ya'ɨwa;* in Wayãpi (a Tupí-Guaraní language of French Guiana), *yã*. One can hypothesize that the original word for Brazil nut in the mother tongue, called Proto-Tupí-Guaraní, was **ña'ɨß* (cf. Shepard and Ramirez 2011). Regardless of phonological details, Brazil nut trees have been around the Amazon for a long time, and they would have been in the aboriginal homeland of the Proto-Tupí-Guaraní people, just as field corn and a long list of other crops were. Because the Brazil nut tree has a patchy distribution in Amazonia, that homeland could not have been just anywhere. Most likely, it was in the southwest.

Another great field botanist, and friend of mine, João Murça Pires (1917–1994) of the Goeldi Museum, had told me in 1987 that the Brazil nut tree was a millenarian species—one of the few known in Amazonia that could live a thousand years. In fact, the oldest one known is only about five hundred years old, based on C-14 dating. In any case, the family of Brazil nut tree, Lecythidaceae, includes the oldest-lived species in the Amazon, *Cariniana micrantha,* with an estimated age of fourteen hundred years. Other massive Amazonian trees living to hundreds of years of age, according to a study carried out in an eighty-thousand-hectare forest being logged near Manaus (Chambers et al. 1998), suggest the oldest trees have high long-term growth rates. These trees include the towering coumarin tree (*Dipteryx odorata*), the large piquiá tree (*Caryocar* spp.), the massive sapote tree or *maçaranduba* (*Manilkara huberi*), and a gargantuan mulberry-family tree, *tatajuba*

(*Bagassa guianensis*). I have also recorded enormous specimens of these trees in the eastern Amazon (Balée 1994).

These are to the Amazon what giant redwoods are to California. The ancient people, the Proto-Tupí-Guaraní, could have lived near ancient Brazil nut groves, and whereas the individual trees must be different, the groves could be the same. If they did inhabit Brazil nut groves, or even transform landscapes that favored Brazil nut trees, these would have been in the Madeira River basin, since Brazil nut groves and the closest families of languages related to Tupí-Guaraní are still encountered there today. They are the languages that didn't migrate out of the Madeira basin but evolved on their own in situ from about twenty-five hundred years ago to the present.

Brazil nut trees are of great economic importance. To my Araweté consultants, whom I had first met in November 1985 (before I met the Assurini), the Brazil nut itself is considered *bïde mukaru-te* ("truly good food, fit for people to eat"). Whereas the old Brazil nut trees do not provide decent firewood, young Brazil nut trees have bark with "high tensile strength" (*eyɨ* in Araweté; *hayɨk* in Ka'apor) that when cut in strips from the tree may be used, like rope, for straps and lashing material. That kind of useful information—about a species that is both important as food and in technology—has probably not been reinvented often. Rather, like manioc flour, açaí palm syrup, and tapioca, Brazil nuts in people's diets have been around for thousands of years. But they may have been less important as food before the dawning of agriculture in the Amazon. That is because where Brazil nut groves are found today, so too does one encounter evidence of long-term, ancient human occupation.

It is common in the vine forests around the Araweté village to see an enormous, solitary Brazil nut tree, with no seedlings growing nearby. A Swiss botanist working at the Goeldi Museum in the first decade of the twentieth century, Jacques Huber (1909), had talked to some Tembé Indians near Belém about Brazil nut trees. They told him that agoutis "planted" the trees. Huber hypothesized from this indigenous knowledge that agoutis, which are seed-scattering and seed-hoarding rodents the size of house cats, would bury caches of seeds, including caches of viable Brazil nuts, across relatively limited areas of forest. Over time, an agouti would find some of its seed caches and eat the contents. But because of its small, rodent brain, it would "forget" the location of all its Brazil nut caches. These forgotten caches were the source, according to Huber, of new Brazil nut trees, which would spring forth thanks to rodent forgetfulness. That may be part of the reason seedlings of Brazil nut trees are rarely if ever found beneath the presumable mother trees in the forest.

It is also true, however, as modern planters know, that the woody shell of Brazil nuts must be cut with a sharp knife down to the endosperm so that the seed can sprout. Perhaps agoutis' sharp incisors (which are also used by many Amazonian Indian groups in ritual scarification and tattooing practices) serve that purpose. On the other hand, it is likely that many Brazil nut groves were planted by the ancients. Brazil nut groves are strictly associated with a past human presence (Shepard and Ramirez 2011). But I had no clue on that in 1986 other than Gentry's cryptic comment spoken at a cocktail party, that is, until I studied the habitat of the Araweté.

The Igarapé Ipixuna is well named. *I* means stream and *pixuna* means black, in Língua Geral Amazônica (General Amazon Language), the trade and creole language based on ancient Tupinambá. Tupinambá was the language first encountered by the Portuguese along the coast of Brazil in 1500. It is also in the Tupí-Guaraní language family. Many place names in Brazil today are derived from it, and from Língua Geral Paulista (a Tupian creole spoken in southern Brazil for a couple hundred years).

The Ipixuna is a blackwater creek that flows into the right bank of the Xingu River, some forty-five miles (about 72 km) south of the current Assurini village. The Araweté village is located on the left bank of the Ipixuna about twenty miles (about 32 km) upstream from its mouth. I took a high-speed aluminum boat up the Ipixuna the second time I visited the Araweté and it was then, on the way, that I stopped briefly in the Assurini village for the first time. The first time I visited the Araweté, in November 1985, I had arrived in a single-engine Piper on the airstrip built by FUNAI near the village—I came out of the plane to be greeted by a colorful, mostly reddened throng of men, women, and children. I say mostly reddened because the skirts of the women are dyed red with urucu pigment and most of the people dye their hair red in it, too. The Araweté have a Hollywood sense of color, and seem to maximize it, for the redness of the people and their clothing stands out but does not conflict with the greenness of the forest (most of the time) all around them. That first time, I carried out a forest inventory of one hectare of vine forest, just upstream from the Araweté village. The second time, in March 1986, coming by boat when the water was high, I came to check earlier determinations I had made of trees on that plot, to re-collect some I had missed before, and to carry out a more general study of the environs. One objective was to determine what, if any, presence there had been of earlier people in the area, and whether this could be determined by looking at the distribution of plants in the forest.

Another Reality

Late in March 1986, I was traveling with two young, recently married Araweté men in a heavy wooden canoe equipped with an outboard engine. Kamaracī (whose name means "white Brazilian") was at the tiller and Tatuavī ("giant armadillo") took the prow. I sat in the center seat as we made our way slowly against the current of the Ipixuna. Like the river into which it drains, the Xingu, the blackwater creek called the Ipixuna ran its course in a relatively fixed channel, thanks to granitic outcrops throughout. That fixity is unlike the Amazon River and others of the region that meander, like the Mississippi River, changing course from time to time, and drowning islands that in previous years had sustained agriculture and even settlements. The Ipixuna, in contrast, tended to be predictable. We were going upstream to the old village of the Araweté; it was also the old village of the Assurini. It was where the Assurini were living on the Day of the Bloody Leaves. After the attack, some victorious Araweté settled there; all of them, from diverse villages and camps, would eventually settle there in about 1977, after they were in peaceful contact with FUNAI. Then there was an attack by the Parakanã raiders from the south in 1982, in which a FUNAI agent was shot and nearly killed by a Parakanã arrow. He had decided to be Roman like the Romans, or go a little bit native, and was dressed quite convincingly as an Araweté man, replete with beaded and feathered ear piercings and all.

After that incident, FUNAI agents moved the Araweté downstream on the Ipixuna to their present location, to get them out of harm's way. That meant moving farther north; the stream flows south to north, as does the Xingu itself. But the village site of the Araweté people's earlier choosing, the one they visited first on the Day of the Bloody Leaves, when they uprooted the Assurini, was where we were heading this morning. After navigating the current of the Ipixuna for several hours, we came to a violent stretch of rapids. It was a time one remembers distinctly, including the sights as well as the sounds of the moment, since the relatively dull daily routines of village life (and those of the ethnographer) are occasionally punctuated by a few seconds or minutes of intense, authentic danger. This was one of those moments. The only way to get past the rapids was to pull the canoe along the extreme edge of the left bank, in a shallow channel of fast-moving, chilly water. The bottom of that channel was granite; a soft, slippery green bryophyte grew spread out over the rocks in the water. The canoe could not be portaged because we were in the steep, rocky cut bank of the creek now. All three of us pushed and pulled the canoe, standing in chest-deep water. If one of us had let go, it would have been nearly impossible not to be pushed by the force of the water into the rocks now below us and facing us. The water was cool but seemed to be boiling—indeed, the Ka'apor word for rapids (on the Gurupi River) is *ita-pupur* ("boiling rocks"). A former director of the Goeldi Museum, Walter Egler, had been killed when his canoe capsized in rapids on the Jari River, north of the Amazon River itself, in 1961. A vague thought of what I had heard about that expedition might have passed through my mind right about then. I don't remember. I do remember that at one point, my foot slipped on the bryophytes covering the rock below and I nearly went under the canoe, but I recovered my footing quickly and held on. We laughed nervously as finally we pulled the canoe free and got above the rapids some one hundred yards distant. We got into the canoe, assuming our original stations, and the Ipixuna suddenly seemed wider and much calmer. We traveled up it for about an hour more, then we came to a swampy bay on the left bank; we traversed the little swamp and reached the shore, now covered with a thicket of vines and small trees. We hacked our way through it and climbed up the embankment, until I could see a grove of dark, green-leaved lime trees. We had made it to the old village, scene of the Day of the Bloody Leaves.

In short order, we shook free some juicy limes, cleared some debris away on the ground (to avoid being stung or bitten by some arthropod or reptile), sat down, and took a lime break. As we reviewed our little journey, I noticed that the ground we had cleared where we now sat was quite dark. Curious, I hacked away at it with my machete and picked up a clump of liberated dirt. It was loamy, charcoal black soil. Indeed, it was Indian black earth (*terra preta do índio*), which my Araweté interlocutors, Kamaracī and Tatuavī, immediately recognized as *iwi-howɨ-me'e* (literally "blue soil," but the word is understood to mean "black soil" also—Chapter 2). It is the best soil for planting corn, their caloric staple, they claim. The other alternatives are *iwi-pirɨ-me'e* ("red sand") and *iwi-ci-me'e* ("yellow clay"). Potsherds were jutting up in the dirt all around us. The Assurini and subsequently the Araweté had lived atop an ancient village site, chosen in part for its superior soil. And that was a soil that only past humans could have helped to make. By the simple fact of long-term occupation, people's garbage—burnt wood, organic refuse from meals, human wastes—all accumulate over time where people live. Imperceptibly, if enough time goes by, the soil changes in quality and color. It gets darker and darker. And it grows in organic richness—in nitrogen, carbon, calcium, phosphorus, magnesium, manganese, and other elements and exchangeable bases essential to plant growth. It becomes a fertile soil horizon of its own that subsequent occupants of the land, if they have agriculture, typically recognize as superior.

Corn, like all grasses, is highly demanding of soil nutrients. Manioc, in contrast, can perform well in relatively depauperate sand. The Araweté have little manioc; rather, they are corn growers above all. Whereas the Ka'apor have lots of manioc and no true black earth (though the soils in *taper* are darker and richer than elsewhere), the Araweté prefer to grow their corn in *iwi-howɨ-me'e* (Chapter 2).

We returned to the village that afternoon, as we had little time left for further exploration of the old village; going downstream, we shot the rapids and for the most part stayed dry. The next day, Tatuavī and I visited a large cornfield about five miles (about 9 km) downstream from the village. There I found Indian black earth to a depth of sixteen inches (40 cm). It was not uniform in its coverage of the site, though, and some surface areas only had a one-inch (2.5 cm) layer of black soil on top. Another Araweté cornfield, which was about twenty acres in extent, also had black earth in it. On the edge of the field, there was an imposing forest with many Brazil nut trees visible as emergents above the canopy mostly riddled with vines. The juxtaposition of black earth and a Brazil nut grove is not unusual, and is no doubt repeated in many other parts of Amazonia. A series of connections between nature, culture, and history make it logical. People live near Brazil nut trees and help them expand into groves. Brazil nut trees help keep people in the vicinity, because the seeds are good to eat, and long-term human occupation helps form black earth. Later generations of people can appreciate the concentration of resources, of which black soils and Brazil nuts are only two, afforded by ancient agriculturalists in the Amazon Basin.

The Araweté and the Assurini know a good village site when they see one. But it is intriguing that they don't attribute its richness to people, like them, who lived long ago. Rather, the creators of old villages, both to the Assurini and to the Araweté, are spiritual beings. Later in March 1986, I inspected a garbage pit on the edge of the Araweté village. Therein, at a depth of about two feet, and protruding from the edge of the blackened pit (called *iwi-ku*), I spotted part of a large stone axehead. The diorite stone from which it was made is not otherwise found in this vicinity, and thus it would have been a trade item imported from thousands of miles away. It was something the ancients would have needed for the purpose of felling trees, chopping firewood, and the like long ago. Kamaracī called the axehead *mai-itãpe* ("spirit's axehead"). The garbage pit itself was of black earth, and potsherds as well as recent waste products, such as tortoise shells, whose former occupants had been eaten by the local people, abounded in it. They call the potsherds "spirits' pottery." If, as the Araweté believe, they received the gift of corn from the gods, and people become godlike upon death, it is true also that people who would now be godlike had much responsibility for making this Amazonian environment of vine forests, black earth, cornfields, and Brazil nut groves. The Araweté I knew were fascinated by the concept of "other realities" (*amute*), which they associated with spirits of the dead. I would here direct the interested reader to an exhaustive account of Araweté religion by Eduardo Viveiros de Castro (1992).

The habitat the Araweté enjoyed today provided a window onto one such "other reality," that of an ancient past perhaps even before the time when there was an Araweté language, which if it could be reconstructed might not seem so strange and ghostly to the living of today after all. I then realized that I had come as close in a cognitive and conscious sense as I would ever come to those ancestors whose labor long lost had built this verdant landscape around me, as I stood inside a blackened garbage pit on the edge of a remote village in the vine forest of the Xingu River basin. I could see then that the forest and the dirt itself were cultural remains, as worthy of comprehension and study as any potsherds or stone axeheads, the usual stuff of archaeology, the principal way of getting at the Amazonian past.

Who Are "They"?

The "other" so often referred to in anthropological discourse on alterity is the people and society scrutinized in any work written by an ethnographer. Every field of scholarship utilizes an "other" concept, wherein the other is merely the object of inquiry, whether it can also have speech and be an interlocutor or not. In ethnology, it is clear too that the other possesses classifications of peoples—that is, "others"— that are at once similar but at odds with scientific, anthropological assessments of ethnic and linguistic differentiation. These classifications are intensely local and cannot be divorced from their rainforest contexts.

The Ka'apor have a host of epithets for surrounding peoples they regard as being "Other." What constitutes the principal difference is in the first place language. The Others do not speak the "good talk" (*ye'ẽ katu*) but rather they utter speech sounds that fall under the linguistic rubric of *ye'ẽ ga'i* ("wayward speech"). There is also a right/left hand dichotomy involved (*katu* when referring to one's arms indicates the right arm; *ga'i* can denote the situational aspect of the left arm—right is proper, left is deviant, associations found in diverse cultures, both in the Old World and the New). But, in another sense, the Ka'apor show knowledge of how the others are different by means of cultural formulae, specifically related to residence choice, food habits, marriage rules, dress codes, and the like. The Ka'apor denote the Canela to their south by the term *Moi-yuwar* ("snake eaters"), an allusion to the documented fact that the Canela on occasion eat anacondas (as do numerous other, but not all, Amazonian groups). It is not terribly dissimilar from the folk English terms "Frogs" for the French and "Krauts" for Germans (*kraut* is the German word for cabbage, a staple for poor Germans especially in the nineteenth century, when grain was expensive). But unlike the Ka'apor, who never eat snakes, folk English speakers sometimes eat frogs and cabbage, too. The Ka'apor call the Guajá, who are one of the last hunting-and-gathering peoples known in eastern Amazonia, "forest dwellers" (*Ka'a pehar*). Although Ka'apor villages and fields are surrounded by high forests, they do not consider themselves to be restricted to forest life because of the very fact that they do have fields and do sleep in villages, that is, cleared spaces with freestanding residences. The Guajá, in contrast, traditionally slept inside huts that involved tying living saplings together and covering these from top to bottom with babaçu palm leaves. The roof and walls are continuous; the structure is a lean-to insofar as it is not freestanding, but it depends on living vegetation, and therefore it also was not designed for long-term occupation. The term "forest dwellers" does not connote a happy life situation in the Ka'apor tongue. It is a pejorative term loaded with repugnance for people who do not live in villages. This ethnocentric bias extends to Guajá food habits.

One of my principal Ka'apor consultants on the *taper* forest that I studied near the village of Urutawi was Kakuri, a man of about forty-five years of age in the late 1980s. Kakuri and I saw an interesting plant along the edge of a trail one morning, while taking a breather on the way to the *taper* forest. As we crouched on our heels, we could make out a stand of *warumã* (*Ischnosiphon* sp.) shrubs a few feet away. The long stalks of these plants—which are reeds related to arrowroot—are used to make the manioc strainer, that is, the cylindrical basket used to squeeze out the liquid cyanide from manioc tubers, much in the manner of a Chinese finger-squeeze, only much larger. Kakuri pointed at the plants and remarked to me that when the Guajá get hungry, they will stoop to eating these. He pulled one out of the ground by the roots and pointed to the little rhizomes at its base. He bit one off, and offered the rest to me. I ate it, and the watery, starchy taste was not unpleasant. Kakuri indicated to me that the Ka'apor virtually never eat such food, but that is the nature of Guajá food, especially when they're hungry.

What struck me as intriguing, though, was Kakuri's knowledge of Guajá knowledge of edibility, and his simultaneous contempt for it. The event also suggested that if a Ka'apor hunter ever needed a morsel of food, he could find it readily on the forest floor. Ka'apor hunters can also snack on nutty-tasting rhizomes of smaller kinds of arrowroot, a preferred food, they say, of white-lipped peccaries also. You only need to wipe off the grit first; the taste vaguely resembles peanuts. I asked Kakuri about the babaçu palm—its durable fruits to the Guajá were like manioc tubers to the Ka'apor, their main source of starch and energy. The Ka'apor rarely eat babaçu fruits, but the Guajá make flour from the mesocarp and they eat the protein-rich kernels. The Araweté eat the beetle larvae that live inside rotting babaçu fruits on the ground. Yet the Guajá do not touch these.

The point came home to me dramatically if not amusingly when, as I sat on a fallen log with some Guajá informants in 1986, I popped a few, raw, into my mouth and munched down on them. I had learned to eat raw beetle larvae with the Araweté, who eat them much as some people in our society, while watching a movie, eat popcorn. Beetle larvae are a healthy snack that takes a little getting used to, since they are insects, they are eaten raw (while still decidedly alive), and they taste like creamy coconut dissolved in vegetable oil. That day as we inventoried an old fallow forest—former home to Ka'apor Indians of many years ago—I had picked up several rotten babaçu fruits that contained these larvae. As I later took the fruits out of my backpack, cracked them open on the log we were resting upon, and ate the larvae inside, my Guajá consultants looked visibly sickened. I learned then that they never eat these nutritious bugs and consider the habit to be revolting (Balée 2012, 19–20). Yet on that earlier day, with Kakuri by the trail, he had told me the Ka'apor believe that eating the beetle larvae from babaçu palm fruits is good for one's teeth, preventing caries. So the Guajá—the supposedly lowly hunter-gatherers who were historically despised, persecuted, and sometimes feared by their agricultural neighbors, whether Indian, black, or white, had something to be disgusted about, too, as concerns those other forms of human culture, that to them were alien and, at times, revolting and disheartening.

Food habits among these diverse human groups are cultural signatures, what cultural anthropologists often refer to as ethnic boundary markers. There are many more such markers; to name only a few: language or dialect, pottery-making style (if any), house-making style, clothing and body ornamentation style, feather work style, arrow and bow–making style, crop genotypes (if any), the incest taboo and the range of relations to which it applies, marriage and kinship patterns, hair tonsure patterns, leadership and political structure, sex roles, gender relations, parent/child interactions, and so on. The differences of culture in a tropical rainforest seem great when one begins to appreciate the details. But in a larger sense, with greater time perspective, we can also begin to acknowledge that these diverse peoples—Araweté, Assuriní, Guajá, Ka'apor, Tembé, and, yes, Aurê and Aurá—share in a history with common roots. They are all Tupí-Guaraní speakers, even if they can only understand a little of what each other says. In the times of the great Atlantic chiefdoms, just before the arrival of the Portuguese, there was no Tupi nation per se. But the language they speak has been around for a long time. And that language tells a story.

2
An Estimate of Anthropogenesis

Scientists and laymen alike most often perceive Amazonia as being one of the primordial cores of earthly nature. Remarkably few studies by scientists in any field have embraced the possibility that large portions of Amazonian forests manifest cultural histories. Many students of cultural ecology, which constitutes the branch of anthropology concerning relationships between human beings and the land, have assumed Amazonian forests to be the special purlieu of environmental factors as opposed to cultural ones. For culture to survive, according to this perspective, it simply adjusts to its environment. I undertake here to expose the fallacies of this reasoning and to offer an alternative point of view.

First, I show that adaptationist theories in cultural ecology ignore, for the most part, the proven capacity of indigenous Amazonians to manage and manipulate critical resources, rather than merely "adapt" themselves to preestablished limits of these. Second, I propose that one should avoid considering indigenous Amazonians as merely "responding" to natural exigencies, since they have transformed much of Amazonia over the millennia. I synthesize widely scattered data that indicate that at least 11.8 percent of the *terra firme* forest in the Brazilian Amazon alone is of archaic cultural origin. As such, one can logically argue that modern indigenous peoples who depend on, manage, and use the resources of anthropogenic forests display an adaptive orientation toward past Amazonian cultures, not merely toward Amazonian nature. In other words, one cannot reasonably explain Amazonian culture simply in reference to seemingly "natural" environments. On the contrary, culture seems to be an independent variable in explicating much of the forested Amazonian mosaic.

This chapter contains two central sections: (1) a critique of adaptationist theories in Amazonian cultural ecology to date, with specific emphasis on the naturalistic bias inherent in these, and (2) a presentation of the evidence for major anthropogenic forests of the Amazon, with emphasis on the *terra firme* of Amazonian Brazil. Discussion and conclusions follow.

Adaptationist Theories

The most problematic term in cultural ecology remains *adaptation,* which has been variously defined (cf. Alland 1975; Hames and Vickers 1983; Hays 1982; Mulder 1987). Hames and Vickers (1983) pointed out that *adaptation* often refers to efficiency in the exploitation of resources. Depending on one's point of view, such efficiency achieves equilibrium with resources, differential reproductive success (based on Darwinian fitness), group fitness (based on kin selection), or group survival and well-being (also see Mulder 1987; Sponsel 1986). Regardless of point of view, most Amazonian specialists who employ an adaptationist scheme differentiate, at least implicitly, between culture (learned behavior) and environmental conditions (sometimes referred to as the ecosystem—see Sponsel 1986). Environmental "constraints" present "challenges" and "problems" that particular human behavioral traits are presumably designed to meet and resolve (e.g., Mulder 1987). Critical environmental resources, according to this view, decline because of human exploitation or are indefinitely maintained because of human avoidance. That resource depletion rates can be mitigated and that resources can be modified, augmented, and even created by cultural practices for human welfare, however defined, are hardly scrutinized in most adaptationist theories.

In contrast, what may be called the "managerial" point of view postulates that presumably natural resources are not a priori givens, that is, unmodified by human interference, past or present. Since pre-Columbian times, cultural management and disturbance of pristine forests may have contributed to the present diversity and distribution of Neotropical forests and species (Anderson and Benson 1980; Nations and Nigh 1980; Sponsel 1986). Huber (1909) long ago noted that prehistoric Indians had disturbed great tracts of Amazonian forest for horticultural ends, a point that Sombroek (1966) later affirmed, implying that such ancient disturbances had left lasting traces in the living vegetation itself.

The point is not to reify but to reject the nature/culture dichotomy, as scientists in all areas of Amazonian studies have employed it to date. Although nature clearly influences certain cultural things, culture also recasts tangible phenomena that, to the naive student of Amazonian ecology, often appear to be wholly "natural."

Perspectives that stress human adaptation to the depletion of critical natural resources, as opposed to the management of resource depletion rates, may be suitably divided into two groups: (1) the limiting factor theory, which posits demographic, sociopolitical, and ritual adjustments to environmental soils and/or protein, and (2) the optimal foraging theory (sometimes called optimal diet breadth theory), which presupposes a human effort to maintain work efficiency (effort/yield), regardless of local, environmental depletions. Although numerous specific differences exist between these two groups of theories, both hold that indigenous peoples tend not to manage environmental resources, either by avoiding them entirely or exploiting them mercilessly.

Limiting Factor Theory

Although Julian Steward, who singlehandedly founded the school of cultural ecology, did not write of "limiting factors," he was the first to suggest, however impressionistically, that environmental soils (Steward 1949; Steward and Faron 1959) and protein (Steward and Faron 1959) may constrain the sociopolitical expressions of indigenous Amazonian groups (also see Hames and Vickers 1983). This hypothesis, as employed by Steward's intellectual heirs in one or more of its incarnations, has been criticized for being reductionist (Hames and Vickers 1983; Neitschmann 1973; Vickers 1980). Since Steward, many cultural ecologists have continued to cite soil and/or protein densities as limiting factors that help account for why so many interfluvial, *terra firme* Amazonian groups evince low population densities, endemic warfare, small and dispersed settlements, relatively short settlement occupation spans, peculiar food taboos, and "simple" sociopolitical organizations (Gross 1975, 1982; Harris 1974, 1984; Lathrap 1970; Meggers 1954, 1971; Roosevelt 1980; Ross 1978). Invariably, the evidence used both to support and refute this hypothesis is indirect.

Protein Limitation. Replying to Chagnon and Hames (1979), whose data suggested that many indigenous groups of Amazonia consume well above minimal protein requirements, Gross (1982) pointed out that these data, in fact, merely support the contention that indigenous groups have adapted to otherwise low environmental protein densities. In other words, Amazonian polities and societies are, according to this view, constrained by strictly autochthonous, environmental facts.

Using this axiom, Ross (1978; cf. McDonald 1977) argued that Achuarä-Jívaro taboos on large, unconfiding game, such as tapirs, deer, and capybaras, contribute to hunting effectiveness and help maintain local densities of these species, which presumably would be most threatened by hunting pressure. This hypothesis, however, appears to lack general applicability to Amazonia. In contrast to the Achuarä, for example, the Ka'apor Indians of eastern Amazonia taboo only the capybara among these large game (Ribeiro 1976); mammalogists Eisenberg et al. (1984, 200) showed that the capybara, because of its prolific reproductive capacity, is a species "that can sustain high predation levels and expand into new habitats." Instead of tabooing other large game, such as deer, tapirs, and peccaries, Ka'apor culture ritually prescribes hunters to capture yellow-footed tortoises, which are small game, for their menstruating and postpartum women. Over the long term, compliance with this food prescription seems to rhythmically focus hunting far away from settlements and swiddens (Balée 1985). This is because tortoises are rapidly hunted out locally while women continue to menstruate and reproduce. Thus, large game, such as deer and peccaries, which are important in the Ka'apor diet, and which are attracted to plants found in settlements and swiddens, are not so quickly depleted in the local habitat, as these probably would be were there no such controls. Ross (1978) suggested that the Achuarä-Jívaro avoid most major faunal resources: in contrast, the Ka'apor seem to be managing these.

Cultural effects on local fauna preclude direct evidence for low environmental protein in Amazonia. No reasonable estimate of the standing crop of animal protein in "undisturbed" Amazonian environments appears to exist (Emmons 1984; Redford and Robinson 1987). The earliest well-documented measure (Fittkau and Klinge 1973) of the standing crop animal biomass in Amazonia was taken in a region that had been hunted out not by Indians but by non-Indians, mostly hailing from the nearby city of Manaus (Beckerman 1979). In contrast, Glanz (1982) indicated that many large herbivores on Barro Colorado Island, Panama, seem to be unusually abundant. But he also noted that such seeming abundance may be misleading, since the current protection of these animals from hunting may induce them to be more confiding than animals in unprotected regions. In other words, local cultural factors in the Neotropics interfere in the measurement of "environmental" protein. Until they weigh cultural factors, it is highly unlikely that cultural ecologists can resolve the question of protein limitation in any Amazonian region.

Soil Limitation. Meggers (1954, 1971) most effectively made the case: the commonly nutrient-poor oxisols and ultisols that cover the great majority of Amazonian *terra firme* limit successive cultivation and hence habitation and population growth in typical locales over the long term. Roosevelt (1980) perceived poor soils, in addition to sparse game and only seasonally productive fishing, as an overall limiting factor that helped explicate low population densities, small and dispersed settlements, trekking, short settlement occupation spans, and simple sociopolitical organizations (i.e., the absence of ranked and state level societies) in the Amazonian interfluves.

Yet the question of poor soils as a limiting factor remains as problematic as that of protein, partly since the empirical data on which this hypothesis is based are exceedingly sparse. Although about 75 percent of the Amazonian soils appear to be nutrient-poor ultisols and oxisols (Cochrane and Sanchez 1982; cf. Moran 1989), which supposedly place severe limits on successive crop production, absolute fertility measures for staple crops seem to depend partly on human manipulations. Even were one to assume that Amazonian soils constituted an absolute limiting factor on indigenous populations and polities (as opposed to environmental protein, for example), one would have to ignore the at least moderately fertile alfisols, mollisols, and vertisols (Hill and Moran 1983; Nicholaides et al. 1983), which account for about 8 percent of all Amazonian soils. One would also have to exclude considering the extremely fertile anthropogenic soils—Indian black earth, or *terra preta*—which are scattered throughout the Amazon Basin and the exact areal extent of which is unknown (Beckerman 1979; Eden et al. 1984; Smith 1980).

As Lathrap (1970) noted, Amazonian soils are only fertile or infertile in terms of specific crop nutrient demands; indigenous groups depending on swidden horticulture tend to possess a wide variety of crops, which they manipulate depending on specific soil conditions. Indigenous cultures, moreover, are eminently equipped to manipulate the soil itself (Hecht and Posey 1989). For example, Carneiro (1983) showed that the Kuikuru and other groups of the upper Xingu River mound soil, which increases friability. Other indigenous groups protect woody pioneer species, such as *Trema micrantha* (Ulmaceae) and some species of *Inga* (Fabaceae),[1] which appear to fix soil nitrogen (e.g., Denevan et al. 1984; Frikel 1959; Hecht and Posey 1989).

No direct links between supposedly poor soils and/or sparse environmental protein and the indigenous demographic, sociopolitical variables, which these presumably limit, have been demonstrated yet. Extremely low population densities in the

Amazonian interfluves (at a modern average of 0.2 persons/km² — Denevan 1976; Hames 1983) may represent artifacts of introduced disease (Beckerman 1979) rather than demographic adjustments to environmental resources. It may even be incorrect to conceive of uncontacted Indians as virgin-soil populations with respect to Old World diseases. Charles Wagley (1977) wrote that the Tapirapé, prior to experiencing contact with Brazilian society, suffered from Old World diseases they had evidently acquired from the neighboring Karajá, who did associate with Brazilians.

In a detailed study of the demography of the Yanomamö, one of the largest, least acculturated interfluvial groups in Amazonia, it could not be demonstrated that all study groups were, in fact, unexposed to Old World diseases (Melancon 1982). Although the Yanomamö (who like most interfluvial groups seem to average 0.2 persons/km², with notable exceptions in the Parima River highlands — Hames 1983; Smole 1976) and other Amazonian groups practice infanticide, it is unclear whether this controls population growth. Selective female infanticide would suggest a cultural curb on population growth rates (Divale and Harris 1976; Harris 1974; Neel 1977) but statistics on the rate of female versus male infanticide are unknown, probably in part because Yanomamö women are reluctant to discuss infanticide of any sort with outsiders (Melancon 1982). These objections to a cultural curb on population growth rates may not explain the highly skewed male/female sex ratios (more males to females) in the Yanomamö junior age groups (Divale and Harris 1976; Harris 1974; Neel 1977), but other possible explanations, such as genetic factors and/or sampling error, do merit further investigation (Albert 1985; Melancon 1982).

It is unlikely that population densities in the aboriginal, predisease interfluves were as low as they are at present; in interfluvial zones of *terra preta,* such as many of those in the basin of the Xingu River, aboriginal population densities were probably considerably higher than they are in modern times (Smith 1980). Indeed, the principal modern groups of the region — the Xikrin-Kayapó, Assurini, and Araweté — have all been reduced by Old World diseases (Müller 1985; Vidal 1977; Viveiros de Castro 1986).

Small and dispersed settlements, short settlement occupation spans, and trekking have also been seen to be adaptive "responses" to environmental limiting factors in the interfluves, but a direct link has yet to be established. Just as indigenous population densities were far greater in some interfluves in pre-Columbian times, so were individual settlements, as has been shown for several Gê groups (Gross 1979; Moreira Neto 1980; Posey 1979b). Settlement occupation spans seem to be conditioned not merely to general soil and/or protein depletion but to numerous other factors, including endemic warfare, epidemic disease, and the expansion of national frontiers (Balée 1984; Beckerman 1979, 1980; Vickers 1980). These factors are often not only post-Columbian, but they exhibit cultural origins as well.

The objections here raised to limiting factor theory should not be read to mean that Amazonian soil and protein resources are inexhaustible. On the contrary, given the importance of these to human welfare, indigenous groups may directly manage the availability, specific compositions, and distribution of these resources.

Optimal Foraging Theory

Optimal foraging theory represents another significant adaptationist theory that is currently being employed in Amazonian cultural ecology. Its practitioners (e.g., Beckerman 1983; Hames 1980; Hames and Vickers 1983; Hill and Hawkes 1983; Vickers 1980) utilize linear programming models to predict what choices of prey a predator will make in given patches and to evaluate the predator's hunting efficiency. With the possible exception of the work of Hill and Hawkes (1983) on the foraging, non-horticultural Aché of Paraguay, the theory as it has been used in Amazonia diverges from the original as applied to nonhuman animal populations in one critical assumption: the populations to be studied are foragers who exploit natural habitats (e.g., Charnov 1976; Charnov et al. 1976; Pulliam 1974). As optimal foraging theorists themselves would readily admit, of course, most Amazonian groups are not mere foragers; they also obtain substantial proportions of their calories from swidden horticulture. This means that portions of the habitat exploited by these groups will have been affected in noticeable ways by cultural activity.

In a series of publications, Raymond Hames and William Vickers (Hames 1980; Hames and Vickers 1983; Vickers 1980) applied optimal foraging theory to the horticultural Yanomamö, Ye'kwana, and Siona-Secoya. Their findings suggest that as people hunt out large game (which usually has a higher caloric yield/effort ratio than small game, all else being equal) from a given patch, small game species, which previously had been neglected, become increasingly important in the hunt. Hames and Vickers argued that as a settlement grows older, large game are taken usually far from the settlement whereas small game are captured mainly near the settlement. As such, their conclusions broadly reflect the predictions of optimal foraging theory.

Yet this hypothesis seems to pay undue attention to post-hoc behavioral responses. Hunting patch rotation and widening of diet breadth (to include previously neglected small game) are seen to occur only *after* the depletion of large game. The hidden assumption is that the Yanomamö, Ye'kwana, and Siona-Secoya make no attempt to control the harvest of large game in any zone. They, like the Bari Indians to Beckerman (1983), are seen to take an opportunistic, "carpe diem" approach to critical environmental resources.

In Hames' (1980) view, the shifting of hunting patches by the Yanomamö is analogous to shifting cultivation. In other words, just as successive cultivation of a patch is presumably impractical (given technoenvironmental constraints), so is the successive hunting of a given patch. On the other hand, just as garden lands in Amazonia tend to be productive of food and other useful items even after "abandonment" (Balée and Gély 1989; Denevan et al. 1984; Hecht and Posey 1989; Irvine 1989; Posey 1984a, 1985a), hunting patches may also remain productive of meat, assuming proper management. The game productivity of Yanomamö and Ye'kwana hunting patches seems not to be a mere result of hunting effort and nearness to settlement of those patches over time. Hames (1980) himself showed that garden land and "abandoned" garden lands, which are presumably no farther from the Yanomamö/Ye'kwana settlement that he studied than other hunting patches, are more productive of game than supposedly primary *terra firme* forest. Optimal foraging theory, as it has been formulated up to the present, therefore, evades a significant cultural fact: the human predator can not only hunt out a patch but also can manipulate and increase the resource availability of certain patches. Most Amazonian Indians are not mere resource foragers. They are resource managers.

I would not deny the major stimulus to research and ecological thought that these adaptationist theories—the limiting factor theory and the optimal foraging theory—have given. In fact, I endorse one hypothesis that both share, namely, that indigenous Amazonians may optimize yield/effort ratios in their subsistence endeavors; one objective of all ecological research in Amazonia should be to test this hypothesis. I would merely recall Street's (1969) point that carrying capacity is difficult to determine, because technology is no a priori constant. Just as changes in technology can improve (or diminish) yield/effort ratios in human work, so can human manipulation of another factor, often assumed to be a constant one—the environment.

Anthropogenic Forests

Indigenous cultures had a significant impact, apparently, on the distribution of forest types and vegetation in contemporary Amazonia (Sponsel 1986). The classical evidence for prehistoric manipulations of the land comes from raised, drained, ridged, ditched, and mounded fields in Amazonian Bolivia, Colombia, Venezuela, and the Guianas, which seem to represent responses to poor drainage (Denevan 1966; Denevan and Zucchi 1978). Other observations of past manipulations of Amazonian environments refer to extensive trenches in the basin of the upper Xingu, which may have been defensive fortifications (Dole 1961/1962; Oberg 1953). A small but growing body of evidence indicates that other, perhaps less dramatic, cultural manipulations of the environment also occurred in Amazonia in the remote past, the effects of which are still observable.

Problems in identifying these manipulations concern distinguishing between horticultural and natural fires in the archaeological record and in estimating the time that elapses before a burned forest recovers. It should be noted that Indians burn forest for a variety of effects and that forest recovery takes place much more rapidly if burning does not occur, since burning destroys the ability of tropical trees to sprout (Prance 1975; Uhl 1982; Uhl and Buschbacher 1985). Although the study of forest recovery is in its infancy, it seems likely that such recovery tends to require far longer than the once frequently cited figure of one hundred years (e.g., Goodland and Irwin 1975), especially if a seed bank is not nearby. Recent research (Saldarriaga and West 1986; Saldarriaga et al. 1986), for example, showed that several plots of *terra firme* dense forest in the upper Negro River basin required, on average, more than two hundred years to recover after burning, based on radiocarbon dates. Although the oldest fires (more than 6000 years B.P.) may have been due to lightning strikes, the more recent dates (3750 years B.P. to the present), coinciding with potsherds, could easily have been produced by human beings (Saldarriaga and West 1986). Uhl (1982) astutely hypothesized that long-disturbed sites (which would have been common in a pre-Columbian Amazonia with far denser indigenous populations than today—Beckerman 1979; Denevan 1976) may never return to pristine forest. Hence, many vegetational associations heretofore considered to be "natural" may, in fact, represent arrested successional forest on archaeological sites, including prehistoric swiddens as well as settlements and camps.

The possible anthropogenic forests I now broach encompass some palm forests, bamboo forests, Brazil nut forests, forest islands of the Central Brazilian Shield, low *caatinga,* liana forests, and three lesser formations (under the subheading "other forests").

Palm Forests

Palms (Arecaceae) are among the most frequently noted disturbance indicators on Amazonian archaeological sites. For example, the peach palm (*Bactris gasipaes* H.B.K.), when encountered in forest, indicates prior human occupation (Balick 1984); the inajá palm (*Attalea maripa* [Aubl.] Mart.) is known to appear on sites of previous human disturbance (Pesce 1985; Schulz 1960; Wessels Boer 1965); and it has been suggested that some monospecific stands of the moriche palm (*Mauritia flexuosa* L.f.) in the Orinoco delta owe their origin to the economic activities of Indians (Heinen and Ruddle 1974). I am specifically concerned here, however, with palms that form monospecific stands or are somehow ecologically important on *terra firme.* The peach palm and inajá, although of *terra firme,* do not usually appear to be dominants; moriche palm stands, even those of cultural origin, are almost entirely confined to swamps. Palms that do seem to be prominent in the vegetational cover of undisturbed archaeological sites in the *terra firme* include *Astrocaryum vulgare* Mart., *Acrocomia aculeata* (Jacq.) Lodd. ex Mart., *Elaeis oleifera* (H.B.K.) Cortes, and *Attalea speciosa* Mart.

The vegetational cover of some of the prehistoric mounds on Marajó Island, which were excavated in the 1950s, seems to reflect prehistoric disturbance and manipulation of the "natural" environment. At site J-10 of the Ananatuba Phase (the earliest pottery-making culture on the island), Meggers and Evans (1957) found the surface thickly covered with a spiny, undetermined palm. This palm was possibly *Astrocaryum vulgare* (tucumã), which is common in some of the advanced successional forests of the Ka'apor habitat in extreme eastern Amazonia (Balée and Gély 1989; also see Roosevelt 1989). This species is also common on a prehistoric site of the Tucumã Phase of coastal Pará state (Corrêa 1985). *A. vulgare* seems to have, moreover, dispersal strategy similar to that of the babaçu palm (*Attalea speciosa*) (Kubitzki 1985), the dominance elsewhere of which is often associated with past human activity (see below). Wessels Boer (1965) described *A. vulgare* as being a species that is "never" encountered in undisturbed forest and that is a "good" indicator of previous human occupation in Suriname. *A. vulgare* (2173) fruits,[2] according to Ka'apor informants, attract tapirs and agoutis. The Ka'apor claim that these animals, especially the tapir, are dispersal agents as opposed to merely seed predators of this species. Cultural uses of *A. vulgare* include fiber for skirts, hammocks, and infant-carrying straps for the Guajá Indians (3481), and the joint between arrow shaft and steel point, made from the seed, for the Ka'apor Indians. The aboriginal inhabitants of Marajó probably also used *A. vulgare* in some way.

Other mound sites of Marajó Island are notable for the frequent presence of another palm, *Acrocomia aculeata* (Roosevelt 1989), known as *mucajá* or *macaúba*, which is commonly encountered also on *terra preta* sites near Santarém in the Lower Amazon (Anderson 1987). This species is planted as an ornamental in Suriname (Wessels Boer 1965) and is often seen in clearings in and near the city of Belém, Brazil. *A. aculeata* fruits appear to have been a very important dietary item to the foraging Hetá Indians of extreme southern Brazil (Kozák et al. 1979). The significance of palms in the vegetation of archaeological sites is not unique to Marajó Island; although the specific palm disturbance indicators may vary, palms are present on other archaeological sites in Amazonia.

Andrade (1983, 23), for example, found a "strict association" between *caiauê* (*Elaeis oleifera*), a native Amazonian oil palm with uses similar to those of its congener, *dendê*, the African oil palm widely used in Brazilian cuisine, and *terra preta*. This association was encountered at numerous sites in the basin of the Madeira River and along the left bank of the Amazon River itself, slightly downstream from Manaus.[3] Andrade (1983) suggested that there had been prehistoric cultural influence on the dispersion of these palms. Wessels Boer (1965) hypothesized that this useful species arrived in northern South America via Central America in pre-Columbian times. Other trees growing in association with *E. oleifera* at Andrade's sites include *Spondias mombin* L. (Anacardiaceae) (Andrade 1983), a light-demanding species frequently found in advanced successional forests elsewhere in Amazonia. The Ka'apor name for this species, *taperiwa'ɨ* (1006) literally means "fruit tree of advanced successional forest."

Babaçu (*Attalea speciosa*) palm forests, which cover 196,370 km² of Brazilian Amazonia (May et al. 1985), also seem to be usually an artifact of intensive disturbance and use of previously "primary" forest. The babaçu palm can dominate burned forest clearings because of its cryptogeal germination. Upon germinating, the apical meristem grows downward (rather than upward, as with most plants); upward growth occurs only years later. Burning of a forest, therefore, does not eliminate stemless babaçu palms, which remain protected underground (May et al. 1985). Although fire kills off other trees and seedlings above ground, young babaçu palms may easily emerge and subsequently dominate and/or become among the ecologically most important species in a forest (Anderson and Anderson 1983;

May et al. 1985). With an estimated life span of 184 years (Anderson 1983), babaçu is a long-lived disturbance indicator.

Although as a result of recent, increased colonization, babaçu forests are probably more extensive now than in aboriginal times, many such forests are still encountered in remote, primarily indigenous areas in the Xingu River basin, Rondônia state, and northwestern Maranhão state, Brazil (cf. May et al. 1985). Numerous enclaves of two to three hectares of babaçu forests exist in the habitat of the Ka'apor, Guajá, and Tembé Indians of northern Maranhão and eastern Pará states; these are always associated with old occupation sites (Balée 1984).

The foraging Guajá, in particular, depend on the kernels and mesocarps of babaçu (3376) for much of their dietary protein and calories (Balée, unpublished data; Mércio Gomes, unpublished report). At present, one of the only two Guajá bands (at a current population of thirty-six persons) in continuous contact with FUNAI (Brazilian National Indian Foundation) exploits babaçu forests, bamboo forests, and what seems to be otherwise "primary" forest on the upper Turiaçu River in Maranhão state. FUNAI workers recovered the remains of a ceramic manioc griddle of the Ka'apor Indians from the site of this band's current settlement. Ka'apor informants indicated to me that their ancestors had occupied this Guajá settlement as an old village site of their own on and off since well before 1928, which marked the end of hostilities between them and Brazilian society. Another babaçu forest enclave, which is situated 5 km downstream from this site, was formerly a Ka'apor settlement also. The late headman of this site, Ori-ru, and his people abandoned the settlement around 1965, because of the increasing presence of bands of Guajá, their traditional enemies; these Ka'apor now live approximately 30 km downstream on a tiny affluent of the Turiaçu River. The Ka'apor and Guajá maintained only hostile relations with each other until 1974, when FUNAI and some of the Guajá entered into peaceful contact with each other.

The Guajá historically tended to locate their camps in babaçu forest enclaves, occasionally raiding Ka'apor swiddens and settlements (Balée 1984). Ethnohistorical evidence suggests that the Guajá once lived in settled villages and practiced horticulture, but because of endemic warfare, they subsequently adopted a nomadic, foraging lifestyle (Balée 1988). Clearly, they have adapted their culture in modern times to anthropogenic forests and, as such, to the material vestiges of an earlier culture.

Bamboo Forests

The foraging Guajá also exploit stands of bamboo (*Guadua glomerata* Munro, Poaceae) (3463) for arrow points and hunting; these bamboo forests are also associated with Ka'apor settlement sites of the past and show an especially high frequency along the Igarapé Jararaca, which drains into the Gurupi River on its right bank, and in the Pindaré River basin to the south (David Oren, personal communication, 1987). Numerous Ka'apor settlements existed in the Jararaca basin until shortly after 1949, when all settlements were abandoned, partly because of a measles epidemic that eliminated one-fifth of the total Ka'apor population (Balée 1984). In the Pindaré River region, the former inhabitants of contemporary Guajá camps would have been the Guajajara Indians. Bamboo forests in the Jararaca and Pindaré regions are quite possibly a consequence of past horticultural activities by the Ka'apor and Guajajara.

Evans and Meggers (1960) reported that Wai-wai Indians of the upper Essequibo River in Guiana believed that a stand of "large-diameter bamboolike cane" (probably *Guadua glomerata*) had been actually planted by the Taruma Indians, whose society had vanished by around 1925. Bamboo forests on *terra firme* cover about 85,000 km² of Brazilian Amazonia (Braga 1979). The evidence here presented supports Sombroek's (1966) claim that bamboo forests are anthropogenic.

Brazil Nut Forests

Forests dominated by stands of Brazil nut trees (*Bertholletia excelsa,* Lecythidaceae), called *castanhais,* occur over an area of 8,000 km² in the basin of the lower Tocantins River, near the city of Marabá (cf. Kitamura and Müller 1984). *Castanhais* of as yet unknown extent also exist in Amapá state of eastern Brazilian Amazonia, in the Jari River basin, and in the Brazilian state of Rondônia in southwestern Amazonia (Manoel Cordeiro, personal communication, 1986). Near Marabá, this forest formation frequently appears on or near *terra preta* sites, in association with archaeological remains, especially along the Itacaiunas River, an affluent of the left bank of the Tocantins (Araújo-Costa 1983; Simões and Araújo-Costa 1987; Simões et al. 1973). *B. excelsa* is a light-demanding species (Pires and Prance 1977) that tends to colonize clearings. It is dispersed mainly by agoutis (Huber 1909), which are, in turn, strongly associated with swiddens (Linares 1976), various types of successional forests, and babaçu forests (Smith 1974). The Kayapó of Gorotire actually plant *B. excelsa,* reflecting its importance as food for human beings and for the game they hunt (Anderson and Posey 1989; Posey 1985a). *B. excelsa* is, moreover, a long-lived disturbance indicator, some individuals of which live for many hundreds of years (João Murça Pires, personal communication, 1987).

Forest Islands

Although it is often difficult to determine whether anthropogenic forests are the result of aboriginal burning, planting, species protection practices, or combinations of these, direct evidence obtains for purposeful planting of forest "islands" (*apêtê*) by the Kayapó of Gorotire in their *campo/cerrado* habitat (Anderson and Posey 1989; Hecht and Posey 1989; Posey 1984a, 1984b, 1985a). The species distribution of these forest islands appears to be consonant with that of *terra firme* dense forest in other Amazonian milieus (Anderson and Posey 1989). If the Kayapó create forests in the ethnographic present, no reason exists to suppose that aboriginal manipulations of the environment, when population densities were much higher, were any less significant.

Low Caatinga

Radiocarbon dates of potsherds suggest that some of the low Amazonia *caatinga* forests, which occupy a total of about 226,000 km² in Brazilian Amazonia, may be a result of repeated burning by aboriginal groups, which arrested succession more than a thousand years ago (Anderson 1981; Pires 1973; Prance and Schubart 1977; Smith 1980; cf. Denevan 1966). Although these savanna-like forests may not be heavily exploited, except for hunting, by modern Indians, they are of great importance for any theory of anthropogenic forests, because of their relative antiquity.

Liana Forests

The Araweté and Assurini Indians inhabit liana forest in the basin of the Xingu River. Liana forest (called *cipoal* or *mata de cipó*) occupies about 100,000 km² in Brazilian Amazonia (Pires 1973), principally between the lower Tocantins and Xingu Rivers, but with outliers as far west as the Tapajós River. This formation is notable for a low to medium basal area (between 18 m² and 24 m² per hectare), high density of lianas and vines (which obstruct tree canopy formation), and a few emergent tree species (Pires and Prance 1985). Several attempts have been made to explain this forest formation by using natural variables. It was originally thought that liana forest resulted from some particular (negative) quality of the soil, even though it was encountered on *terra preta* also (Heinsdijk 1957). Brazilian farmers of the region, moreover, "consider the cipoal parts [of the forest] preferable for shifting cultivation" (Sombroek 1966, 195).

Evidence against soil as a limiting factor in liana forest formation came from Falesi (1972), who noted this forest growing on a variety of soil types, including the relatively fertile *terra roxa estruturada eutrófica* (orthoxic rhodic paleustalf). Other natural variables, such as rainfall and water table, do not seem to account for the origins of liana forest either, since liana forests "never catch fire without being felled" (Pires 1973, 187). In other words, these forests are not stressed by lack of water. The possibility that liana forest is a kind of disclimax, a consequence of millennia of human interference in "natural" successional processes, has been recently proposed (Alwyn Gentry, personal communication, 1985; Dodson 1985; Smith 1982; cf. Sombroek 1966).

In 1985–1986, I carried out two tree inventories of 1 ha each, similar to that of Boom (1986), among the Araweté and Assurini Indians. The Araweté number 156 persons in one settlement on the Igarapé Ipixuna, a small tributary of the right bank of the Xingu, which because of shoals and rapids becomes virtually unnavigable in the dry season. The Assurini have a population of fifty-five persons and occupy a site on the right bank of the Xingu itself, somewhat below the mouth of the Ipixuna. Both settlements are located in the *município* (county) of Senador José Porfírio, less than 400 km upstream from the town of Altamira. Most of the *terra firme* habitat of their region is decidedly liana forest. The most common families and genera of lianas and vines in these forests, based on my collections, include

Araceae (*Philodendron*), Bignoniaceae (*Arrabidaea, Memora*); Fabaceae (*Bauhinia, Machaerium, Mucuna, Acacia, Mimosa*); Dioscoreaceae (*Dioscorea*); and Sapindaceae (*Paullinia*).

FUNAI has maintained peaceful contact with the Assurini since 1971 and with the Araweté since 1974. The basic ethnography on the Araweté is found in Viveiros de Castro (1986) and on the Assurini in Aspelin and Coelho dos Santos (1981) and Müller (1985). The two groups speak different languages of the Tupí-Guaraní family. Until the end of belligerence between the Assurini and Brazilian society in 1971, which was facilitated by the fact that the Assurini had suffered a major defeat and loss of population resulting from an Araweté raid shortly before, the Assurini and Araweté were at war.

The Assurini and Araweté inhabit prehistoric sites. At both settlements, one finds potsherds of the relatively simple Tupí-Guaraní tradition,[4] which is characteristic of the Brazilian littoral. To a much lesser degree, one also encounters sherds of the often incised, *caraipé* (*Licania* spp., Chrysobalanaceae) tempered, decorated pottery so typical of the well-known Amazonian traditions, such as Tapajoara and Marajoara. The different ceramic traditions and their rough distributions seem to accord with findings near Altamira, the Fresco River, the Itacaiunas River, and the Tocantins River (Araújo-Costa 1983; Palmatary 1960; Simões et al. 1973; Smith 1980). I have also recovered andesite axeheads from the Araweté site that are quite similar to some of those found in the basin of the Fresco River (Carapanã Phase), almost 400 km upstream from the Assurini settlement and 300 km upstream from the Araweté settlement (F. Araújo-Costa, personal communication, 1986; Simões et al. 1973). Simões and Araújo-Costa (Simões and Lopes 1987) reported C-14 dates from the Itacaiunas and Tocantins archaeological materials of A.D. 1000 and A.D. 1580. The antiquity and great areal extent of these cultural remains, which to a fair degree correspond with the principal distribution of liana forest, suggest that two major indigenous societies, perhaps paramount chiefdoms, occupied these regions in prehistoric times.

Terra preta is encountered in both the Araweté and the Assurini settlements as well as in nearby swiddens. I collected *terra preta* to a depth of 39 cm in the center of an Araweté maize field. The results of chemical analysis of this sample are shown in Table 2.1 in comparison with the aggregate results of Smith's (1980) study of numerous *terra preta* sites along the Trans-Amazonian Highway.[5] The high carbon and nitrogen content of the *terra preta* from the Araweté site indicates positive conditions for plant growth. Phosphorus at 31.8 mg/100 g at the Araweté site is relatively high, although lower than Smith's (1980) average of 40.1. According to Smith (1980), high phosphorus content is due to animal bones and shells, feces, urine, and ash from many ancient fires.

Calcium levels at the Araweté site average 10.9 mg/100 g of soil, which are considerably higher than Smith's average of 4.7 mg/100 g (Smith 1980). Elevated calcium levels, as a consequence of deposits of animal bones, demonstrate high

Table 2.1. Chemical comparison of *terra preta* from Araweté maize field with average *terra preta* values from Smith

Site	% Carbon	% Nitrogen	P₂O₅ (mg/100 g)	Al (mg/100 g)	H₂O pH	Exchangeable Bases (mg/100 g) Ca	Mg	K	Sum
Araweté	2.97	0.25	31.8	0.20	5.8	10.9	1.6	0.06	12.6
Smith's sites	1.73	0.15	40.1	0.40	5.4	4.7	0.9	0.15	5.8

base saturation, a property especially propitious for maize cultivation. The Araweté dump shells of *jabotis* (*Geochelone* tortoises), which are important objects of hunting and trekking, as well as bones from other game animals and fish in pits along the edge of the settlement. The process of shell and bone accumulation in their garbage pits at present, with concomitant soil enrichment, probably occurred in a similar manner in the prehistoric past.

The pH of the Araweté sample is 5.8, slightly more than the average of 5.4 from Smith (1980), which suggests a relatively high availability of nutrients for plant growth. Aluminum levels at 0.2 mg/100 g in the Araweté sample are lower than the average of 0.4 mg/100 g (Smith 1980); levels higher than 1.0 mg/100 g can be deleterious to plant growth (Smith 1980). The carbon content of the Araweté sample at 2.97 percent is high compared to the average of 1.73 percent from Smith, implying the long-term existence of low-energy hearths (Smith 1980).

These results prove that the *terra preta* near the Araweté is of extremely high fertility. It is no coincidence that maize (*Zea mays* L., Poaceae), which many consider to be an "absolute gauge" of Amazonian soil fertility (Lathrap 1970), is grown by the Araweté on these soils. The Araweté describe *terra preta* as *iwi-howi-me'e* (lit., "blue soil"); they claim that it "makes the corn grow" and/or "makes the corn happy" (*awaci-huri-hã*). One of the four varieties of Araweté maize, *awaci-puku* (lit., "long maize") (2145) has an opaque phenotype (the spike is completely purplish-black). This variety is currently being analyzed to determine whether it contains the opaque-2 gene, which results in high lysine content, an amino acid that is usually low in maize cultivars (Mertz et al. 1964). Other indigenous maize varieties with opaque phenotypes in lowland South America—from the Kaingáng (Xokleng) and Guaraní—have been shown to possess the opaque-2 gene and high lysine content (E. Andrade, personal communication, 1987).[6] This suggests that environmental protein (i.e., from game and fish) may not be a limiting factor for the Araweté, given the possibly high protein content of their maize and large quantity of land they devote to maize cultivation.

Insofar as the average surface of Araweté swiddens covered by maize alone is about 82 percent, according to my measurements, it is likely that they exploited *terra preta* at previous settlement sites for many years prior to their contact with FUNAI; the same is also probably true of the Assurini. At the site where the Araweté last attacked the Assurini, which is now unoccupied, I measured *terra preta* to a depth of 55 cm. If 1 cm of *terra preta* represents ten years of human occupation (Smith 1980), then this site was occupied, perhaps intermittently, for 550 years. The utilization of *terra preta* by these modern groups represents an adaptation to past human activity, an adaptation to culture itself.

The pottery-making peoples who long inhabited the present sites of the Araweté and Assurini probably exploited forest within a range of 5–10 km (Balée 1985) for swidden cultivation. Tree inventories were carried out on *terra firme* liana forest 5 km from the Araweté and 4 km from the Assurini settlements. Although I did not collect charcoal samples on these sites and the soils, although dark, are not *terra preta,* it is possible that these sites were cleared for swidden cultivation in the remote past. Table 2.2 shows the ten ecologically most important tree species ≥10 cm dbh (diameter at breast height) from the Araweté plot, giving the importance value, indigenous name, and indigenous use of each; Table 2.3 shows the same information from the Assurini plot.[7]

Several of the ten most important species occurring on both plots are known to be disturbance indicators. These include *Attalea speciosa* (babaçu), *Alexa imperatricis* (Fabaceae) (see Fanshawe 1954), and *Theobroma speciosum* (Malvaceae) (see Frikel 1959). On the Araweté plot alone, two other known disturbance indicators also occur among the ten most important species: *Inga* sp. 1 (Frikel 1959; Huber 1909; Schulz 1960) and *Bertholletia excelsa* (Brazil nut tree). Species of *Trichilia* (Meliaceae) obtain among the ten most important species on both plots; in Maranhão, with the Ka'apor Indians, I have found *Trichilia* sp. to be a most important species only on plots of known successional forest (*taper* in Ka'apor—see Chapter 1). It seems likely that further research will show many other species of these liana forests to be disturbance indicators as well.

In addition to the high ecological importance of several known disturbance indicator trees on both plots, the nearby presence of stone axeheads and potsherds, and the depth of *terra preta* deposits in the general vicinity, another datum that supports prior manipulation of these liana forests concerns basal area. Both 1-ha plots exhibit very low basal areas for all trees ≥10 cm dbh (22.0 m^2 for the Araweté plot and 22.4 m^2 for the Assurini plot) (Balée and Campbell 1990). One hectare of *terra firme* dense forest, in contrast, may have a basal area of more than 40 m^2 (Pires and Prance 1985); at the same time, such low basal areas may further reflect past human disturbance of a "primary" forest (Boom 1986).

Although the Indians consider the liana forests to be "primary" forest (*ka'a-ete* in Assurini and *ka'ã-hete* in Araweté), in the sense of not having been felled in human memory,[8] one can hypothesize that these sites, if they were not prehistoric swiddens, were at least influenced by nearby burning. This is because the depth of *terra preta* at the present settlement sites suggests long periods of human occupa-

Table 2.2. Ten most ecologically important species from 1-ha forest inventory near the Araweté Indians, with importance values, local names, and uses

Species	Collection No.	Importance Value	Araweté Name	Uses to the Araweté
Alexa imperatricis (R.H. Schom.) Baill.	1634	21.83	*iwira'i*	Wood for pestle; *Cebus* monkeys eat fruit
Cenostigma macrophyllum Tul.	1680	19.73	*komani'i*	Lathe handle made from wood
Euterpe oleracea Mart.	1833	19.22	*acai'i*	Exocarp edible to people and game
Theobroma speciosum Willd. ex Spreng.	1670	8.43	*aka'awi'i*	Pulp edible to people and game; firewood
Inga sp. 1	1864	8.41	*ñā-pχkχ'i*	Pulp edible to people and game; firewood
Attalea speciosa Barb. Rodr.	1776a	8.37	*nata'i*	Edible kernels and mesocarps; agoutis and hyacinthine macaws eat kernels; roofing thatch
Acacia polyphylla DC.	1696	7.43	*tariri'i*	No reported use
Bertholletia excelsa Humb. and Bonpl.	B1770	7.07	*ya'i*	Edible seed to people and agoutis; lashing material from bark of young specimens
Sterculia pruriens (Aubl.) K. Schum.	B1740	6.71	*tapidopaimi'i*	Lashing material from bark; toucans and monkeys eat fruit
Trichilia sp. 1	B1700	6.22	*pia'i*	Excellent firewood; toucans eat fruit

tion; it is likely that the aboriginal horticulturalists of these sites would have extended their swiddens at least to distances of 5 km from their settlements (Balée 1985), that is, to ranges equidistant with the plots of liana forest I surveyed.

If we consider disturbance indicator trees and liana forests to be archaeological resources, the infrastructures of Araweté and Assurini societies thrive, therefore, on the living artifacts of long-extinct cultures. Tables 2.2 and 2.3 show the uses to the Araweté and Assurini, respectively, of the ten ecologically most impor-

Table 2.3. Ten most ecologically important species from 1-ha forest inventory near the Assurini Indians, with importance values, local names, and uses

Species	Collection No.	Importance Value	Assurini Name	Uses to the Assurini
Cenostigma macrophyllum Tul.	2135	62.23	*kumandu'ɨwa*	Cebus monkeys eat fruit; firewood
Attalea speciosa Barb. Rodr.	None	18.58	*marita'ɨwa*	Insect repellent from oil; roofing thatch; agoutis eat kernels
Alexa imperatricis (Schom.) Baill.	2391	15.95	*añigirána*	Bark used to relieve toothache; firewood
Trichilia lecointei Ducke	2323	12.12	*iwɨrapitik'ɨwa*	House post; game eat fruit; firewood
Neea oppositifolia Ruiz & Pav.	2320	10.35	*tepečikurɨp*	Tortoises eat flower; firewood
Matisia sp. 1	2428	10.11	*murure'ɨwa*	Game eat fruit; firewood
Unonopsis guatterioides (A.DC.) Fries	2324	8.52	*pina'ɨwa*	House post; lashing material from bark; game eat fruit; firewood
Theobroma speciosum Willd. ex Spreng.	2338	8.44	*akawɨwa*	Edible flower and fruit to people and monkeys; firewood
Eschweilera coriacea (DC.) S.A. Mori	2329	4.23	*iwitirɨ'ɨwa*	House post; lashing material from bark; bark used in curing skin infections; firewood
Chamaecrista xinguensis (Ducke) H.S. Irwin & Barneby	2415	4.15	*tirɨrɨ'ɨwa*	Toucans eat fruit; firewood

tant trees on each plot. Various categories of use can be discerned. These trees supply food for people (e.g., babaçu kernels and mesocarps, fruit pulps of *Theobroma speciosum*, and Brazil nuts); food for game animals (such as fruits of *Eschweilera coriacea* [Lecythidaceae] and *Euterpe oleracea* [Arecaceae], eaten by numerous species in the hunt); construction material (e.g., *Unonopsis guatterioides* [Annonaceae], the trunks of which serve as house posts, and babaçu, the fronds of which are used by both groups as roofing thatch); raw material for tools (such as *Ceno-*

stigma macrophyllum [Fabaceae], from the wood of which the Araweté make a lathe handle, and *Eschweilera coriacea* and *Sterculia pruriens* [Malvaceae], from the barks of which are obtained lashing material); medicine (e.g., *Alexa imperatricis,* the bark of which the Assurini apply directly to relieve toothache); insect repellent (the oil of babaçu kernels is used by the Assurini to repel noxious blackflies); and fuel (many of the species, such as *Trichilia* spp., supply firewood).

These are the uses of only the ecologically most important species from both plots; the number of useful species, of course, is far larger than this in either habitat for both groups. The Araweté and Assurini manifest, therefore, an extraordinarily high economic dependence on soils and forests that evidently were modified by human cultures of the past.

Other Forests

Three other possible anthropogenic forests on Amazonian *terra firme* merit discussion. These are forests of bacuri, cacao, and pequi.

In the center of the Ka'apor Indian reserve in Maranhão, I surveyed a very old occupation site, which Ka'apor informants of the nearby village of Gurupiuna call the *pakuri-ti* (bacuri grove). Potsherds are found on the surface and to an indeterminate depth at this site, which occupies approximately 4 ha. The most common tree species is bacuri (*Platonia insignis* Mart., Clusiaceae) (3026). Many individual bacuri trees attain greater than 50 cm in diameter at this site; bacuri appears to be one of the only species on this site that reaches such size. Charcoal at a depth of 55–60 cm was collected here and dated at 3690 ± 140 years B.P.[9] Although this charcoal could have been produced by an ancient lightning strike, it is likely that horticulture in this region is as old or nearly as old as the C-14 date, given the potsherds.

Although the age of the bacuri grove is unknown, bacuri as a species appears to be a long-lived disturbance indicator. Even the oldest Ka'apor of Gurupiuna simply do not know the history of this site, having called it "bacuri grove" since their earliest memories; yet, they do recognize its origins as being cultural. The bacuri fruits (a most important plant food of the Ka'apor) from this site are more phenotypically diverse than those that come from *terra firme* dense forest elsewhere in the Ka'apor reserve. The Ka'apor of Gurupiuna name three varieties of bacuri that occur on the old occupation site; in contrast, Ka'apor in other villages distinguish only one bacuri variety, which one finds in *terra firme* dense forest. Prehistoric human manipulation of bacuri may account for its present diversity at this archaeological site.

Meggers and Evans (1957, 260) found the surface of mound number 1 on the upper Anajás River of Marajó Island to be covered with mature trees, especially cacao (*Theobroma cacao* L.), "which the present inhabitants believe to be of Indian origin." It is notable not only that Meggers and Evans' informants believed these cacao trees to be of indigenous origin but also that the last Indians to inhabit the region, the Aruã, were no longer present by about 1800 (cf. Nimuendaju 1948), that is, about 150 years prior to Meggers and Evans' excavations.

Table 2.4. Areal extent of possible anthropogenic forests of Brazilian Amazonian *terra firme*

Forest Type	Estimated Area (km²)	Source of Estimate
Babaçu forest	196,370	May et al. 1985, 115
Other anthropogenic palm forests	?	—
Bamboo forest	85,000	Braga 1979, 55
Brazil nut forest near Marabá	8,000	Kitamura and Müller 1984, 8
Other Brazil nut forests	?	
Apêtê forests (forest islands)	?	
Anthropogenic low *caatinga*	?	
Liana forest	100,000	Pires 1973, 182
Other forests on *terra preta*	?	
Bacuri, cacao, pequi forests	?	
Total known area	389,370	

Finally, Myazaki (1979) reported that groves of *Caryocar villosum* (Aubl.) Pers. (as *C. butyrosum*), the pequi, which is an important dietary item for many indigenous peoples, indicates ancient settlement in the basin of the upper Xingu.

The Extent of Anthropogenic Forests

Recent research has shown that very high proportions of tree species in primary forests of *terra firme* are useful to Amazonian Indians (Balée 1986, 1987; Boom 1989; Carneiro 1978; Prance et al. 1987). This raises an adaptational question. Are so many species useful because human beings over the millennia "discovered" their uses in nature; or did the human beings tamper with successional processes by planting, transplanting, and/or protecting useful species vis-à-vis useless or harmful species (cf. Alcorn 1989)? In either case, large portions of Amazonian forests appear to exhibit the continuing effects of past human interference.

One will probably never know what the original peak distribution of anthropogenic forests of the Amazon was, since many have been already destroyed. Were one to sum the already known areas of possible anthropogenic forests, however, one would perceive that such forests account for 11.8 percent of the *terra firme* in Brazilian Amazonia (see Table 2.4).[10] The total area of anthropogenic forests in Brazilian Amazonia is no doubt much higher than this figure, given that the areas of several anthropogenic palm forests, Brazil nut forests, *apêtê* forests, and other forest types on *terra preta* remain unmeasured. Many anthropogenic forests tend to be felled and burned by colonists, either for swidden cultivation or cattle pasture. As Hilbert (1955) noted, *terra preta* is practically synonymous with *roça* ("swidden"), since it is a so highly preferred soil type for horticulture. Liana forests also harbor economically useful vestiges of the cultural past and are sometimes even felled to take advantage of these. My estimate of at least 11.8 percent of Brazilian Amazonian *terra firme* forest being of cultural origin should be read literally, with emphasis on "at least," and should be further considered as the first attempt at such an estimate.

Given anthropogenic forests of such scope, most indigenous groups, even foraging peoples, seem to exploit "natural" environments that display cultural components. Even the foraging Aché of Paraguay, who arguably were once horticultural (Clastres 1968), trek through liana forest and "large stands of bamboo" (Hill and Hawkes 1983, 143), which are probably anything but "primary forest." Anthropogenic forests do not constitute, of course, the universe of vegetation that Amazonian Indians utilize and manage. Were one to study all anthropogenic zones in indigenous Amazonia, one would have to consider also modern manipulations of the environment.

Discussion

Although the assumption that resource management practices of pre-Columbian and early post-Columbian Indians are somehow comparable to those of contemporary Indians would surely be mistaken (Roosevelt 1989), given denser populations and more complex indigenous cultural infrastructures of the past, one does note the long-term management and use of certain critical resources up to the present. Since the Amazonian Neolithic, for example, cultural influences on Amazonian environments probably altered the distribution of what are, at present, nearly universally useful resources among Amazonian indigenous groups. These resources include soils, plants, game, and fish.

A clear example of such continuity concerns hunting in and near swiddens, which has taken place since pre-Columbian times (Donkin 1985; Linares 1976). Neotropical groups that regularly hunt in swiddens include the Achuarä-Jívaro of Peru (Ross 1978); the Yanomama (Smole 1976) and Ye'kwana of Venezuela (Hames 1983); the Lacandon Maya of Mexico (Nations and Nigh 1980); the Yukpa of the Venezuelan/Colombian frontier (Ruddle 1974); the Miskito Indians of coastal Nicaragua (Neitschmann 1973); the Sirionó of Bolivia (Holmberg 1969); the Kayapó of Gorotire (Posey 1985a); the Ka'apor (Balée 1984a, 1985); and many other groups discussed in this volume and elsewhere. In addition to hunting in swiddens, some groups, such as the Karajá of the Central Brazilian Shield (Krause 1911), burned grassland to stimulate new growth that attracted game. Many of these groups explicitly recognize that swiddens, swidden/forest ecotones, and other manipulated zones lure important game species.

"Wild" peccaries (*Tayassu* spp.), which rank among the largest and most frequently taken game animals in Amazonia (Hames 1980; Kiltie 1980; Sponsel 1986), feed in and near swiddens as well as in other managed vegetational zones (Donkin 1985). "Wild" but poorly describes their habit in relation to human beings; in many Circum-Caribbean, Central American, and Amazonian cultures, peccaries were and are practically semidomesticated (Donkin 1985). Upon slaughtering anywhere from six to twenty adult white-lipped peccaries in a communal hunt, Ka'apor hunters frequently take several live juveniles home to raise. The explanation they give for this practice is that the distress cries of the young animals in captivity will alert and draw other herds of white-lips to within close range of the village.

Small and dispersed swiddens probably produced ecotonal environments that favored higher densities of peccaries, in addition to other important game species, such as tapir, deer, paca, and agouti, than would have otherwise been so. Semidomesticated animals were lured not only to domesticated crops but also to semidomesticated (and sometimes deliberately planted) tree species. Indigenous peoples often plant such species with the declared intention of baiting game animals (e.g., Posey 1984a, 1984b, 1985a). Elsewhere, Indians protect such species of plants (Denevan et al. 1984). Such game management strategies represent the opposite of what optimal foraging theorists call "exploitation depression" of prey (Charnov et al. 1976).

In addition to game management, Chernela (1982, 1989) has shown that fish-dependent populations in the Negro River basin protect flooded forests and their associated tree species, which nourish the fish and without which many important food fish could not survive in the otherwise nutrient-poor water (Gottsberger 1978; Goulding 1980). Likewise, many of the tree species, some of which proffer edible fruit for human beings, depend, in part, on ichthyochory.

Conclusions

Complex interactions between Amazonian soils, plants, and animals require equally sophisticated strategies of resource management, the explication of which is the objective of many chapters in this volume. Human manipulation of critical resources—by relocating, attracting, protecting, planting, transplanting, semidomesticating, domesticating, and using the resources—constitutes the cultural factor in Amazonian adaptations. As Alcorn (1984, 1989) noted, people manipulate not only species and other "things," but processes as well. Human adaptation to the environment is a process, too. Given past and present human management of Amazonian forests, many inhabitants thereof also adapt to and depend on cultural vestiges that are only seemingly masked by nature. Indigenous resource management strategies that have persisted since the aboriginal Amazonian past have done so against incalculable odds. As such, these strategies should be considered as being of potential value to humankind, in general.

3
Comparison of High and Fallow Forests

A scheme that pigeonholes Amazonian forests as being somehow pristine—the "wilderness" or *selvas*—has dominated the Western scientific as well as popular imagination since at least the nineteenth century. Most theories in cultural ecology tend to evade whether indigenous societies and technologies, in fact, might have transformed the Amazonian wilderness permanently. This evasion may be comprehended upon considering that demonstration of such transformation would undermine a key principle of cultural ecology, namely, environmental determinism (often euphemistically called environmental "conditioning," "possibilism," and "limiting factors") with respect to stateless societies (cf. Moran 1990a, 9–10).[1] However, these same stateless societies, with regard to Amazonia, appear to have transformed a "rich realm of nature" (apologies to David Sweet) into an even more analytically and empirically complex biocultural domain than had ever been conceived in the nineteenth-century naturalistic mentality that, to a certain extent, still prevails in ecology. At least 12 percent of the *terra firme* forests of Brazilian Amazonia appears to be anthropogenic (Balée 1989a)—that is, of a biocultural origin that would not have existed without past human interference. In addition, at least twenty-four perennial crop plants, including trees, were domesticated in Amazonia alone (Smith et al. 1991, 8; also Clement 1989). Management of domesticated plants is probably more than five thousand years old in Amazonia (e.g., Bush et al. 1989; Plucknett 1976). These and related findings (various articles in Denevan and Padoch 1988; Posey and Balée 1989), however recent, have already led to speculation concerning possible effects such past intervention by indigenous peoples may have had on regional biodiversity (qua existing numbers of plant and animal species).

Biologist Kent Redford (1991), for example, has argued that evidence for environmental intervention by past Amazonian societies suggests deleterious effects on "virgin" forests and plant and animal species therein: "These people behaved as humans do now: they did whatever they had to to feed themselves and their families"; as for modern indigenous peoples, he wrote, "They have the same capacities, desires, and, perhaps, needs to overexploit their environment as did our European ancestors" (Redford 1991, 46, 47; see insightful critique by Sponsel 1992). This speculation harbors an implicit theory of human nature (one not confined to Redford—see, e.g., Johnson 1989 and Rambo 1985), specifically, that it has been encoded to *reduce* biological and ecological diversity. It is ultimately based, however implicitly, on the doctrine of the psychic unity of humankind. In principle, human beings, regardless of displaying a diversity of sociopolitical types (such as foraging bands, horticultural villages, agricultural chiefdoms, and high civilizations dependent on intensive and/or mechanized agriculture), as is transparent in the ethnographic knowledge of the world, are seen *as a species* to be responsible for the extinction spasm of our times (e.g., Myers 1988; NSB 1989; see discussion in Balée 1992b). Such pessimism, were one to adopt a sociobiological point of view, is perhaps justified from the abysmal environmental record of biologically modern, upper Pleistocene hunters (e.g., Martin 1984). In other words, if humanity in a state of nature (i.e., not environed by domesticates) was anathema to nature's economy, subsequent generations that enjoyed the fruits of various Neolithic revolutions could not have been any more conservationist.

Even if one grudgingly admits, as did Redford (1991, 47), that some indigenous Amazonian societies of the past may have managed their heritage of natural resources, their modern descendants, in this view, are inevitably to be drawn into the vortex of expanding Western society, and to a certain degree voluntarily (also see Ribeiro 1970). This bespeaks a not so hidden metaphysic in an emerging dialogue between anthropologists and biologists. In fact, it represents a late twentieth-century version of unilinear progress, perhaps the only significant difference between L. H. Morgan's and the latter-day acculturationists' schemes being the mechanisms of such progress. For Morgan, these were tools; for the latter-day acculturationists, it is the supposedly undeniable and self-evident allure of the material abundance to be appropriated in Western society itself (the seemingly natural expansion of Western society onto indigenous lands is thus explained not as conquest, but as seduction, with organized crime conveniently put on a back burner). In any case, both theses, however empirical their predictions, rely ultimately on a metaphysical proposition, that is, in relation to the diversity of their natural, infrahuman environments, people must be everywhere the same. The issue to be evaluated is whether a comparative view of human history, with special regard to Amazonia, genuinely supports such a conclusion.

Avoiding predictions, this chapter seeks to clarify past relationships between certain indigenous societies and regional biodiversity in Amazonia. On the one hand, the findings presented here do not conflate indigenous resource management (which need not be conscious and deliberate, and which most certainly has existed in Amazonia ever since the appearance of domesticated plants) with defilement of the natural environment; such a conflation would be a mistake that, I believe, could have further disastrous implications for remaining indigenous societies in Amazonia. On the other hand, these findings should be relevant to anthropologists and biologists now undertaking research in the growing and extremely important field of restoration ecology in Amazonia. In particular, I suggest that indigenous peoples displaying an agroforestry complex may have enhanced, rather than diminished, regional biodiversity, at least concerning the domain of plants.

My specific aims are fourfold: (1) to demonstrate the existence of anthropogenic forests in the phytogeographic region known as pre-Amazonia; (2) to compare the similarities and differences between these forests and undisturbed forests of the same region; (3) to propose that these anthropogenic forests, produced by indigenous agroforestry practices of the past, represent net *increases* in plant biodiversity; and (4) to discuss the possible extent to which indigenous deliberation and ecological knowledge were involved in the formation of such forests and possibly increased plant diversity in the region.

Forests of Pre-Amazonia

The region in Maranhão state, Brazil, that is inhabited by the Ka'apor, Guajá, Tembé, and Guajajara Indians, all of whom pertain to the Tupí-Guaraní language family, has been called the pre-Amazonian forest (Daly and Prance 1989, 421; SUDAM 1976) or Amazonian Maranhão (Fróis 1953). Its limits are roughly understood to be the Gurupi River on the west; the Atlantic Ocean to the north; the upper courses of the Grajaú, Pindaré, and Gurupi Rivers on the south; and the left bank of the Mearim River on the east (Fróis 1953, 99, map). The region appears to represent an easterly extension of or penetration by the high forests drained by the Amazon River to the west (Ducke and Black 1953; Fróis 1953).

Various sources further divide pre-Amazonia into areas of savanna, palm groves (especially nearly pure stands of the babaçu palm—*Attalea speciosa* Mart.), and "high forest" (called *hiléia* or *floresta densa*) (Ducke and Black 1953; Fróis 1953; Projeto RADAM 1973; Rizzini 1963), understood, in context, to be pristine. According to one source, the high forest lies between latitudes 3°18′ S and 6°21′ S and longitudes 44°35′ E and 48°20′ W (SUDAM 1976), its southern and easternmost limits being the upper Gurupi River and the left bank of the Mearim River. Ducke and Black (1953, 6–7) were more conservative, indicating that the true high forest (*hiléia*) of Maranhão lies between the Gurupi River and the Turiaçu River and upper Pindaré River. According to Rizzini (1963), who was in agreement with Ducke and Black (1953), this region is part of the phytogeographic province called "Amazon forest." It lies within the southeastern sector of the subprovince "Tertiary Plain," which runs east from the foot of the Andes. This southeastern sector includes the basins of the lower Tocantins River; the area around Belém; lower Xingu and Gurupi Rivers; and upper Pindaré and Turiaçu Rivers (Rizzini 1963, 45 [map], 51). The part of this sector lying in Maranhão corresponds well with the present habitat of the Ka'apor, Guajá, and Tembé and is essentially definitive of the phytogeographic limits of the region under study.

At least two kinds of *terra firme* forest occur in the region: old fallow (or simply fallow) and high forest. Fallow refers to sites that were used long ago for agriculture but that are now forested. The age of disturbance is between forty and more than one hundred years ago. High forest is not necessarily "higher" (in terms of height of the canopy) than fallow; it is, rather, a forest of the *terra firme* that displays a primary character. It seems not to have been disturbed for agriculture (or, by implication, fairly large-scale fires as distinguished from incidental lightning strikes) within the past two or three hundred years, if ever. Regardless of the distinctions between these forest types, interpretations of radar and satellite imagery to date, as reflected in mapping, portray both as being high forest (*floresta densa*) (e.g., Projeto RADAM 1973). In part, this may reflect a problem in terms of scale (see Moran 1990b), but it also points to serious, perhaps epistemological, problems in the interpretation of what is pristine and what is not. Even if scale were to be considerably reduced from the usual 1:250,000 (Moran 1990b), for example, the aerial signatures of the different forest types remain to be clarified. The ground truth, as presented here, supplies the basis for a much more refined interpretation of the forests of pre-Amazonian Maranhão, which may someday be readable from and falsifiable by remote sensing (if the forests do not disappear first). This interpretation admits of substantial heterogeneity between stands of remaining forest, depending on whether these were once the object of an indigenous agroforestry complex or not.

Between 1985 and 1990, I carried out inventories of eight separate hectares of forest in the habitats of the Ka'apor, Guajá and Tembé Indians. The methods used were comparable for each of the eight hectares: (1) all trees ≥10 cm in diameter at breast height (dbh) (using a standard of 1.7 m) on each plot were measured, tagged, collected (except in a few isolated instances), and identified; (2) all plots were subdivided into forty sampling units (or subplots) of 10 m × 25 m, in order to sample relative diversity (see below); (3) all plots were narrowly rectangular in dimension, being either 10 m × 1,000 m or 20 m × 500 m; and (4) all plots were situated near indigenous villages, but none were the objects, at the time of study, of agricultural activity. These inventory methods are identical to those of many other recent phytosociological studies in Amazonia (e.g., Boom 1986; Campbell et al. 1986; Gentry 1988; Salomão et al. 1988) so the results of this study are comparable with those other studies.

The eight hectares span a linear distance of about 150 km, from the left bank of the Pindaré River to the left bank (and Pará side) of the Gurupi River, which is often considered to be part of the pre-Amazonian forest (e.g., Projeto GURUPI 1975). Analysis of these hectares permits one to calculate several important aspects of the pre-Amazonian forest, such as floristic composition, species richness, and physiognomy. For all eight hectares, and any combinations thereof, one can calculate total basal area (i.e., the number of square meters at breast height [1.7 m] occupied by individuals ≥10 cm dbh), basal area of individual species, relative frequency (the number of individuals of a species/all individuals of the plot × 100), relative diversity (the number of sampling units in which a species occurs/all occurrences of all species × 100), and relative dominance (the basal area of a species/total basal area of the plot). In addition, on the basis of these measures, one can calculate the ecological importance value for each species. The ecological importance value of a species is a derived measure involving the sum of relative frequency, relative density, and relative dominance (see Campbell et al. 1986; Greig-Smith 1983, 151; Salomão et al. 1988).

These inventories were carried out exclusively in indigenous areas, near cur-

Table 3.1. Summary of eight hectares of *terra firme* forest in pre-Amazonia

Hectare	Location	Type	Individuals	Families	Species	Basal Area (m²)
1	P.I. Awá/Rio Pindaré	Fallow	506	38	157	22.1
2	P.I. Guajá/Rio Turiaçu	Fallow	563	41	125	21.1
3	"	High forest	521	45	145	27.2
4	Urutawi/Rio Turiaçu	High forest	519	41	126	25.3
5	"	Fallow	451	36	95	30.3
6	Gurupiuna (Rio Gurupi basin)	High forest	467	43	123	30.3
7	"	Fallow	497	43	141	23.3
8	P.I. Canindé/Rio Gurupi	High forest	475	41	144	34.5
Summary				*Averages*		
		High forest	486/ha	43/ha	135/ha	29.3
		Fallow	504/ha	40/ha	130/ha	24.2
		All plots	500/ha	41/ha	132/ha	26.8

rent occupation sites of a traditional foraging people of the region (the Guajá) and of three societies that have always displayed an agroforestry complex (the Ka'apor, Tembé, and Guajajara). A synoptic description of each of the eight hectares follows (also see Table 3.1; related data appear in Balée 1992b; more complete phytosociological data on all sites appear in Balée 1994).

Hectare 1. Location: near P.I. (Posto Indígena) Awá, occupied by fifty-two Guajá Indians, within Reserva Indígena Caru, left bank of Pindaré River, approximately 46°2′ W, 3°48′ S. Forest type: fallow. The dimensions of this site were 500 m × 20 m. The site had 506 individuals, 157 species in 38 families, and a total basal area of 22.1 m². The ecologically most important species was the babaçu palm (*Attalea speciosa*). The Guajá of the site described it as *ka'a-ate* (high forest), but it is clear that the site is a fallow, based on species distributions, much surface pottery, and charcoal. This fallow would have resulted from agricultural activities of Guajajara Indians, probably more than a hundred years ago. These activities altered what was high forest on the site; a preliminary archaeological survey (shovel test) of the plot yielded (in addition to many potsherds) a large charcoal sample from the wood of *Dinizia excelsa,* an enormous fabaceous tree typically associated with high forest (Kipnis 1990).

Hectare 2. Location: near P.I. Guajá, in region occupied by thirty-six Guajá Indians, right bank of upper Turiaçu River, within Reserva Indígena Alto Turiaçu (now called Terra Indígena Alto Turiaçu: see Chapter 1), approximately 45°58′ W, 3°6′ S. Forest type: fallow. This site was 500 m × 20 m. It had 563 individuals, 125 species in 41 families, and a total basal area of 21.1 m². The ecologically most important species was also babaçu palm. This site had been a Ka'apor village in the 1940s, according to oral testimony from the Ka'apor Indians of the village of Urutawi (which is 19 km to the east) as well as from a non-Indian visitor, Major, a longtime employee of the SPI (Serviço de Proteção aos Índios, or Indian Protection Service) and later FUNAI (the Brazilian National Indian Foundation). Potsherds and remains of Ka'apor ceramic manioc griddles are found here. This was near the site where the Guajá were first peaceably (and officially) contacted by the federal government during 1973–1975. Guajá informants describe this forest as *wa'ï-'ï-tu* (babaçu grove); nearby Ka'apor informants call it either *yetahu-'ï-tï* (babaçu grove) or *taper* (fallow)—see Chapter 1.

Hectare 3. Location: about 1.5 km southeast of Hectare 2. Forest type: high forest. This site measured 10 m × 1,000 m. The site had 521 individuals, 145 species in 45 families, and a basal area of 27.2 m². The ecologically most important species was matamatá (*Eschweilera coriacea*). There was no evidence for disturbance on this site and it was classified as high forest (*ka'a-te*) by Ka'apor informants as well as by Guajá informants.

Hectare 4. Location: near village of Urutawi, in region inhabited by about eighty Ka'apor Indians, within Reserva Indígena Alto Turiaçu, on minor tributary of right bank of Turiaçu River, 19 km due east of P.I. Guajá. Forest type: high forest. This site measured 500 m × 20 m. It had 519 individuals, 126 species in 41 families, and a basal area of 25.3 m². The ecologically most important species was matamatá (*E. coriacea*). The site was classified as *ka'a-te* (high forest) by Ka'apor informants and there was no evidence of disturbance.

Hectare 5. Location: about 2 km east of Hectare 4, in region inhabited by Ka'apor Indians. Forest type: fallow. This site measured 500 m × 20 m. It had 451 individuals, 95 species in 36 families, and a total basal area of 30.3 m². The ecologically most important species was wild papaya (*Jacaratia spinosa*). This site was a Ka'apor swidden in the late 1940s to early 1950s, according to oral testimony. Of the four fallows, it is the most recent. It is classified, nevertheless, as *taper* (fallow), not *taperer* (old swidden or young secondary forest) by the Ka'apor. The basal area of 30.3 m² is artificially high because of a large number of caespitose individuals, especially the armed palm tucumã (*Astrocaryum vulgare*), which alone occupied a basal area of 5.9 m², or about 20 percent of the total basal area. As this fallow ages, it is likely that caespitose palms and multistemmed young trees (such as *Gustavia augusta*) will become less dominant and the basal area will actually diminish. Of all eight hectares, at 95 species this was least diverse—its species/area curve (not shown), unlike those of all others, is already becoming asymptotic after subplot (sampling unit) 20. This confirms the widely held assumption that the younger a fallow (or secondary forest), the lower its diversity.

Hectare 6. Location: near village of Gurupiuna, in region occupied by about 120 Ka'apor Indians, drained by tributary of the Igarapé Gurupiuna, which empties into right bank of Gurupi River, approximately 46°20' W, 2°40' S. It is about 60 km north of P.I. Guajá and exactly 280 km south-southwest of Belém (based on measured air distance). Forest type: high forest. This site measured 10 m × 1,000 m. It had 467 individuals, 123 species in 43 families, and a basal area of 30.3 m². The ecologically most important species was breu manga (*Tetragastris altissima*). The site was classified as *ka'a-te* (high forest) by the Ka'apor of Gurupiuna and no evidence for human disturbance was encountered.

Hectare 7. Location: about 2.5 km west-northwest of Hectare 6. Forest type: fallow. The site measured 10 m × 1,000 m. It had 497 individuals, 141 species in 43 families, and a basal area of 23.3 m². The ecologically most important species was bacuri (*Platonia insignis*). The site was classified alternately as *taper* (based on age of disturbance) and *pakuri-tɨ* (bacuri grove) (based on the relative abundance of the bacuri tree). The site was disturbed most likely in the 1870s, which was when the ancestral Ka'apor expelled Afro-Brazilian refugee slaves from here (Balée 1988a). Considerable quantities of potsherds and charcoal are found on this site. The oldest Ka'apor of the village of Gurupiuna remember the site from their childhood (ca. 1920s) as having been a bacuri grove. In other words, this fallow has not been significantly disturbed (certainly not by fire) probably since the Ka'apor arrived in the region in the 1870s. The importance of the site to the Ka'apor today lies in its being a source of highly prized bacuri fruits.

Hectare 8. Location: about 5 km west of P.I. Canindé, in region inhabited by about 150 Tembé Indians, left bank of Gurupi River (state of Pará), 26 km west-northwest of Hectare 7, and 254 km east-southeast of Belém (based on measured air distance). The dimensions of this site were 10 m × 1,000 m. Forest type: high forest. The site had 475 individuals, 144 species in 41 families, and a basal area of 34.5 m². The ecologically most important species was matamatá (*E. coriacea*). The Tembé classified the site as *ka'a-ete* (high forest) as did Ka'apor informants (*ka'a-te*) who accompanied me to the site. No evidence for prior human disturbance was encountered.

The high forest hectare samples (3, 4, 6, and 8) compare favorably in species diversity to other eastern Amazonian primary forests. It is clear, moreover, that any one hectare is insufficient for sampling diversity of *terra firme* forests, since the species/area curves for all hectares except one (5) are still steeply rising after one hectare (not shown). For ten *terra firme* forest one-hectare samples, using identical methods to those used here, from Breves (Marajó Island), Belém, the middle Xingu basin, and Serra dos Karajás, the average number of species per hectare is 126 (Campbell et al. 1986; Salomão et al. 1988). The three hectares of high forest I sampled in Maranhão (3, 4, and 6—excluding 8) have an average of 131 species per hectare. Adding Hectare 8, from the left bank of the Gurupi River in Pará state, this average climbs to 135 species per hectare. The total average for all eight hectares, including fallow, is 132 species per hectare (Table 3.1). It may be concluded that although pre-Amazonian forests are less species rich than upper Amazonian high forests in general (see Gentry 1988), these forests are among the richest in species diversity of eastern Amazonia.

Fallow versus High Forest

Although current interpretations of radar and satellite imagery (because of scale and/or signature problems) do not yet distinguish between fallow and high forest in pre-Amazonia, certain ground truth criteria exist for maintaining a distinction between the two. On the surface, these ground truth criteria include basal area, floristic composition, and species richness. An important physiognomic difference concerns basal area. The data from the fallow forests, with the exception of Hectare 5 (which is abnormally high in basal area because of very high frequency and abundance of large but nonwoody caespitose species—see above), show a consistently lower basal area than the high forests. The fallows, including Hectare 5, average 24.2 m^2; excluding Hectare 5, this average is 22.1 m^2. The high forest plots, in contrast, average 29.3 m^2. (These differences are not significant statistically, but a larger sample size might show significance.) It seems that fallows are typically within a range of about 18–24 m^2, whereas high forests range from 25 to 40 m^2 (Balée and Campbell 1990; Boom 1986; Pires and Prance 1985; Saldarriaga and West 1986, 364). These differences in basal area between high forests and fallow appear to be at least partly diagnostic.

In identifying a plot of forest as fallow, from a strictly botanical point of view, a great deal also typically hinges on the presence of "disturbance indicator" species. It would be inaccurate to consider disturbance indicators as synonymous with "pioneer" species (cf. Brown and Lugo 1990), since many pioneer species are short-lived. Disturbance indicators, in contrast, may also be long-lived, as with the babaçu palm and Brazil nut tree (see Balée 1989a). Many fallow species are light demanding, yet in comparison to some pioneer species (e.g., *Trema*), they are relatively shade tolerant (i.e., prosper in small light gaps) (see Denslow 1987, 441–442). As for the hectares in this sample, and for reasons possibly related to human disturbance, typical light-gap genera of the high forest, while also present in fallow, tend to be represented by different species in the fallow. For example, only one (*Cecropia sciadophylla*) of five species of *Cecropia* collected in fallow hectare inventories is also represented among the four species of *Cecropia* collected in high forest hectares; in fact, of a total of fifty-four individual *Cecropia* trees occurring on all eight hectares, merely three pertain to the common species *Cecropia sciadophylla*. For Van Steenis (1958), disturbance indicators were biological nomads. They occurred as isolated or even rare individuals in a primary forest until a disturbance, such as fire in the service of agriculture, opened space for them. Van Steenis proposed no quantitative measure to determine when a biological nomad becomes an indicator of human activity—that is, how diverse, frequent, abundant, and/or ecologically important must a nomad plant species become to qualify as an indicator of agricultural perturbation on a given plot of forest? In fact, to date, no solid measures for disturbance indicators have been proposed (see Brown and Lugo 1990).

There are good phytosociological reasons for the separation of high forest hectares from the fallow hectares and for considering these to represent two different composite forest types in pre-Amazonia. The eight hectares of fallow and high forest in this sample yield twenty-eight pairs of hectares (see Table 3.2). The similarity of these pairs can be systematically compared using the Jaccard coefficient, which is simply the number of species/total number of species in the sample (i.e, the sum of the total number of species of each plot minus the shared species) expressed as a percentage (Greig-Smith 1983, 151). On average, the coefficient of similarity for pairs of hectares of high forest/fallow is only 10.9 percent. In contrast, the average coefficient of similarity for pairs of hectares of high forest/high forest is 22.8 percent, which is very significantly higher. Also, the average coefficient of similarity for pairs of hectares of fallow/fallow is 17.2 percent which is also very significantly higher than the fallow/high forest average but not significantly lower than the high forest/high forest average.

In addition, no tendency exists for nearby plots of high forest and fallow to be more similar than more distantly separated hectares of the same type (i.e., fallow/fallow, high forest/high forest). For example, if one compares the high forest hectare near Gurupiuna (6) to the fallow hectare (7) near Gurupiuna (the two are separated by about 2.5 km), the coefficient of similarity is only 10.5 percent. Likewise, comparing the high forest hectare (3) with the fallow hectare (2) near P.I. Guajá in the Turiaçu basin (the two are separated by only about 1.5 km), the coefficient of similarity is only 11.6 percent. The Gurupiuna sites on one hand and the P.I. Guajá sites on the other are separated by an aerial distance of about 60 km. It is interesting, therefore, that the fallow hectares of P.I. Guajá and Gurupiuna (2 and 7, respectively) have a similarity coefficient of 19.3 percent, which is very significantly higher than those of two pairs of nearby hectares. The high forest hectares of Gurupiuna and P.I. Guajá (6 and 4, respectively), with a similarity coefficient of 25.1 percent, are also much more similar to each other than either is to its nearby fallow forest. As would be expected, the pairings Gurupiuna high forest/P.I. Guajá fallow (6 and 2) and P.I. Guajá high forest/Gurupiuna fallow (3 and 7) have low similarity coefficients, respectively, of 11.2 percent and 8.7 percent. In other words, it is clear that the overriding factor that accounts for divergence in floristic composition between forest stands in this eight-hectare sample of pre-Amazonia is not distance between stands, but rather past perturbation by an agroforestry complex.

When plotted on a species/area curve (best-fit curve), as separate forest parcels, the fallow forest and the high forest accumulate diversity at similar rates. These plots are organized along a continuum; the fallow curve represents increasing diversity from hectares 1, 2, 5, and 7, in that order (i.e., from the fallow plot of P.I. Awá [Pindaré River basin], through the Turiaçu River basin fallows, to the basin of the

Table 3.2. Jaccard coefficients of similarity for all twenty-eight pairs of hectares

Pairs of Hectares	No. Species in Common	Total No. Species	Coefficient of Similarity (as %)
High forest			
3 and 6	49	219	22.4
3 and 4	53	218	24.3
3 and 8	48	241	19.9
6 and 4	50	199	25.10
6 and 8	50	217	23.0
4 and 8	49	221	22.2
Fallows			
1 and 2	44	238	18.5
1 and 5	35	217	16.1
1 and 7	40	258	15.5
2 and 5	37	183	20.2
2 and 7	43	223	19.3
5 and 7	28	208	13.5
Fallow and high forest			
1 and 3	30	272	11.0
1 and 6	25	255	9.8
1 and 4	27	256	10.5
1 and 8	24	277	8.7
2 and 3	28	242	11.6
2 and 6	25	223	11.2
2 and 4	28	223	12.6
2 and 8	29	240	12.1
5 and 3	29	211	13.7
5 and 6	25	193	13.0
5 and 4	21	200	10.5
5 and 8	23	216	10.6
7 and 3	23	263	8.7
7 and 6	25	239	10.5
7 and 4	28	239	11.7
7 and 8	23	262	8.8

Gurupi River). The forty 10-m-×-25-m sampling units of each hectare are, moreover, seen as a continuum from 0 to 159, as with the high forest plots. The high forest plots are represented in this order: hectares 4, 3, 6, and 8 (i.e., from the Turiaçu to the Gurupi). Incidentally, in my survey of the forests of the Pindaré, no evidence suggests that any plot of *terra firme* was high forest. This is not altogether surprising, considering a probably much higher density of Guajajara settlements in the Pindaré than there ever was for Ka'apor settlements in the Turiaçu—this does not imply necessarily a regional extinction of high forest species, since these still occur to the north, despite disturbances.

These curves and floristic composition data support the familiar notion that for secondary forests, "within a span of 80 yr or less, the number of species approaches that of mature forests" (Brown and Lugo 1990, 6). The data from the pre-Amazonian forests, however, also show that not only does fallow "approach" high forest in species diversity, but also the plant diversity between the two forest types is statistically insignificant. This is *conceptually* very significant proof that these fallow forests represent a kind of indigenous reforestation, insofar as species richness of high forests is being replaced by an equivalently rich secondary forest through cultural mediation, although the most important species, indeed, are different between the two forest types (see below). While no evidence has been yet presented for Holocene (i.e., during and after the rise of indigenous agroforestry complexes) extinctions of fauna and flora in Amazonia, one can argue that, on the other hand, an indigenous agroforestry complex, such as that displayed by the Ka'apor, Tembé, and Guajajara, may have actually increased the abundance of certain desirable plant species.

The most astounding difference, in terms of species, concerns ecologically important species. In comparing the thirty ecologically most important species between the four fallows on one hand and the four high forest plots on the other, the two forest types share but a single species, *Eschweilera coriacea* (matamatá) (see Table 3.3; EIV = ecological importance value [sum of relative density, relative frequency, and relative dominance]). This yields a coefficient of similarity of only 1.7 percent for the thirty most important species (number of shared species [1]/ total number of species on the two plots [(30 + 30) − 1] × 100). This difference is extremely significant. The average coefficient of similarity for the thirty most important species of pairs of high forest is 16.4 percent and the average index of similarity for the thirty most important species of pairs of fallow is 11 percent; the average index of similarity for the thirty most important species of pairs of high forest/fallow, however, is consistently less than 2 percent. This permits one to conclude that the important species between fallow and high forest are very significantly and consistently different. In this quantitative sense, the thirty most important species of fallow (minus *E. coriacea*) may be considered to be indicators of disturbance; similarly, the thirty most important species of high forest (minus *E. coriacea*) may be considered to be indicators of nondisturbance. Although from this vantage point *E. coriacea* is a facultative species, in three of the four high forest hectares (3, 4, and 8) it is *the* ecologically most important species, whereas in any fallow forest hectare it does not attain a rank higher than fourteenth ecologically most important species. In fallow forests the babaçu palm (*Attalea speciosa*) is the ecologically most important species on two hectares (1 and 2), yet it does not approach the extremely high value that *E. coriacea* does on high forest plots.

For reasons that remain unclear, and that do not accord with received wisdom, the fallow forests are actually less dominated by a few species than are the high, presumably primary, forests. For example, the average total of the two ecologically most important species (with a total possible value of 300) of the four fallows is only 40.3, whereas the comparable figure for the four high forest hectares is 60.3, which is significantly higher. The data here presented offer but partial support to the statement "[a] large number of species in mature forests is due to the presence of rare species. In contrast, secondary forests are usually composed of common species" (Brown and Lugo 1990, 7). For analytical purposes, one may consider a species to be rare if it occurs only once, regardless of whether on fallow or high forest. By this criterion, the high forest has 199 species (or 59 percent of the total) that are rare; fallow forest has 139 species (or 39 percent of the total) that are rare. Both forest types, in other words, harbor significant quantities of rare species. Aside from differences in age, basal area, overall floristic composition, and ecologically important species between fallow and high forest, another difference to which I have alluded, being strictly related to utilitarian concerns, deserves mention. Fallows are indigenous orchards (e.g., Denevan et al. 1984), whether consciously planted or not. Of the thirty ecologically most important species of fallow, fourteen are significant food species for one or more of the indigenous peoples of the region, whereas for the thirty ecologically most important species of high forest, there are only six important food species. Some significant food species of the fallow include babaçu palm (*Attalea speciosa*), hog plum (*Spondias mombin*), tucumã palm (*Astrocaryum vulgare*), inajá palm (*Attalea maripa*), bacuri (*Platonia insignis*), and copal trees (*Hymenaea* sp.). These orchards exist because of the past presence of an indigenous agroforestry complex, yet that they are cognized results of this complex is doubtful.

What Do They Know and When Did They Know It?

The modern Ka'apor are, minus historical borrowings from other societies, heirs to the agroforestry complex of their Proto-Tupí-Guaraní–speaking forebears who, if only on the basis of linguistic evidence, lived in societies that were certainly associated with domesticated plants (Balée and Moore 1991; Lemle 1971). Yet the extent to which this ancient, socially transmitted expertise is cognized by Ka'apor adults today remains questionable (e.g., see Parker 1992 with respect to the Kayapó).

Rindos (1984, 99) pointed out that early agricultural human beings may have been aware, in a retrospective sense, that their dependence on plant management represented a different lifestyle from that of other, still foraging societies. This is certainly applicable to the modern Ka'apor, who readily distinguish the neighbor-

Table 3.3. Comparison of the thirty ecologically most important species from high forest (4 ha) and fallow (4 ha)

High Forest Species	EIV	Fallow Species	EIV
Eschweilera coriacea	37.83	*Jacaratia spinosa*	11.4
Lecythis idatimon	14.53	*Gustavia augusta*	10.41
Sagotia racemosa	12.67	*Attalea speciosa*	9.37
Tetragastris altissima	11.60	*Astrocaryum vulgare*	7.76
Protium trifoliolatum	7.76	*Spondias mombin*	6.53
Protium decandrum	7.07	*Neea* sp. 1	6.26
Protium pallidum	6.78	*Pisonia* sp. 2	6.25
Carapa guianensis	5.69	*Pouteria macrophylla*	5.71
Couepia guianensis	5.07	*Attalea maripa*	5.40
Pourouma minor	4.54	*Platypodium elegans*	5.02
Taralea oppositifolia	4.51	*Platonia insignis*	4.32
Mabea sp.	4.06	*Simaba cedron*	4.26
Pourouma guianensis	3.28	*Hymenaea parvifolia*	4.17
Dodecastigma integrifolium	3.10	*Trichilia quadrijuga*	4.06
Couratari guianensis	2.77	*Lecythis pisonis*	3.56
Oenocarpus distichus	2.72	*Dialium guianense*	3.32
Sterculia pruriens	2.65	*Astrocaryum gynacanthum*	3.31
Bagassa guianensis	2.65	*Eschweilera coriacea*	3.19
Cecropia obtusa	2.60	*Theobroma speciosum*	3.11
Newtonia psilostachya	2.47	*Lindackeria latifolia*	3.05
Chimarrhis turbinata	2.40	*Handroanthus impetiginosus*	2.85
Simarouba amara	2.39	*Myrciaria obscura*	2.75
Euterpe oleracea	2.37	*Neea* sp. 2	2.64
Lecythis chartacea	2.25	*Hymenaea courbaril*	2.61
Parkia pendula	2.23	*Protium heptaphyllum*	2.59
Protium polybotryum	2.22	*Tetragastris panamensis*	2.56
Apeiba echinata	2.19	*Apuleia leiocarpa*	2.53
Fusaea longifolia	2.18	*Mouriri guianensis*	2.49
Protium giganteum	2.12	*Cupania scrobiculata*	2.40
Tachigali myrmecophila	2.11	*Pouteria bilocularis*	2.34
Total EIV values	166.81		136.22

ing, foraging Guajá Indians as a people who do not "swidden" (*kupiša moú 'ɨm*) (Chapter 7). Ka'apor informants indicate that the Guajá are *purara* ("poor") because they depend mainly on babaçu nuts and other nondomesticates, instead of on the domesticated tubers, rhizomes, corms, and fruits found, for example, in Ka'apor swiddens (see Balée 1988b, 1992a for discussion of Guajá ethnobotany; also Chapter 5). The Guajá themselves readily admit "we do not plant" (*nɔɔ tum awa*). To paraphrase Rindos (1984, 99), the Ka'apor and the Guajá know who they are, as distinguished from each other, at least on the basis of their radically different means of associating with plants.

It seems unlikely, however, that they know exactly who they were. Guajá informants, for example, show no historical memory of ever having planted swidden fields, yet linguistic and other evidence suggest that their ancestors lost control of an agroforestry complex during a process of regression induced by disease, depopulation, and colonial/indigenous warfare (Balée 1992a). In the same way that Ka'apor informants cannot often "remember" grandparents' names (as with many other Amazonian societies; see Murphy 1979), do not conceptually distinguish more than a few of the many internally poisonous principles in their habitat, avoid consuming some fruits that are otherwise edible, and believe that stone axeheads in their forests are "thunder-seeds" (*tupã-ra'ĭ*) that were never used for felling trees (Balée 1988a), it cannot be said that they intend the effects of their environmental interventions to be somehow beneficial to more distant, ensuing generations of their own kind. They do not themselves, moreover, make such claims.

Although Ka'apor informants called all fallow forests in this sample *taper* and all high forests *ka'a-te,* they do not encode a succession between fallow and high forest (Balée and Gély 1989). The Araweté and Assurini, moreover, respectively called old fallow forests of their region by the terms *ka'ã-hete* and *ka'a-te,* which mean "high forest" (Balée and Campbell 1990). The Guajá of the P.I. Awá on the Pindaré River also described the fallow hectare that was inventoried (Hectare 1) as *ka'a-ate* ("high forest"). The implication is that if a site is left fallow long enough, many peoples will consider it to be high forest, not recognizing that it was once occupied by people bearing an agroforestry complex (instead of, for example, by divinities). In other words, the successional processes responsible for the formation of fallows do not appear to be cognized in many indigenous cultures.

In contrast, a very influential case has been made for incorporating indigenous knowledge into rational development schemes for Amazonia (Posey 1983a, 1984a, 1985b). This is partly based on the transparent fact that Amazonian Indians, unlike many state societies, have not, by and large, converted their forests, polluted their streams, and otherwise defiled natural ecosystems (cf. Redford 1991), yet they still exploited the flora and fauna and survived in these for thousands of years. It is also based on the notion that this custodial care of nature is deliberate, conscious, and easily elicited in speech. Specifically, in the case of the Kayapó, Anderson and Posey (1985) and Posey (1983a, 1984a, 1984b) have argued that the "forest islands" (*apêtê*) of fruit trees and other utilitarian, nondomesticated plants were actually planted partly with the *intention* of benefitting future Kayapó generations (cf. Parker 1992). Yet the principal long-term beneficiaries of forest management by the Ka'apor Indians, on the other hand, have been their traditional enemies, the foraging Guajá, who rely heavily on the palms and other disturbance indicator plants found in Ka'apor (and Guajajara/Tembé) fallows, not the Ka'apor of today (Balée 1988b, 1992a).

In spite of a universal and immediate sense that they possess an agroforestry complex, which affects the distributions of several hundred plant species, including both domesticates and semidomesticates, the Ka'apor exhibit no rationalistic knowledge concerning the remote acquisition and/or domestication of Neotropical crops found in their swiddens today. If the ancestral Ka'apor (and earlier Tupí-Guaraní forebears) had transmitted an entirely rationalistic knowledge of Amazonian plants to their descendants, one would expect this to be reflected in modern Ka'apor speech and behavior, especially concerning the origins of plants that to continue surviving and reproducing must remain under relatively constant human supervision and interference. Yet even the most significant traditional domesticates are not perceived as having been derived from any related plant, even though closely related undomesticated congenerics of these domesticates are to be found throughout the Ka'apor habitat and one need not employ a microscope to note salient, overall resemblances between them (see Grenand 1980, 43 for similar data with regard to the Tupí-Guaraní–speaking Wayãpi).

For example, although at least two species of the genus *Anacardium* (cashew) (*A. giganteum* and *A. parvifolium*) occur in mature forests of the Ka'apor habitat, both of which are closely related to their domesticated congener, cashew (*Anacardium occidentale*) (see Mitchell and Mori 1987), Ka'apor informants say domesticated cashew trees came into being when the culture hero *Ma'ir* planted branches of *yašiamɨr* (*Lecythis idatimon*), a profoundly unrelated tree in the Brazil nut family and denizen of the high forest. Sweet potatoes (*Ipomoea batatas*) are said to have first sprouted and bloomed when *Ma'ir* planted *iwɨ pu'a* ("round soil," a clayey loam), yet at least four species of nondomesticated morning glories (*Ipomoea*), closely related to sweet potatoes, are to be found in the Ka'apor habitat, usually on the edges of swiddens. Finally, it is said that manioc (*Manihot esculenta*) originated when *Ma'ir* planted the (unspecified) branches of high forest trees, yet three closely related nondomesticated species of *Manihot* are common in Ka'apor swiddens (see below). In fact, these species, as with all other known wild manioc species, can be intercrossed with *M. esculenta* (Jennings 1976, 81; Rogers and Appan 1973).

These explanations for the origins of traditional domesticates are probably associated with an unconscious lexical and cognitive dichotomy between traditional domesticates and nondomesticates in Ka'apor ethnobotany (Balée 1989b). In any case, they evince a lack of lexically encoded knowledge regarding hybridization in plants. They also indicate lack of an explicit concept of "semidomestication," which is also not encoded linguistically in Ka'apor, yet many nondomesticated species of Ka'apor swiddens and fallows occur nowhere else and are perhaps best understood as being semidomesticated; names for semidomesticates, moreover, appear to be retained at a very significantly higher rate than names for plants of the high forest (Balée and Moore 1991). In other words, although over the centuries the Ka'apor have been certainly breeders of plants (testimony to this is to be seen in numerous landraces of certain domesticates and semidomesticates in Ka'apor swiddens and fallows that are not to be found elsewhere), this was not by conscious design.

Most Ka'apor adults, to be sure, exhibit a sophisticated knowledge of the body parts and life processes of and techniques for manipulating and harvesting many individual plants; the plant lexicon, moreover, contains more than four hundred generic names for local plants (Balée 1989b). Yet numerous plants display life spans far shorter than that of an average human being. A knowledge of these plants, their habits, and requirements does not perforce imply a rationalistic (i.e., empirically falsifiable) knowledge of *long-term* ecological and successional processes, involving several human generations. Although one can live to see that, as the elders (*tamũi*) say, red brocket deer eat wild manioc leaves, no one lives long enough to observe a young swidden grow into an old fallow, that is, upwards of one hundred years since initial clearing and burning. Likewise, elsewhere in Amazonia, no one lives long enough to see yellow clay become transformed into typically deep (>70 cm) horizons of anthropogenic black earth, which, if Smith (1980) is correct, accumulates at a rate of only 1 cm per ten years. To expect such knowledge would be tantamount to ignoring significant limitations on the oral transmission of information about events, people, and places (Goody 1977). This does not deny to the Ka'apor, however, a material role in the management of their habitat for more than a hundred years (the first Ka'apor to occupy the habitat arrived in the mid-1870s; Balée 1988a). Rather, the Ka'apor agroforestry complex, as well as that of the Guajajara and Tembé, appears to be incidental to developmental processes associated with semisedentary, egalitarian Amazonian societies—it is, simply, not a product of long-term design, indigenous or otherwise.

Although some Ka'apor do plant consciously on occasion nondomesticated tree species, such as ingá, wild cacao, *jutaipororoca* (*Dialium guianense*), and wild soursop (*Annona* sp.) (Balée and Gély 1989) in their dooryard gardens, most of the ecologically important tree species of the fallow are *not* present because they were planted. When queried about fallow formation, informants strongly tend to agree that many of these species were introduced and/or dispersed by game animals attracted to human settlements. For example, agoutis disperse babaçu palm, inajá palm, and copal trees; deer disperse bacuri and hog plum; and *Cebus* monkeys disperse wild cacao and ingá, according to informants—such knowledge is readily confirmed, moreover, in the biological literature. With a few individuals of certain species, some human planting was involved (wild cacao, ingá, hog plum). Informants claim, nevertheless, that groves of these species were not entirely planted, but rather dispersed from an originally planted tree. Many palms, such as bacaba (*Oenocarpus distichus*) and inajá, are said to be present because humans first scattered (*omor*) the seeds (while throwing them away on the edge of the dooryard garden or village), but animals subsequently dispersed them. As for babaçu, one of the most important fallow species, I have found no evidence for deliberate human intervention in its life processes and distribution—the Ka'apor eat this only as snack food, do not return to the village with it, and do not plant it. Rather, its presence near long-abandoned human settlements is probably due to spread by agoutis originally attracted to the settlements because of certain plant species common to old and new swiddens (a similar argument has been proposed for the association of Brazil nut groves and prehistoric sites in lower Amazonia; Balée 1989a).

In other words, human involvement in fallows dominated by babaçu palms most likely took place indirectly. This kind of fallow management is fundamentally unlike that described for the Gorotire Kayapó by Posey (1983a, 1985b) and Anderson and Posey (1985, 1989), which has been most recently questioned by Parker (1992). In fact, many of the species planted on "forest islands" by the Kayapó and that are also present in Ka'apor fallows are never planted (although some may be protected) by the Ka'apor. These include *Tapirira guianensis, Himatanthus sucuuba, Schefflera, Handroanthus serratifolius, Tetragastris altissima, Maytenus, Casearia, Sacoglottis, Mascagnia, Cecropia palmata, Neea* spp., *Coccoloba paniculata, Simarouba amara,* and *Vitex flavens* (cf. Anderson and Posey 1989, 162–168) (Table 3.1). The Kayapó forest island (*apêtê*) does manifest species that in the Ka'apor habitat would indicate fallow (*taper*); it exhibits 75 percent "plantable" species according to Kayapó informants interviewed by Anderson and Posey (1989, 169; cf. Anderson and Posey 1985). Yet in Ka'apor fallows, the percentage of tree and vine species occasionally planted by the Ka'apor is only about 1 percent (or 4 divided by 360 [the total number of tree and vine species collected in fallow inventories] times 100; the four species are *Dialium guianense, Rollinia exsucca,* hog plum [*Spondias mombin*], and wild cacao [*Theobroma speciosum*]). Whereas the Kayapó may be consciously planting and propagating nondomesticated tree species, a point questioned by Parker (1992), Ka'apor fallows represent one of the unintended (yet expected) results of human/animal interactions (Balée and Gély 1989). Many of the plant species therein are present, in other words, because of animals attracted to the *deliberately* and intensively managed domain of dooryard gardens and producing swiddens, not because of having been planted by human beings. One need not plant what will be predictably "planted" and in sufficient quantity by infrahuman animals or what will germinate and grow upon scarification by swidden burning (as may be the case with copal trees; Balée and Gély 1989). The vast majority of the time the Ka'apor spend gardening is not directed to planting nondomesticated trees, moreover, but to the manipulation of domesticated starchy tuber plants and other domesticated herbs, vines, and shrubs, the management of which is partly determined by normal social life.

Conclusion

Despite the relative nondeliberation involved in management of nondomesticated tree species, it is still logical to conclude that fallows are human, specifically stateless human, creations. Regardless of a relative lack of human attempts to transform actively the composition of old swiddens and thereby directly contribute to the development of forest fallow, these fallows would not be present were it not for the Ka'apor management of other plants (especially domesticates) and concomitant rise of diverse anthropogenic zones, including old and new villages as well as old and new swiddens and dooryard gardens, which by themselves attract many of the principal dispersal agents of important fallow tree species. Fallows are indigenous orchards, but to a large extent they represent unintended artifacts of the manipulation of herbaceous domesticates, that is, nontree vegetation. At the same time, they represent a measurable indigenous contribution to regional biodiversity. As such, past and present indigenous agroforestry complexes of pre-Amazonia, and probably many of those elsewhere in Amazonia, do not a priori merit the charge of being degrading, but rather should be perceived in terms of their enhancing effects on the environment, regardless of the actors' relative lack of deliberation and foreknowledge of such.

PART II

CONTACT AND ATTRITION
Overview

By the beginning of the twenty-first century, a definitive picture was beginning to emerge of indigenous occupations of diverse landscapes in Amazonia. It was becoming clear that the concept of forests that had been altered in terms of their soils and biota by indigenous peoples over hundreds of years was not only plausible, but even likely. But when I first set foot in the Amazon, at age twenty-four in July 1979, which is when I also first visited a village of Ka'apor Indians for one day only, to begin a feasibility study for doctoral dissertation research, I took with me what I had learned in anthropology while a graduate student at Columbia University, namely, that indigenous Amazonian populations were small, scattered, and lacking in centralized leadership and planning. The reason they were like that was that they were putatively subject to limitations of Amazonian nature. These limitations were widely thought to be found in the mostly poor and infertile soils, inappropriate for intensive agriculture, and hence incapable of supporting large populations that otherwise might have had noticeable effects on their surroundings, and that would have begotten truly built environments. In addition to inhibitive soil conditions, the environmental limitations on human society were at one point thought to be based on protein scarcities (Gross 1975, 1979, 1982; cf. Beckerman 1980), intrinsic to a dense tropical rainforest. The theory seemed appealing, since because people had not domesticated any fauna to speak of and because their crops were mainly starchy, not protein rich, they had to keep moving their small settlements in search of new game supplies. Such movements, either because of impoverished soil or low protein supplies, were not conducive to the rise of complex sedentary society, urbanism, and civilization, as we knew it, and that is why none of these were found in Amazonia, at least not in the Amazonia viewed in the present time, or the ethnographic present. Environmental limitation was also a convenient way of explaining why one found hunter-gatherers (egalitarian, mobile, low-density, nonagricultural people) and horticultural village societies only, as the native populations of that region.

The view that Amazonian Indians were essentially constrained in their social and political organization by environmental limitations came from a point of view known as *cultural ecology,* an influential paradigm in American anthropology during the second half of the twentieth century. The problem was—as first noted by a luminary who had been schooled in it, Eric Wolf—it could not well explicate the empirical findings about the origins of Amazonian forests that were beginning to emerge by the late 1980s and were flourishing by the end of the 1990s.

These findings, as the preceding chapters suggest, had demonstrated indigenous societies that actually changed forests, and that forests once classified as primary and pristine were, in fact, nothing more than artificial landscapes that had resulted from human interventions of the past. Cultural-ecological understandings of Amazonian forests, peoples, and societies were part of a broader perspective in both archaeology and cultural anthropology that has come to be called the adaptationist or standard model (Stahl 2002; Viveiros de Castro 1996). It is blind to human agency, if the agency is egalitarian, hunter-gatherer, horticultural, or in some other way not complex or stratified.

If people could have altered the structure and composition of native forests, and turn them into cultural forests, why could they not do the same with poor soils and limited protein? The evidence on soils is that of course they did modify these, sometimes significantly, in diverse patches across the Amazon Basin, creating new soil types known collectively today as Amazonian Dark Earths (Glaser and Woods 2004; Lehmann et al. 2003; Smith 1980; Sombroek 1966; Woods and McCann 1999; Woods et al. 2009). Protein limitation was a moot point in light of the fact that people could grow proteinaceous crops, such as maize, on these soils, as originally noted by Stephen Beckerman (1979); in addition, their very gardens attracted rather than repelled game, including deer and other important mammals, in the context of garden hunting, an insight pioneered by Olga Linares (Linares 1976) (see Chapter 2).

These forests also contained evidence of many useful tree species. In fact, so useful were cultural forests, which were the artifacts of agricultural settlements, of the sort described in Chapter 1, that hunters and gatherers actually preferred to make their temporary campsites in these forests. And because these forests were not actually pristine, it could not be any longer sustained on empirical grounds that hunters and gatherers based their livelihoods on wild species from untamed nature—they were not savages in the wilderness.

In fact, in a sense, they were adapted to culture, but not their own culture: the culture of the agricultural people who preceded them on the landscape, and who prepared it, so to speak, for them, though not having intended necessarily to do so. Some of the species in them, such as babaçu palm, which supplies a starchy mesocarp, and tucumã palm with its fibrous sword-leaf, serve as substitutes for the domesticated crops of the horticultural villagers, manioc and cotton, respectively, in these two cases. That issue of the historical contingency of the foraging lifestyle, and its dependence on culture, as much as nature, is the subject of the first chapter in this section, Chapter 4. I have called the steps leading up to the contingency of being without agriculture as essentially orderly, though not planned or predictable necessarily from events themselves. I also firmly believe these were not desirable by the people who initially underwent the changes.

The term I use in Chapter 4 for what must initially have been a devastating loss of the ability to maintain sedentary society and food production from cultivated crops is *agricultural regression.* It has been somewhat controversial, and that is mostly because I think the term "regression" has been conflated with "devolution." The two terms are very different in my dictionary. It's hard to get around "regression," for those who don't like that word, partly because, regrettably, I can't find a better term for what is usually a *process,* even if a contingent one. It is contingent on the destructive effects of European colonialism and its sequelae, though I would certainly not exclude the possibility of pre-European, hierarchical, expansionistic societies causing some groups to flee from them, avoid them, and perhaps become less sedentary and technologically dependent on agriculture (e.g., Fausto 1997). I therefore agree with Loretta Cormier (2006, 357), who, in writing of the Guajá, cogently observed: "I would argue that *agricultural regression* and *cultural devolution* are not synonyms. The term *agricultural regression* suggests a loss of a particular type of cultural knowledge. However, the term *cultural evolution* might be misread to suggest unilinear evolutionism—that a group has taken a step backward on an inevitable chain of cultural development. For the Guajá, the argument that they have lost agricultural knowledge is not meant to suggest that they themselves are devolved" (italics in original). I am convinced the Guajá and many other foragers of Amazonian tropical forests lost agriculture—they did not give it up willingly, even if they have developed remarkably competent, efficient, satisfying technologies and societies in the long aftermath of that initial loss. They have sometimes even lost fire-making technology, and these losses of crops and fire occurred systematically and unwillingly, at least at first, as described in the first two chapters of this section (Chapters 4 and 5), not by choice. Southeast Asian specialist James C. Scott (2009), in his reading of some of the authors I have cited in this book, especially Allyn Stearman (1984), Charles Mann (2005), and Pierre Clastres (1989), has eloquently proposed that the Sirionó and Yuquí (discussed in Chapter 6), together with other groups of trekkers and foragers, who arguably once had more sedentary lifestyles, had eventually become more nomadic and less tied to agriculture because they did not wish to be governed, as he puts it. Of agriculture, he writes, they had "purposely given it up to evade subordination" (Scott 2009, 188). Scott suggests they attempted to evade subjection to hierarchically organized forces that are more operational in success at conquest and the like when people are sedentary along rivers and bottomlands, not mobile and more or less free of such sociopolitical restraints. It is abundantly clear from the experience of the Assurini that their temporary period of living without agriculture was not desirable, but forced on them (Chapter 1).

Scott's position recalls somewhat but not entirely archaeologist Donald Lathrap's (1968, 1970) thesis that hunter-gatherers in Amazonian tropical forests, such as the Amahuaca of Peru, were "devolved" from highly agricultural and sedentary societies of the Ucayali River and other major rivers. According to Lathrap, they had abandoned the riparian settlements and taken to the woods on the high ground, before the arrival of the Europeans, due to population growth and attendant social pressures arising in the fertile floodplains, richer in foodstuffs from aquatic wildlife and more arable than the *terra firme*. Either they did so willingly, of course, or they did not. Lathrap suggests they were pushed out of these highly productive riverine zones. His theory of a prehistoric connection between the people of the highland forest and those of the Ucayali bottomlands is supported by the fact that the Amahuaca speak a Panoan language, as do the populous, agricultural, settled Shipibo of the banks of the Ucayali River itself. The difference between the highland and lowland Southeast Asians seems more based on class and stratification; that has yet to be demonstrated in the Amazon. The attempt at comparison is inherently interesting though, and merits further investigation.

Certainly people choose some aspects of their lifestyle; they are not all determined by contingent factors of history and events, such as conquest and warfare. I nevertheless find it hard to believe that people would give up fire-making technology, above all, by choice: it is not rhetorical merely to ask, therefore, who could want to survive in any landscape without having access to fire, that most useful of instruments? Chapters 4 and 5 develop an argument that in addition to foraging people's preadaptation, so to speak, to the culture of agricultural societies of the past, which transformed natural landscapes into orchards of fruit trees, fiber plants, and other useful commodities in lieu of domesticated crops, they also were not always hunter-gatherers.

Evidence in their languages, discussed for the Guajá in detail in Chapter 5, suggests that many hunter-gatherers of Amazonian landscapes are no more pristine than the forests themselves, and here I mean especially the cultural forests, which are the ones they built their shelters in and where they extract their livelihoods. Chapter 6 continues this thread of thinking, focusing on the mechanisms of change, in a discussion of how the Sirionó people of the eastern Bolivian Amazon gradually modified their vocabulary and classification of plants and animals due to migrations, adoption of new methods of making a living, and variable contacts and interactions with diverse groups over hundreds of years across disparate landscapes, including ones engineered in the past.

4
People of the Fallow Forest

Ecological studies often presume that the habitats of lowland South American foragers are somehow "natural" or pristine. Modern foragers (hunter-gatherers) are etched in anthropological minds as being the few remaining people of the earth who use no agriculture. The underlying assumption is that hunter-gatherers exploit "wild" resources over whose reproduction and distribution human beings have little influence (Foley 1987, 750). Since hunting and gathering was the only means of subsistence throughout most of human evolution, research on modern foragers putatively can help elucidate patterns of resource use by preagricultural hominids (Foley 1987, 76–77; Lee and DeVore 1968).

These assumptions about foragers derive from cultural ecology and evolutionary ecology. In both paradigms, a strategic ahistoricity has prevailed. Such guiding assumptions about foragers in lowland South America are at best partly mistaken and at worst gross distortions of historical processes. In addition, historical ecology is a more powerful tool for explaining patterns of resource use by lowland South American foragers than is either cultural ecology or evolutionary ecology. In anthropology, widely held views on resource use by tropical South American foragers date from the work of Julian Steward, if not earlier. In his major theoretical work, Steward (1955, 30) envisioned cultural ecology as being "an heuristic device for understanding the effect of environment upon culture." In his writings on South American Indians, not only was Steward uninterested in the possible effects of culture upon the environment, but he also showed no concern with the possible interpenetration of the two. On these questions, modern evolutionary ecology offers no quibble with Steward. Evolutionary ecology accepts the primacy of environment over culture and the treatment of the two as alienated subject and object (Balée 1989a). This ultimately derives from reductionism in evolutionary biology itself, wherein "organism" and "environment" represent segregated, noninteractive entities of research (Levins and Lewontin 1985, 134).

Steward's adherents codified the primacy of environment over culture in terms of "environmental limiting factors." Steward (1949; Steward and Faron 1959, 293, 453–454) specified that soil exhaustion "required" village movement and that the "need" periodically to relocate swiddens and/or the village itself precluded "community stability" of tropical forest Indians. This anticipated, of course, Meggers' (1954, 1971) theory that Amazonian soils had restrained cultural development in lowland South America, given that community stability is a prerequisite for the development of complex society. Meggers interpreted extinct prehistoric Amazonian chiefdoms in the archaeological record, for example, as being the relics of introduced civilization, which could not have survived as such, in the tropical forest (Roosevelt 1987, 161).

Steward (1947, 1949) proposed an essentially evolutionary typology of South American Indian means of economic production (Willey and Sabloff 1980, 175). Steward classified foragers as "marginal tribes" who mainly inhabited zones that were unfavorable for agriculture (Steward 1947, 1949; Steward and Faron 1959). Steward (1947) suggested that foragers (or trekkers) living in lands favorable to agriculture (those of the "tropical forest tribes") had borrowed certain cultural traits from horticulturalists. I would not deny that lowland South America, including Amazonia, supplies evidence for preceramic, prehorticultural society. But Steward's "marginals" were mostly known from colonial and postcolonial times, and many were coinhabitants of the tropical lowlands with horticultural peoples. The presence of foragers in arable habitats probably may be best explained by nonenvironmental considerations (Lyon 1985, 3). Such considerations belie an evolutionary typology of modern lowland South American Indians.

The alienation of culture from environment in cultural ecology and evolutionary ecology, in attempting to establish a historical approach to foraging in the tropical lowlands of South America, obscures the fact that foragers use not merely pristine forests, but agricultural zones, especially old fallows and the culturally dependent biological resources therein encountered. Some foragers of the tropical lowlands, moreover, apparently regressed from an earlier phase of horticultural society, because of historical and sociopolitical forces. It seems unlikely, hence, that study of their patterns of resource use would reflect pre-Neolithic patterns, even in South America. In addition, since these foragers depend on the resource concentrations of old fallows, the habitats they exploit must be qualitatively and quantitatively distinct from those of pre-Neolithic peoples of the tropical lowlands.

Agricultural Regression in Lowland South America

Lowland South American Indian societies affiliated with the Tupí-Guaraní language family are remarkably unalike (Viveiros de Castro 1986, 106). The Tupinambá and Guaraní societies of late prehistoric and early colonial times displayed paramount chiefdoms (Balée 1984b; Clastres 1973; Dean 1984). But many others such as the Hetá, Aché, and Avá-Canoeiro exhibited no social units more complex than nomadic bands (Clastres 1972; Kozák et al. 1979; Toral 1986). The Tupinambá occupied villages with an average population of 625 people. The modern Tupí-Guaraní–speaking Ka'apor, in contrast, have average village sizes of only thirty-three (Balée 1984a, 254; 1988a, 167), while Guajá bands number only between five and fifteen persons (Balée 1988b, 48). Although the Ka'apor prescribe the ritual consumption of tortoise meat for girls at puberty, menstruating women, and men and women in the couvade (Balée 1985), the linguistically related Guajá actually forbid this meat to such ritually restricted persons. The list of infrastructural, structural, and superstructural differences between Tupí-Guaraní societies, therefore, is indeed long. But, in addition to broadly defined activities like "gathering" and "hunting," these otherwise disjunct societies share, or shared in the past, a key characteristic: the intensive cultivation of plants.

Food plants aboriginally cultivated by Tupí-Guaraní peoples of the Atlantic littoral, for example, included manioc, bananas, maize, peanuts, capsicum peppers, pumpkins, sweet potatoes, and pineapples (Benitez 1967, 27; d'Évreux 1864, 74; Léry 1960, 113, 115, 158, 160, 162; Lisboa 1967, 97, 99, 100, 101, 115, 121; Soares de Sousa 1974, 89, 95, 105–106, 110; Staden 1930, 141, 176; Vasconcellos 1865, 131, 133). The inventory of cultivated species was high, with numerous varieties of many of these crops. There were some twenty-eight named varieties of manioc (Métraux 1928, 65–67). The Tupinambá also had access to semidomesticated tree species such as *Hymenaea* and *Inga* (Soares de Sousa 1974, 101; Vasconcellos 1865, 133).

Given that no chronicler reported a Tupí-Guaraní society of early colonial times to be lacking horticulture, it is curious that several Tupí-Guaraní societies of the present lead exclusively foraging existences. These include the Hetá, Aché, Avá-Canoeiro, Guajá, and Setá. Yet historical and "inferential linguistic evidence" (Sapir 1949, 430–432) suggests that each of these groups regressed from an earlier phase of horticultural society (Clastres 1973, 271; Lathrap 1968; Martin 1969). The Aché of Paraguay, for example, possess a Tupí-Guaraní cognate for maize (*waté*) (Clastres 1968, 51–53). Evidently, the Aché cultivated maize in the eighteenth century (Clastres 1972, 143; Métraux and Baldus 1946, 436). The Avá-Canoeiro, who number some thirty-five people, did at one time cultivate maize, the term for which, *avaši,* is cognate in Tupí-Guaraní (Rivet 1924, 177; Toral 1986). Despite their mischaracterization as foragers (Holmberg 1969), the trekking Sirionó of Bolivia also planted some maize, their word for which *ibaši,* also a cognate (Allyn Stearman, personal communication, 1987). The Hetá, who spoke a language phonologically and lexically derived from Guaraní (Rodrigues 1978), evidently "practiced some plant cultivation" before they became total foragers (Kozák et al. 1979, 366). The Guajá language has cognates for cashew (*akayu*), annatto (*araku*), pineapple (*nana'i*), yam (*kara*), chili peppers (*kiki*), and maize (*wači*), all traditional Tupí-Guaraní cultigens, but none of which the Guajá have recently cultivated. The Setá, a now extinct society of southern Brazil, were hunter-gatherers whose language had Tupí-Guaraní cognates for manioc and maize (Loukotka 1929, 374, 392).

Sociopolitical, not evolutionary, forces appear to be responsible for the regression of semisedentary horticultural societies into foraging nomadism (Balée 1988a; Clastres 1968; Martin 1969). These forces, such as epidemic disease, slave raids, and colonial warfare, also devastated the paramount chiefdoms of the Tupinambá of the Atlantic littoral. In the emergence of true foraging societies, these forces were acting on societies already of a much smaller scale. The Guajá (then called Uayás) apparently lived in settled villages in the 1760s in the lower Tocantins River (Noronha 1856, 8–9). The lower Tocantins saw intensive warfare between colonial militia, their Indian mercenaries, and uncontacted Indian societies after this time (Balée 1988a). By 1872, the clandestine Guajá, now of the upper Gurupi to the east, were described as being "persecuted" by all surrounding indigenous groups and as lacking horticulture (Dodt 1939, 177). According to a Ka'apor informant, the Ka'apor would not have permitted the Guajá to establish permanent settlements, Ka'apor warriors also being more numerous than those of the Guajá (Balée 1988b).

The forests of the Hetá and Aché, respectively in Paraná state, Brazil, and eastern Paraguay, experienced a long history of missionization, epidemic disease, and slave raids (Clastres 1972, 143–144; Métraux 1948). In the region of the Hetá (between the Piquiri and Ivai Rivers in Serra dos Dourados), there are a number of archaeological sites dating from thousands of years ago (Chmyz and Sauner 1971). Jesuit missions (*reducciones*) were founded between 1576 and 1630 in the region (Maack 1968, 41), where Guaraní Indians were concentrated. After 1632, however, gold and slave seekers from São Paulo, called *bandeirantes*, destroyed these missions (Chmyz and Sauner 1971, 10; Maack 1968, 42–43). The Guaraní, who fled into Paraguay, abandoned their village sites. Gê-speaking Indians (the Kaingáng) subsequently occupied these settlements (Maack 1968, 44). These newcomers, in turn, later became enemies of the Hetá and may have prevented the Hetá from maintaining a sedentary lifestyle (Kozák et al. 1979, 366). Full-time hunting and gathering does not appear to be an option that a horticultural society freely chooses. Warfare, epidemic disease, and depopulation evidently can make horticulture and semisedentism less rewarding than foraging for some societies.

The process of agricultural regression itself has been fairly unexplored. It does not seem to be often an abrupt but rather a gradual transition from sedentarism through trekking to complete nomadism. Dependence on given plant species diminishes over time, until only one or two cultigens are left and finally none. One of these final cultigens appears to be maize. Although maize cultivation is closely associated with the sedentary pre-Columbian Tupí-Guaraní chiefdoms, it is also ironically seen among modern trekking peoples. The Araweté, who until recently trekked for an average of six months a year (Viveiros de Castro 1986, 271), are highly dependent on maize instead of starchy tubers. Viveiros de Castro (1986, 26) appropriately noted that, with respect to the Araweté, "maize aggregates people; it is practically the only force which does so. Innumerable other forces work towards dispersion." Maize aggregated the Araweté during planting and harvest seasons. In the other seasons, people tended to forage in the forest. Among several macro-Gê peoples such as the Botocudos, Tapuias, central Kayapó, and Akwẽ-Shavante, the maize harvest united society. Otherwise, the people dispersed themselves into small groups for gathering, hunting, and fishing (Galvão 1979, 245; Maybury-Lewis 1967, 47). Maize dependence combined with trekking also occurs with the Panoan-speaking Amahuaca of eastern Peru (Carneiro 1985) and the linguistically isolated Hoti of southern Venezuela (Coppens 1975, 68; Coppens and Mitrani 1974, 136). Several trekking groups traditionally cultivated other food plants, especially bananas and sweet manioc, one or more of which may have been calorically more significant than maize in the diet, but little or no bitter manioc (Coppens and Mitrani 1974, 136; Smole 1976, 119).

Manioc is, of course, the caloric staple of most semisedentary (nontrekking) lowland South American Indians. Brochado (1977, 57) found that of 553 indigenous peoples of the tropical forest who cultivate manioc, 478 (86.4 percent) employ it as a primary food source and 75 (13.6 percent) use it only as a supplementary resource. For 60 (80 percent) of these latter groups, the dietary staple is maize, bananas, peanuts, or sweet potatoes. In particular, trekking peoples seem to choose maize rather than bitter manioc as a staple because of its relatively low start-up costs, simple processing requirements, low transport costs, and fast maturation time (Balée 1985; Galvão 1979, 245). When trekking people do cultivate manioc, it tends to be only the sweet variety, as with the Hoti (Coppens and Mitrani 1974, 136) and the Yanoama of Venezuela (Smole 1976, 119). Sweet manioc demands neither elaborate detoxification nor installations in order to be made fit for human consumption (Balée 1985, 489–490). Brochado (1977, 64) observed that among the 75 groups that employ manioc as a dietary supplement, the nontoxic varieties outnumber the toxic ones almost two to one (Lathrap 1970, 53).

One can reasonably propose that of the two staple cultigens, maize and bitter manioc, maize disappears last from a trekking people's horticultural repertoire, prior to their becoming true foragers. That the Guajá language has retained a cognate for maize but not one for bitter manioc (called *tarəmə*) supports this hypothesis. This is precisely the case, moreover, with the Aché (Clastres 1968, 53).

In the transition from sedentarism through trekking to nomadism, it is logical to assume that there is a growing economic dependence on plants that are not cultivated per se by the groups undergoing this transition. The species that tend to become primary botanical resources, however, are not readily characterized as being "wild." Rather, these plants, in their dominance and/or frequency, appear to be vegetational artifacts of an agricultural past, sometimes of other societies.

Artifactual Resources of Foragers

What is an artifact? Two basic definitions, implicitly or explicitly, are common in the literature. One is based on purposeful human activity, whereby an artifact is any object "made or unmade by *deliberate* human action" (Childe 1956, 11, emphasis mine). The other is broader, entailing no necessary consciousness on the part of human actors. In this second definition, an artifact is "any material expression of human cultural activity" (Spaulding 1960, 438). This definition seems more appropriate in considering use of biological resources by lowland South American Indians. It is true, however, that some contemporary Indians use prehistoric artifacts that were produced by "deliberate human action." In North America, for example, the Navajo use Anasazi ruins for gathering medicinal plants and clay and as sites for curing ceremonies (Holt 1983). In South America, the Yanomamö and Araweté recently felled trees with prehistoric stone axeheads, the origins of which they believe to be divine (Chagnon 1977, 46; Viveiros de Castro 1986, 152). But many groups, including the Araweté, use an anthropogenic resource, *terra preta do índio* (literally, "indigenous black earth," a type of Amazonian Dark Earth), which is better understood as an artifact in the broad, materialistic sense that Spaulding gave it. The formation of *terra preta do índio* in hearths and garbage pits over hundreds or thousands of years is clearly no result of "deliberate human action." The Araweté, Kuikuru, Mawé, Munduruku, Xikrin-Kayapó, and many other Indians prefer to plant nutrient-demanding crops in this most fertile cultural horizon (Balée 1989; Carneiro 1983, 67; Frikel 1959, 1968; Smith 1980, 562).

Another artifact of this sort concerns disturbance indicator and late successional plants. In their dominance and/or frequency, these form artifactual landscapes in Amazonia (Anderson and Posey 1985; Balée 1989; Balée and Campbell 1990; Posey 1985a), Central America (Gómez-Pompa et al. 1987), and no doubt elsewhere. Although not all of these landscapes, such as anthropogenic savannas of northwestern and central Amazonia (Huber et al. 1984; Prance and Schubart 1977; Smole 1976; Whitten 1979), proffer very important plant resources to indigenous peoples, many others do. These old, artifactual landscapes, which often cover archaeological sites, appear to be especially useful to modern foragers of the tropical South American lowlands. These foragers, for sociopolitical reasons, could not remain sedentary horticulturalists, and thereby could no longer produce any artifactual landscapes of their own. But they could use the artifactual landscapes of other societies.

Descriptions of the habitats of the foraging Hetá of Paraná and Aché of Paraguay point to past agricultural management. To begin with, recall that the region was at one time occupied by agricultural Indians, the Guaraní, whose population at contact has been estimated at 1,500,000, with a population density of four persons per square kilometer (Clastres 1973). This is higher than the population density of modern Amazonia and twenty times higher than the average Indian population density (0.2 persons per square kilometer; see Hames 1983, 425) of lowland South America at present. The region was subsequently colonized, became the scene of much colonial warfare, and was later in large part abandoned. It is possible that much of the "primary" forest of the region never recovered or was at least irrevocably modified by past disturbance, in spite of what has been described often as "primary" forest cover (Fischer 1987, 16; Hill and Hawkes 1983, 143; Maack 1968, 217). The forests the Hetá exploited, like the Hetá themselves, regrettably no longer exist. When they did, no systematic inventories or studies of forest recovery were made (Paulo Sodero Martins, personal communication, 1988). General studies of the vegetation of the Serra dos Dourados, the habitat of the Hetá, nevertheless, were carried out. The geographer Maack (1968, 215–217) mentioned some nine species that were "predominant" in the "rain forest" (*mata pluvial*) that typified the region of the Hetá. These included *Acrocomia sclerocarpa* Mart., *Acacia polyphylla* DC., *Cassia multijuga* H. & P., and *Jacaratia spinosa* A. DC., all of which are well-known late successional species. The mucajá palm (*Acrocomia sclerocarpa*) in the central zone of eastern Paraguay, near Aché and less than 200 km from the Serra dos Dourados, is strongly associated with agricultural fallows. Markley (1956, 12–13) noted that this palm germinates best in areas that have been used for horticulture. He added, "It is not surprising, therefore, that the most intensively and thoroughly cultivated farmlands of the Central Zone contain the greatest concentration of palms" (Markley 1956, 13). Evidently, *Acrocomia* is a long-lived late successional species that, through the agency of human beings, can grow in what otherwise appear to be "virgin" forests. Markley is worth quoting further with respect to a seemingly dense forest with *Acrocomia* present, at a place called Curuguaty: "A subsequent study of the history of Curuguaty revealed that the area had been a center of agriculture off and on from pre-Columbian times until after the war of the Triple Alliance (1865–1870). The settlements of the area, its agricultural development, its subsequent decline and the return of the land to what now resembles virgin forest explain the apparently anomalous occurrence of the mbocayá [*Acrocomia*] palm" (Markley 1956, 15).

In the Amazon, *Acrocomia* often occurs in *terra preta do índio* and other fallows. It has been described as "semidomesticated" (Balée 1988b, 48; Huber 1900, 21). *Acrocomia* was an important dietary item for the Hetá Indians. They consumed its proteinaceous kernels, a flour made from the palm heart (the only flour they made from any plant), and the weevil larvae that bored into the fruit (Kozák et al. 1979, 379; Loureiro Fernandes 1964, 40–41).

Continuing with Maack's list, *Acacia polyphylla* is known to be a late successional species elsewhere in lowland South America (Balée and Campbell 1990; Huber 1909, 162). *Cassia multijuga,* according to many herbarium labels I have seen at the Museu Goeldi, is by far most commonly found in secondary forests from eastern to western Amazonia (Lisboa et al. 1987, 55). The bamboo (*Bambusa guadua* H. & P.) stands of the Hetá habitat referred to by Maack (1968, 217) and Kozák et al. (1979, 367) are also usually found in areas previously disturbed by slash-and-burn agriculture (Balée 1989; Sombroek 1966, 188–189). The Hetá used bamboo blades as the plaits for their sieves (Loureiro Fernandes 1964, 41).

Maack (1968, 217) also noted a "wealth" of lianas, epiphytes, aroids, bromeliads, and orchids in the Hetá forest. Liana forests (*mata de cipó*) appear to be late successional in Amazonia (Balée 1989; Balée and Campbell 1990) if not elsewhere in the Neotropics (Al Gentry, personal communication, 1987). The authors of the habitat descriptions of the Hetá Indians observed two other palms, *Euterpe edulis* Mart. and *Syagrus romanzoffiana* (Cham.) Glass. *Syagrus* (called *jerivá* in Brazil) was the "most common" species in the Hetá habitat (Kozák et al. 1979, 367–368; Loureiro Fernandes 1964, 39; Maack 1968, 217). Both these palms are now commonly cultivated as ornamentals in the major cities of southern Brazil (Pinheiro and Balick 1987, 28). *Jerivá* is extremely varied in habit and occurs in a wide variety of climates, soils, and elevations (Barbosa Rodrigues 1898, 15; 1899, 8). Its wide dispersion may, in fact, have been related to human influence, given its frequent occurrence in swiddens and fallows (Barbosa Rodrigues 1898, 16). In the 1980s, *jerivá* was seen almost only in disturbed areas (Dalmo Giacometti, personal communication, 1988). The Hetá used *jeriva* palms for many utilitarian ends. Palm fruits, in general, were described as constituting the "basis of the diet" of the Hetá (Loureiro Fernandes 1964, 40), and those of *jeriva*, which is perennial, were most important (Kozák et al. 1979, 367–368; Loureiro Fernandes 1964, 40). The Hetá consumed the mesocarp, the kernel, and the palm heart of *jeriva*. They used the spathe as a receptacle. All basketry, rope, infant-carrying straps, sleeping mats, and fishing line derived from the fiber and/or sword leaves of this palm (Kozák et al. 1979, 395; Loureiro Fernandes 1959, 25; 1964, 41–43).

The frequent presence of one final species, feral oranges (*Citrus sinensis* [L.] Osb.), supplies incontrovertible evidence for prior human modification of the landscape in which the Hetá were first reported to be living in the 1950s (Maack 1968, 217). The orange trees were planted by Jesuit missionaries in the sixteenth and seventeenth centuries. Although some orange groves appear to have expanded since then, these would not have existed in the habitat of the Hetá were it not for their having been an agricultural introduction of the past. The orange trees, together with the many late successional species, signify that the habitat of the Hetá should hardly be described as "primeval" (Kozák et al. 1979, 357).

Feral orange trees exist as well in the habitat of the Aché, having been also originally planted by Jesuits of the sixteenth- and seventeenth-century missions (Hill et al. 1984, 125; Kaplan and Hill 1985, 229; Métraux and Baldus 1946, 436; Vellard 1934, 240). It seems inaccurate from this evidence alone, even if the orange trees were subsequently dispersed by nonhuman agents, to describe their habitat as "primary forest" (Bailey et al. 1989, 65; Hill and Hawkes 1983, 143). Some question exists as to whether feral oranges were edible to the Aché. Clastres (1972, 156) indicated that they were not, but others (Hill et al. 1984, 112; Kaplan and Hill 1985, 229; Vellard 1934, 240) listed the oranges among the Aché's gathered foods.

In addition to feral orange trees, Vellard (1934, 228) listed numerous trees of the Aché habitat that are well-known late successional species, such as *Cedrela, Piptadenia, Astronium, Cordia,* and *Ceiba*. He also noted the presence of *Jacaratia*, the fruits of which both the Aché and Hetá ate (Loureiro Fernandes 1964, 39, 42; Vellard 1934, 240). This species (*Jacaratia spinosa* A. DC.) is the ecologically most important species on a one-hectare fallow of forty years in the habitat of the Ka'apor Indians (Balée, unpublished data; Balée and Gély 1989). It does not dominate in "primary" forest.

Although the Aché obtained most of their calories from game animals and not from plants (Clastres 1972, 160; Kaplan and Hill 1985, 228; Vellard 1934), the most important botanical resource, both as a food and for making many tools and utensils, was *jeriva* (*Syagrus romanzoffiana*), as it was for the Hetá. *Jeriva* supplied the Aché with edible buds, flour, and weevil larvae in addition to bow wood, bowstring, roofing thatch, matting, fans, and cases for keeping feather art (Clastres 1972, 161; Métraux and Baldus 1946, 436; Vellard 1934, 232, 240). The preferred campsites of the Aché, moreover, were groves of *jeriva* palms (Clastres 1972, 150).

Table 4.1. Economic palms in Guajá society

Scientific Name	Guajá Name	Uses	Part Used
Astrocaryum gynacanthum Mart.	*yu* (3483)	Food	Seed
Bactris acanthocarpa (Mart.) Henderson var. *acanthocarpa*	*wəʔə* (3355)	Arrow shaft	Petiole
Bactris maraja Mart.	*mariawa* (3385)	Food	Mesocarp
Bactris major Jacq.	*kiripiri-hu* (3490)	Food	Mesocarp
Bactris sp. 2	*kiripirim'ɨ* (3525)	Food	Mesocarp
Euterpe oleracea Mart.	*pinuwā-pihun* (3281)	Food	Mesocarp
Geonoma baculifera (Poiteau) Kunth.	*yowo'ɨ* (3347)	Arm band	Seed
Attalea maripa Mart.	*inaya'ɨ* (3377)	Food, bowl	Mesocarp, seed, spathe
Oenocarpus distichus Mart.	*pinuwa'ɨ* (3144)	Food	Exocarp
Attalea speciosa Mart. ex Spreng.	*waʔiʔɨ* (3376)	Food, thatch	Kernel, mesocarp, young leaves

The disturbance indicator *Acrocomia* (mucajá palm), a food resource of the Aché, was also present (Hill et al. 1984).

Other gross indicators of past agricultural perturbations of the Aché habitat include bamboo forests and liana forests (Baldus 1936, 750; Hill and Hawkes 1983, 143; Vellard 1934, 229). Regardless of whether the Aché would have been described as a society of honey gatherers (Vellard 1939) or a society of hunters (Clastres 1972; Kaplan and Hill 1985), it is not unreasonable to hypothesize that they were also a people of old fallows. These fallows harbor artifactual resources upon which they directly and indirectly depended.

The campsites and principal botanical resources of the foraging Guajá Indians of northern Maranhão, Brazil, evoke striking similarities to those of the Hetá and Aché. In particular, as with the Hetá and Aché, the Guajá display a very high dependence on palms for food, shelter, clothing, tools, and utensils (Table 4.1; Balée 1988b, 50–51) (In Table 4.1, the four-digit numbers are collection numbers in the series Balée with vouchers at the New York Botanical Garden and duplicates at the Museu Paraense Emílio Goeldi.) The economically most important palms in Guajá subsistence, moreover, are well-known disturbance indicators. These include babaçu (*Attalea speciosa* Mart.), inajá (*Attalea maripa* [Aubl.] Mart.), and tucumã

(*Astrocaryum vulgare* Mart.). These species often occur in enclaves of two to three hectares, surrounded by dense *terra firme* forest. Babaçu, when present, is usually the most dominant species in such enclaves. The Guajá refer to this forest as "babaçu grove" (*wa'ĭ-tu*). Babaçu groves are the exclusive sites of their camps (Balée 1988b), as are *jerivá* groves for the campsites of the Aché. The babaçu palm (*wa'ĭ*) supplies edible kernels (*ha'ĭ*) and mesocarp flour (*tərəma*). The Guajá extract the kernel by placing the fruit on a small concave rock (*ita*) and then cracking the fruit open with another smooth rock. They eat the kernels raw. To obtain flour, young fruits are placed directly in hot ashes, which soften the exocarp. The exocarp is peeled off and the mesocarp, now doughy, falls off in irregular plates, which are then eaten by hand. Although consuming babaçu mesocarps is not common in Amazonia, the Araweté, Apinayé, and Guajajara do so (May et al. 1985, 125; Viveiros de Castro 1986, 164), especially in the failure of domesticated staples. May et al. (1985, 125) pointed out that for rural families, as well as for some indigenous peoples of the babaçu zone south of the Amazon River, the babaçu mesocarp flour serves as a "substitute" for manioc flour during manioc shortages (Balée 1988b, 52). The Guajá harvest babaçu fruits on the ground. Like the fruits of *jerivá*, so important to the Aché and Hetá, babaçu fruits are available year round, with young fruits appearing at the beginning of the rainy season (December to January). In addition to using babaçu for food, the Guajá use the leaves as roofing thatch.

Babaçu palm groves in the habitat of the Guajá are always indicative of prior agricultural occupations. In the upper Turiaçu River region, where I have done research with the Guajá, all babaçu forest enclaves were former Ka'apor Indian agricultural settlements and fields (Balée 1988b), attested to by Ka'apor oral history and the remains of ceramic manioc griddles.

Fruits of the inajá palm (*Attalea maripa*) provide oil, protein, and calories in the Guajá diet. The Guajá heat the exocarp, which they then peel off and discard. They boil the mesocarp to make a porridge (*mika*) that they serve on an inajá spathe (*hawera-kutai*). They also eat the inajá kernels. Inajá, like babaçu, frequently occurs in fallows in Amazonia (Balée 1988b, 48; Pesce 1985, 66; Schulz 1960, 22; Wessels Boer 1965, 152).

The Guajá make all garments and rope from the young sword-leaf fibers (*tikwira*) of *takamã* (tucumã palm [*Astrocaryum vulgare*]). Infant-carrying straps, hammocks, women's skirts, bowstrings, string for tying the foreskin, rope, and string for armbands all derive from tucumã fibers (Balée 1988b, 50). In the habitat of the Guajá, the tucumã palm occurs only on former sites of Ka'apor agricultural occupation, with or without the presence of babaçu (Balée and Gély 1989). In a forty-year-old fallow of the Ka'apor, also mentioned above, only 20 km from the Guajá of the upper Turiaçu River, tucumã was the second ecologically most important species. Tucumã is one of two species of *Astrocaryum* taken to be a "good" indicator of prior human settlement in Surinamese noninundated habitats; further, it is "never" found in undisturbed forest there (Wessels Boer 1965, 132).

Since babaçu, inajá, and tucumã occur on and indicate prior settlements of the horticultural Ka'apor, the Guajá, who camp within babaçu groves and who intensively exploit these palm species therein, have adapted not merely to a "natural" environment but to the vegetational artifacts of another society.

Discussion

The foraging Guajá, Hetá, and Aché appear to have undergone a kind of parallel historical regression from prior agricultural societies to exclusively foraging ones. Although the specific botanical resources they utilize vary (less so between the Hetá and Aché), it is remarkable that all three groups depend heavily on palms that appear to dominate in old agricultural fallows. In each case, the principal habitats they exploit exhibit plant indicators of past human disturbance. These "hunter-gatherers," in other words, heavily rely on agriculture (Bailey et al. 1989; Headland 1987). This does not mean they use a priori domesticated crops, even though the Aché (Vellard 1934, 241) and the Guajá (Araújo Brusque 1862, 15) were historically known to raid occasionally the swiddens of other Indians and settlers for maize, sweet potatoes, and the like. Such raids entailed high risks, since at least in the Guajá case, when the other Indians caught them in their swiddens, they would "beat or kill them" (Araújo Brusque 1862, 15). For the Guajá, Aché, and Hetá, raiding swiddens was clearly a far less usual occupation than harvesting the fruits of old fallows. Sometimes these fruits were domesticates, as with the feral oranges of the Hetá and Aché and the yams and lime trees that the Ka'apor planted many years ago and that the Guajá now harvest in old fallows (Balée and Gély 1989). But for all three foraging societies the economically most important plant resources were palms.

The foraging Warao Indians of the Orinoco delta demonstrated a major dependence also on a palm, the moriche palm (*Mauritia flexuosa* L.f.) (Heinen and Ruddle 1974). Even though some monospecific groves of moriche palm in swamp forests of the Warao habitat seemed to be anthropogenic (Heinen and Ruddle 1974, 124), many were no doubt natural groves unrelated to human activity. The palm groves of the Aché, Hetá, and Guajá Indians, however, appear to be anthropogenic, considering, in addition, many other late successional species found therein. The Aché, Hetá, and Guajá also underwent agricultural regression, about which no evidence exists with respect to the Warao (Wilbert 1969, 19).

Not a general theory of foraging, therefore, but a general theory of agricultural regression and the utilization by foragers of old fallows in lowland South America follows from this information. Because of epidemic disease and/or defeat in warfare, some sedentary agricultural societies became oriented toward a trekking lifestyle. Some eventually became full-time foragers, with population densities considerably lower than those of surrounding agricultural peoples. This is certainly so with the Aché, Hetá, and Guajá, who are (or were) circumscribed by far more dense agricultural populations, such as those of the Guaraní, Mbyá, Kaingáng, Ka'apor, and Tenetehara. In other words, agricultural regression appears to be associated with a concomitant drop in population density. It is unlikely that very small bands of five to fifteen people could have maintained or opted for agriculture ex post facto. First, they would be more liable to defeat in warfare and the subsequent loss of their swiddens, orchards, and other agricultural investments. Second, with only one or at most two small swiddens, appropriate for such a tiny village size, the group would run the risk of generalized crop failure. Agricultural societies such as that of the Ka'apor tend to spread this risk, insofar as many small families of a village have their own swiddens. If one fails (due to insect pests, overgrazing by peccaries, and so on), this does not ipso facto become a villagewide crisis (Balée 1984a). In addition, no one family keeps the seeds and/or cuttings of all the crops grown by the Ka'apor. The largest Ka'apor settlements, incidentally, tend to be the longest lived, perhaps because of ecological risk management and a well-established system of economic reciprocity between many households (Balée 1984b).

The smaller a society gets, in contrast, the more nomadic it becomes. In late September 1987, an isolated family of Araweté Indians came to the attention of the Brazilian National Indian Foundation (FUNAI). They were living some 200 km distant from the Araweté village of 155 people, having become "lost" in 1976, at the time the FUNAI authorities were first making contact with the Araweté and when introduced epidemic disease killed some sixty Araweté (Viveiros de Castro 1986, 181). Early in September 1987, Kayapó Indians attacked this isolated family of seven people, killing one man and one boy. The Kayapó raiders took captive two women and two children. One Araweté man escaped into the forest. The hostages were taken to the Kayapó village of Cateté (in the basin of the Itacaiunas River, between the Tocantins and Xingu Rivers). Once apprised of the situation, the FUNAI authorities, with two Araweté Indian interpreters from the Araweté village, departed in search of the missing Araweté man. They also entered into negotiations with the Kayapó for the release of the four hostages, which eventually took place.

The search party made contact after a few days with the missing Araweté man, convincing him to return with them. All the surviving Araweté of this isolated group were later returned to live among their far more numerous kinfolk again (the missing man had a sister in the Araweté village). I interviewed these survivors, their interpreters, and members of the FUNAI search party when they arrived in the city of Marabá on the Tocantins River in transit to the Araweté village. At the site of the raid stood a house in a small swidden. The only crop in the swidden was maize. The Araweté traditionally grow a reasonable number of species, including maize, manioc, papaya, cotton, annatto, a bromeliaceous fiber plant (*Neoglaziovia variegata* [Arruda da Câmara] Mez), and yam (Viveiros de Castro 1986, 150–153). The isolated family was subsisting, according to the remains found at the site, from maize, babaçu mesocarps, and Brazil nuts. These last two botanical resources are often present because of past agricultural legacies (Balée 1989; Ducke 1946, 8). Although the Araweté traditionally make their hammocks from cotton, the hammocks of the isolated family were woven from the bark of an uncultivated tree (*Couratari guianensis* Aubl.). Whereas the Araweté usually fashion their bowstrings from the cultivated bromeliad *Neoglaziovia variegata*, the bowstring of the missing man was from *Astrocaryum vulgare* (tucumã palm) sword-leaf fibers. Evidently, this group of seven people had become so small that they could not maintain an inventory of more than one crop, maize. As pointed out above, maize appears to be the last cultigen a trekking society loses before becoming true foragers. These Araweté seemed to have been living on the edge between trekking and foraging.

Even though agriculture becomes increasingly implausible for a group approaching band size (between about fifteen and fifty persons) or even fewer in numbers, a very small number of people can survive exclusively by foraging—even a number smaller than that of a band. I met the "lost" Araweté in Marabá while I was in transit to a hitherto unknown group of Indians consisting of only two men between the ages of thirty and forty who were living some 190 km west-northwest of Marabá on a tributary of the Tapirapé River (Chapter 1). I was participating in a FUNAI mission with the objective of identifying the cultural and linguistic affinities of these Indians and determining whether there were more of their kind in the vicinity (there were none). Their language was determined to be a new Tupí-Guaraní language (Moore and Maciel 1987). Although a substantial list of Tupí-Guaraní cognates for cultivated plants was elicited from the two men (Balée 1987; Moore and Maciel 1987), their campsites evinced no horticulture. Based on material encountered at their only two known campsites, the botanical resources they were subsisting on were babaçu fruits and Brazil nuts. Evidently, they had lived in a horticultural village in the past, but after their kinfolk died (apparently in a massacre by other Indians), they alone, who were probably teenagers at the time, could not keep horticulture viable (Balée 1987). The two orphans turned instead to a major reliance on plant species that dominate in very old fallows, depending on artifactual resources. Later, the two men were transferred to the Assurini do Xingu village, where they were living as of October 1988.

It is probably not merely the small size to which a trekking people can be reduced that leads to the inevitability of a foraging lifestyle. An intervening variable is, no doubt, the frequent unevenness of adult sex ratios in such tiny populations, given statistical odds. In a group with few or no women left, as with the orphans of the Tapirapé River, a previously learned sexual division of labor in horticulture would cease to be meaningful, as would horticulture itself.

The resource concentrations of old fallows are, to foragers, substitutes for the caloric vitality, to horticulturalists, of young swiddens. Whereas these old fallows could not alone permanently support a horticultural village without being supplemented by new clearings and plantings, they can maintain for indefinite periods small, nomadic foraging bands. The orphans of the Tapirapé, like the Hetá, Aché, and Guajá, relied on agriculture without raising any crops. A cultural past, in other words, was embodied in the food they gathered.

The principal problems concerning the use of biological resources by lowland South American Indians therefore are not evolutionary, but historical. The artificiality of culture/environment dichotomies and of evolutionary typologies with regard to lowland South America should by now be apparent. Cultural ecology and evolutionary ecology offer no theories for explicating three seemingly paradoxical facts: (1) the natural habitats of various foragers are cultural; (2) many foraging societies began in a historical process of agricultural regression, buoyed by sociopolitical forces; and (3) many modern foragers (or "hunter-gatherers") use agriculture, as is readily documented in their profound dependence on old fallows. These facts are not paradoxical, however, within the framework of a historical ecology.

5
Vanishing Plant Names

We must be very cautious in assuming that elements of culture are so useful or so important that they would never be allowed to disappear.
—William H. R. Rivers (1926, 206)

In Italic (i.e., Latin) the birch reflex shifted to "ash." . . . The shift to "ash" in Latin, like the total loss in Greek, is often thought to have been motivated by the absence of birch in these climes.
—Paul Friedrich (1970, 29)

Although a transition from horticultural to foraging society may be seen as far less common than the development of horticulture in a slow process that began with foraging as the exclusive means of human subsistence (Clastres 1989, 201; Gellner 1988), an increasing amount of empirical work is suggesting that such a transition has occurred repeatedly in the forested lowlands of South America. It has probably happened in other parts of the world as well, such as southern Africa, the Philippines, and equatorial Africa (Bailey and Headland 1992; Headland and Reid 1989). This observation does not imply, however, that all contemporary and recent foraging peoples were once horticultural. Many Australian Aborigines, the Inuit Peoples, the California Indians, peoples of the Great Basin, and the Selk'nam and other peoples of Patagonia no doubt lacked domesticates (with the possible exception of the dog) altogether before contact. Some contemporary, pristine foragers lived in environments in which early Neolithic technology could not function; in other words, the lands were habitable but not arable. Antarctica was both unarable and uninhabitable and has been until quite recently for the human species as a whole. It is not an example of environmental determinism or cultural ecology to make the rather trivial point that, under given technological conditions, agriculture and perhaps even human occupation were impossible in certain parts of the globe until quite recently. Rather, environmental determinism has tended to fail in explaining differential modes of production where, as in the humid tropics, technological and geophysical conditions do not present severe and absolute constraints. Here, historical ecology would seem to offer a foundation for more penetrating insights into the relations of peoples, cultures, biota, and regions. Historical ecology essentially holds that people and their landscape can be understood as a dialectical, and dynamic, whole (Crumley 1994).

Lévi-Strauss (1950, 466–467) postulated that many of the differences in plant usage among native societies across the Amazon Basin, such as the disjunct boundaries of the use of tapa cloth, could not be explained in terms of the plant materials at hand—wild fig trees used for making tapa cloth are found throughout the region. And numerous fruit trees, hallucinogenic vines and trees, and even domesticated crop plants that constitute part of a pan-Amazonian wealth of common plants are paradoxically used, if used at all, in different ways by different sociocultural groups (also see Balée 1994, 99–100; Milton 1991). Lévi-Strauss (1950, 467) aptly noted that such unlike utilizations of the same plant species result from cultural differences, what one may in several empirical instances now consider to be differences of historical ecology. Differences of plant utilization, some of which can be explained by fundamentally unlike modes of production, such as the distinction between foraging and horticulture, do not connote differences in the human potential to perceive, classify, name, exploit, manage, and even domesticate the biota and landscapes of given environments. It is important to understand, rather, in historical-ecological terms, how and why the relations between peoples and plants may vary quite substantively, especially in the same region (Balée 1994, 99–100; Milton 1991). Contemporary foraging (or hunting and gathering) as a mode of production has tended to be associated with low population density as well as a generalized absence of political institutions other than age and gender hierarchies; also, little surplus food tends to accrue from gathering, fishing, and hunting, and permanent or semipermanent settlements are uncommon. Some apparent exceptions to this pattern included certain Northwest Coast societies (as is widely known), the Ainu of Sakhalin Island and environs (Ohnuki-Tierney 1984; Watanabe 1972, 59–64, 80), and perhaps a number of late prehistoric societies of southeastern North America (Kidder and Fritz 1993; Marquardt 1992), where the exploitation of marine resources was evidently sustainable and highly efficient. With the possible exception of the estuarine Warao of northern South America (cf. Steward 1977, 135), however, the foraging societies of the humid tropics of South America may have generally regressed from a horticultural mode of production in the past (Balée 1992a, 1994, 1995; Lathrap 1968; Roosevelt 1992; Stearman 1984, 1989).[1] Incidentally, the term *regression* as I am using it is not intended to carry pejorative connotations, nor would I wish to imply that horticulture is better than or superior to foraging as a means of production in lowland South America or anywhere else.[2] Indeed, in certain sociopolitical circumstances, even in Amazonia, horticulture as a lifestyle would probably be maladaptive for the individuals who practiced it.

To argue in principle that horticultural peoples have in some instances become foraging peoples is no less reasonable from an informed historical perspective than assenting to the proposition that prehistoric empires and states, such as Tiwanaku and the Maya, underwent collapse in part associated with a failure of agroecosystems (Culbert 1988; Kolata 1987, 1993, 283–298; Wiseman 1985—for a counterargument, see Graffam 1992). Intensive agriculture, moreover, is not typically found together with nonstratified political systems. Likewise, horticulture is usually not found among fully nomadic, strongly egalitarian societies. Utilitarian items—aside from the complex of horticulture itself—have also vanished from nonstratified societies. So have social and political institutions. As Gertrude Dole (1991, 374) observed, in a remark especially pertinent to lowland South America, "history and the ethnographic record attest to a loss of complexity in the majority of native societies in modern times." W. H. R. Rivers (1926) long ago pointed out that the canoe, pottery, and bow and arrow had disappeared from cultural inventories on different islands of Melanesia due mostly to the dying out of elders who knew but did not pass on certain ritual as well as technical knowledge regarding these crafts. The Polar Eskimo had lost, for a time during the nineteenth century, the bow and arrow, the kayak, and the three-pronged fish spear, so they were at that time forgoing caribou hunting, open-water sealing, and the taking of salmon in lakes and streams (Vanstone 1972, 38, 48). These losses had been attributed to their extreme isolation from other peoples, hence, they were a consequence of what one might analogously term cultural drift.

Yet the complete loss of a "useful" complex, such as horticulture and its associated crops, in which much practical consciousness (Giddens 1987, 63) and habitual behavior over many generations have been invested by an entire society, may seem counterintuitive in certain regions. In arable lands of lowland South America where a developed horticultural people regressed to a foraging mode of production, one encounters the paradox first raised by Pierre Clastres (1968), wherein land is arable, horticulture is to be expected, yet foraging is instead sometimes exclusively present. Absence of horticulture in arable and inhabited lands requires an explanation, one that obviously cannot be based on environmental determinism. Perhaps in the face of the data from Lathrap (1968) that suggested some Upper Amazonian groups had regressed to trekking or foraging from more horticultural beginnings, Steward (1977) (following John Cooper) classified the foragers of arable lands in lowland South America as "internal marginals," a kind of residual category that was clearly situated beyond his explanatory framework. In fact, a process of agricultural regression may be as intelligible and as convergent as many authors have claimed with regard to the origins and development of agriculture. The remainder of this chapter presents some of the evidence that supports such a process of agricultural regression, which would have involved a transition in the mode of production with regard to the Guajá Indians. In addition, I document some of the possible consequences of this transition for botanical vocabulary. The reasoning behind my argument is based mainly on a controlled comparison of the ethnobotanical systems of the Guajá and Ka'apor. I also invoke ancillary evidence collected among other Tupí-Guaraní–speaking peoples in making this comparison. The comparison is essentially controlled because the Guajá and Ka'apor are peoples who speak languages of the same linguistic family known as Tupí-Guaraní and who exploit the same botanical habitat, namely, the extreme eastern part of Amazonia in Maranhão state, Brazil, sometimes called pre-Amazonia.

The Guajá: Former Horticulturalists?

The Guajá number approximately 230 persons. Somewhat more than half the population lives in one of three permanent villages centered around FUNAI posts in the basins of the Turiaçu, Caru, and Pindaré Rivers. Much of the region has been the target of major, illegal invasions by squatters and loggers since about 1989. It now seems unlikely that even a small percentage, perhaps 20 percent, of the forests of this region of about one and a half million hectares, which were essentially intact as of about 1970, will survive much longer.

The majority of the region is occupied by the Ka'apor, a horticultural people whose ethnobotany I have attempted to describe elsewhere (Balée 1994) and the Guajá, a foraging people who seem to have possessed, at least until quite recently, no truly domesticated plants. Although Guajá has a term for "planting" (which may be polysemous with "burial" and other terms related to "digging"), informants claim, when queried about planting and harvesting techniques, that they were never horticulturalists before: *nə tum awa* ("We [the Guajá] do not plant [crops]"). There is some evidence, however, that the antecedents of the Guajá were a horticultural people. Some of this evidence appears in earlier work (Balée 1992a, 1994); additional evidence surfaced in research by Ricardo Nassif and me carried out during 1991–1992. This evidence, as with the earlier data, is mainly of a linguistic sort.

First and most obvious, the Guajá speak a language of the Tupí-Guaraní family. It does not appear that the Guajá were colonized by other Tupí-Guaraní speakers. Rather, Guajá, as with many other Tupí-Guaraní languages of the present, including Ka'apor, is descended from a mother language, Proto-Tupí-Guaraní, that was spoken in the Amazon at about the same time Latin was spoken in Rome. Because this language contained words for domesticated plants (Lemle 1971), it can be assumed that it was once associated with horticultural society. Apart from this truism, it can be argued that the Guajá language exhibits evidence both for and against a horticultural past.

Suggestive evidence of a horticultural past for the Guajá is to be found in the apparently cognate terms that the language exhibits for some traditional domesticates of lowland South America. These mere nine terms are listed in Table 5.1. In contrast, out of a total of sixty-seven Ka'apor generic names, including those referring to both introduced and traditional domesticates of lowland South America, at least twenty-five are apparently cognate terms—see Balée (1994, 352–356). The Guajá terms listed in Table 5.1 are, by inspection, cognate with words in other Tupí-Guaraní languages associated with horticultural society.[3] In other words, although the Guajá have not cultivated any of these plants in historical memory (a point independently confirmed by Ricardo Nassif [1993]), the words for them have persisted in the language, perhaps because the Guajá abandoned these last in a processual loss of domesticates and because one or more of these plants continued to

Table 5.1. Cognate generic names for traditional domesticates in Guajá

Guajá Generic Name	Botanical Referent	English Name
araku-ate	*Bixa orellana*	Annatto tree
kamana'ĩ	*Phaseolus* sp.	Lima bean
kara	*Dioscorea trifida*	Yam
kwi	*Crescentia cujete*	Calabash tree
nana'ĩ	*Ananas comosus*	Pineapple
paku	*Musa* sp.	Banana
urumũ	*Cucurbita moschata*	Squash
wači	*Zea mays*	Maize
waya	*Psidium guayava*	Guava tree

be the object of Guajá raids on other, horticultural peoples' swiddens and fallows until recently (Balée 1992a, 1994).

In addition to the cognate evidence, some patterns of plant nomenclature in Guajá are also similar to those seen in several other Tupí-Guaraní languages of Amazonia (Balée 1989b; Balée and Moore 1991, 1994). One of these patterns concerns names for nondomesticated plants modeled on the names for domesticated plants, of which there are only four such names in Guajá (see Table 5.2).[4] In contrast, the Ka'apor botanical lexicon exhibits forty (or ten times more) such names (Balée 1994, 194–195, table 7.4). The names for nondomesticated plants in Table 5.2 are marked in relation to their domesticated models. A linguistically marked term involves a base (the first element, or the name for the domesticate or model) plus an overt mark (the second element, which is an attribute often translating as "false" or "similar") (see Witkowski and Brown 1983, 569).

In several other Tupí-Guaraní languages associated with horticulture, names of wild plants are sometimes modeled on names for domesticates, usually an indication of the historical primacy of the name for the domesticated plant (Balée 1994; Balée and Moore 1991). This comment needs to be conditioned with the term *usually*, because one may counterargue that in certain instances marking reversals for names of domesticates could have occurred: "'Marking-reversals' are dramatic examples of lexical change resulting from shifts in cultural importance" (Witkowski and Brown 1983, 570). In many Tupian and Cariban languages, the terms for the domesticated dog and the panther or jaguar share the same base or even the same term. In Ka'apor, the word for dog is *yawa*; the panther is called *yawa-pitã*, "dog-red." The word for panther is marked with the postponed adjunct for "red," whereas the term for dog is unmarked. Both species occur in the habitat, the dog,

Table 5.2. Guajá generic names for nondomesticates modeled by analogy on names of domesticated plants

Guajá Generic Name	Gloss	Botanical Referent	Botanical Model
araku-rána	Annatto-false	*Bixa arborea*	*Bixa orellana*
mani'o-'ɨ	Manioc-stem	*Buchenavia parvifolia*	*Manihot esculenta*
waya-ran-'ɨ	Guava-false-stem	*Myrciaria floribunda*	*Psidium guayava*
arapa-mani'i-'ɨ	Brocket deer-manioc-stem	*Manihot* spp.	*Manihot esculenta*

of course, being closely associated with people and the settlement. But the dog arrived in the region probably via early Portuguese colonists, and thus long after the panther. In other words, when the dog was first introduced, the new indigenous term for it was probably marked, whereas the term for panther may not have been, or if it was marked, it would likely have been so in relation to some other native feline of the forest, such as the spotted jaguar.

As for plants, the word for the banana's most closely related nondomesticated relative in lowland South America, *Phenakospermum guyannense* (A. Rich.) Endl. ex Miq. (Strelitziaceae), is a case in point. Although there is some suggestive evidence that a term for banana reconstructs in Proto-Tupi, the mother language of Tupí-Guaraní, which was probably spoken in the southwestern Amazon Basin at the time Italic was spoken in the Italian peninsula (Aryon Rodrigues, personal communication, 1990), the genetic and biogeographic data, with which I have no quibble, strongly support a Southeast Asian origin (specifically, in the area of the Malay Peninsula and major outlying islands, including Borneo) (Harlan 1992, 209; Simmonds 1982, 308–309). As is widely known, Friar Tomás de Berlanga brought bananas from the Canary Islands to Santo Domingo (Hispaniola), where he successfully introduced them, in 1516 (Reynolds 1927, 31–32; Simmonds 1982, 310; Smith et al. 1992, 273). The crop reached Mexico, where it also had been evidently known before, by 1531 (Reynolds 1927, 34). The generic term for the domesticated banana in Ka'apor is *pako*, a seeming cognate with the Guajá term *paku*. Actually, similar terms for banana are found in many Tupí-Guaraní languages. The question is whether even these terms have an indigenous origin. P. K. Reynolds (1927, 30–31) argued (without primary source documentation) that *pako* (or *paco*) was derived from Portuguese *bago,* which he glossed as "fruit in a bunch."[5] In botanical Portuguese, "*bago*" actually refers to a drupe. In Brazilian Portuguese generally, moreover, the term *bago* refers to each individual fruit of a bunch of grapes or to a "fruit or grain that resembles a grape" ("*fruto ou grão que lembre a uva*"; Holanda Ferreira n.d., 174). However titillating the hypothesis, there is no evidence that Portuguese colonists in Brazil of the sixteenth century called bananas by the term *bago,* even if the word *banana* appears to have been borrowed by Portuguese from an unidentified West African language; also, *bago* would not likely have been borrowed by a Tupí-Guaraní language as *pako,* or *pakóßa,* which was the probable Tupinambá term. Even in Munduruku, a macro-Tupian language, the initial consonant for the term meaning banana has been dropped, which does not occur with borrowings (Denny Moore, personal communication, 1995).

The word for the wild banana in Ka'apor is *pako-sororo,* or "hunger-banana," a clear example of marking. If the arguments that bananas did not exist in Amazonia or anywhere in the New World until 1516 are correct (e.g., Nordenskiöld 1931, 25), this may evince a marking reversal. This argument would hold that the original wild banana (*P. guyannense*) was called *pako* (or the mother term that would have been close to it morphophonemically) but the term became marked after the introduction and absorption of the banana into the nucleus of the horticultural lifestyle of the Indians.

On the other hand, other Old World crops, such as the bottle gourd (from Africa), arrived in the New World before 1516 (e.g., Lathrap 1977). By the time of the first descriptions of the cultivated plants of the Tupinambá in the mid-sixteenth century, bananas were widely present at least along the Atlantic coast from Rio de Janeiro to Salvador (Bahia) (Balée and Moore 1991; Smole 1980) if not also in the continental hinterlands, including Amazonia. If the banana was introduced early in the sixteenth century, it would have spread and become well established with comparatively high velocity across northern and eastern South America. In contrast, although manioc has been one of Africa's most important food crops since the late nineteenth century, it spread very slowly after its introduction by the Portuguese in the late 1500s at the latest, near the mouth of the Zaire River, having reached Angola only by 1660 and having still not reached the East African hinterlands by 1800 (Jones 1959, 80), that is, at least two hundred years later.

Generally speaking, it seems that bananas are not as economically important in the lowlands of South America as is manioc in tropical Africa (see Smith et al. 1992). The terms for banana in many Tupí-Guaraní languages are phonologically similar but do not seem to have been borrowed (Balée and Moore 1991). In light of the present data (and the limitations of these), perhaps the most parsimonious inference, from a historical-linguistic point of view, is that while bananas were likely domesticated in Southeast Asia, they may have arrived in the New World and entered the agricultural repertoire of South American Indians before 1516, or even before 1492. Such an inference is not unreasonable, in principle, given the fact that many other individual crops, bottle gourd and coconut being the main exceptions, would have been unlikely to float across any ocean unassisted and still be viable in the new terrestrial environment. It is simultaneously also clear, however, that only palaeoethnobotanical research, including intensive collection of pollen cores, combined with further historical-linguistic and ethnobotanical research can ultimately determine whether bananas were present in pre-Columbian South America (Smith et al. 1992, 272–273).

If *paku* descended from the pre-Columbian mother language of Guajá, it retained its meaning despite the loss of its cultivation, a remarkable but not unique survival. In fact, if the term for wild banana in Guajá, which is *yawa-ka'a*, "spotted jaguar-herb" (also *yawa-ka'a-hu*, "large spotted jaguar herb"; see Appendix I) has retained any of its principal morphemic components, these are quite unrelated to the term for domesticated banana. The reason the word for domesticated banana, *paku*, was retained at all is probably related to the fact that there are no poisonous banana varieties (unlike some manioc varieties) and the Guajá did with regularity raid the producing swiddens, replete with bananas, of surrounding horticultural peoples since at least the 1850s (Araújo Brusque 1862). But raiding the swiddens of others does not a horticultural society make.

The Guajá lexicon for recently introduced domesticated plants also has parallels in other Tupí-Guaraní languages. Names for recently introduced, nontraditional (or Old World) plants may be modeled on names for nondomesticates or other, not closely related, domesticates. In Guajá, rice (*Oryza sativa*) is called *takwari-ake*, "kind of bamboo-untranslatable," and sugar cane (*Saccharum officinarum*) is named *takwar-rána*, "kind of bamboo-false." Both of these crops are from the Old World. As a member of the grass family, native bamboo is phylogenetically and in terms of outward appearances "like" rice and sugar cane. Names for introduced domesticates tend to be borrowed, invented based on some observable quality (as with the term for a citrus species, *tai-hamãi*, "spicy-much"), or marked in relation to some other existing name, as in these two cases, which instantiate patterns observed in other Tupí-Guaraní languages long associated with horticultural societies (Balée and Moore 1991, 1994).

As with several other Tupí-Guaraní languages, moreover, Guajá has evidently retained cognate terms for several semidomesticated plant species. This retention includes several palms (such as *takamã* [*Astrocaryum vulgare*] and *inaya* [*Attalea maripa*]), a genus of large bamboos (*takwar* [*Guadua*]), *Cecropia* spp. (*ama'i*), two species of edible sapotes (*akučiterewa'i*), mountain soursop (*aračiko'a'i*), the araliaceous *Schefflera morototoni* (*maratatowa'i*), and the moraceous *Brosimum acutifolium* (*merere'i*) (see Balée and Moore 1991 for possible cognates in other languages). Terms for semidomesticates in Guajá, as in other Tupí-Guaraní languages, at this stage of analysis, seem to have been retained at a significantly higher rate than terms for wild nondomesticates.

On the other hand, some of the other linguistic evidence suggests that, whereas the original Guajá ancestors may once have been swidden horticulturalists, those of the recent past were not. Whereas in the Ka'apor language and evidently in other related languages of the Tupí-Guaraní family the words for traditional domesticates are never or only rarely modeled on words for wild plants, the Guajá language displays two intriguing exceptions. The domesticated cashew tree, originally from south-central Brazil, is called *akayu-rána*, "cashew tree-false." It is the marked term when compared to the word for the enormous, nondomesticated cashew trees *Anacardium giganteum* and *Anacardium spruceanum*, which are called *akayu* or *akayu-ate,* "cashew tree-genuine." Based on evidence from other Tupí-Guaraní languages, in which the nondomesticated cashew trees receive marked terms and the domesticated generic does not, this probably represents a marking reversal. In other words, at one time the term for domesticated cashew in Guajá was probably *akayu*, or a morphophonemically similar term, as it is today in many Tupí-Guaraní languages (Balée and Moore 1991, 1994). In Guajá, the nondomesticated cashew trees would have been named by this base plus an overt mark, such as *-ránə* (which may be glossed as "false" or "similar"). As the Guajá lost horticulture, perhaps their reliance on the edible fruits of wild cashew trees grew proportionately greater than their dependence on domesticated cashew trees, which now could be harvested only from other peoples' swiddens and dooryard gardens. Likewise, the native Guajá term for the papaya plant, originally from the western Amazon Basin, is *arakači'a-hu*, "*Jacaratia spinosa*-big."[6] The species *Jacaratia spinosa* is typically found in old fallows of the region and elsewhere in Amazonia, although virtually never for its having been planted by humans. It is a semidomesticated plant in the same family as the papaya plant, Caricaceae. *Jacaratia spinosa* produces a smaller and more acidic yet still edible fruit for the Guajá and some other Amazonian peoples. The Guajá name for *Jacaratia spinosa*, *arakači'a*, is, moreover, an apparent cognate for the same species in different Tupí-Guaraní tongues of Amazonia where I have collected data: *zarakači'a'ɨw* (Tembé [Gurupi River basin]) and *arakači'í* (Araweté [Xingu River basin]). Actually, the Assurini of the Xingu River basin may also represent an exception to the Tupí-Guaraní pattern in this case in the same way as do the Guajá (a point I failed to note previously in Balée 1994, 201), perhaps because they did not traditionally cultivate papaya. They called papaya *yarakači'a-ú,* or "big *yarakači'a*" (which could only be provisionally glossed as "*Jacaratia spinosa*-big" at this time because I neither collected nor saw papaya being grown by the Assurini at the time of my fieldwork there in 1986). At least the Guajá term for papaya, if not also that of the Assurini, suggests a marking reversal and a departure from the nomenclatural norm observed in several other languages of the Tupí-Guaraní family.

The Guajá plant vocabulary diverges in another respect from these other languages. A few generic names "lump" domesticated plants with phylogenetically related nondomesticated plants. Regardless of the intrinsic difficulties related to separating domesticates from nondomesticates, especially in Amazonia (Lévi-Strauss 1950, 465), such an artificial dichotomy of biological reality seems to be operative in Ka'apor plant classification at a latent (and perhaps even practical) level of consciousness (Balée 1994; Giddens 1987, 62–63). This dichotomy seems to be extant, moreover, in the Tembé plant classification system and other Tupí-Guaraní systems (Balée 1989b). Whereas domesticates do not tend to be lumped as to life form with nondomesticates in Guajá, exceptions belie what was probably once a more nested hierarchical system of plant classification. Guajá plant classification and vocabulary are seemingly in transition. I suspect that certain distinctions among plant names will emerge in the form of marking and even marking reversals, if the Guajá of today are fully incorporated into local agricultural society, assuming their culture and language are not both soon engulfed and extinguished by the continuing invasions of their lands by settlers and ranchers.

The lumping occurs with two generic names, *nana'ĩ* and *ako'o'ĩ*, the morpheme boundaries having not been indicated. The term *nana'ĩ* refers to the pineapple (*Ananas comosus*), as noted in Table 5.1. But as a word it also denotes at least four nondomesticated bromeliads of three genera (*Aechmea, Billbergia,* and *Bromelia*) found in the fallow forest. Although pineapples are grown by horticultural peoples in eastern Amazonia, these are not typically important food crops like manioc, bananas, sweet potatoes, and yams. Perhaps the ancestral, foraging Guajá did not find it worth their while to harvest pineapples when undertaking the palpably dangerous exercise of invading an enemy people's swidden. (This does not mean they were never eaten.) The term *nana* is the Ka'apor generic and monomorphemic word for pineapple, probably a cognate with Guajá in *nana'ĩ*. It should be considered that the Guajá term seems to have represented two morphemes, *nana* (the proto-term for pineapple proper, which was likely domesticated in the environs of the Gran Chaco; Pickersgill 1976, 16; Schultes 1984, 22) and *'ĩ*, which when in final position of a word means "not." In other words, one plausible gloss for *nana'ĩ* is "pineapple-not," an apt rubric for the nondomesticated bromeliads covered by the term.[7] That it does not, in a morpheme-by-morpheme gloss, well denote the domesticated pineapple may mean that Guajá raiders did not often pluck pineapple fruits from their enemies' swiddens, but sought more nourishing or more transportable crops in the little time available to them. (In this regard, bananas tend to be planted near the edges of Ka'apor swiddens, growing well on soils that are less well drained than required by other crops, such as pineapples and manioc, which perform best when there is no waterlogging, as in the centers of swiddens [which tend to be sloping] or in dooryard gardens. In other words, bananas tend to be more accessible if one is approaching a swidden from the outside, whereas pineapples are either centrally located in the swidden or, more typically, found in dooryard gardens, where an enemy approach would tend to be most risky, especially if Ka'apor dogs and men with weaponry are present.) One may reasonably expect that the Guajá, who are just beginning to cultivate pineapples, as an adjunct to several other crops at the FUNAI posts of their region, may one day lexically distinguish on a habitual basis the domesticated pineapple from its nondomesticated (and generally nonedible) relatives, given that they are beginning to distinguish lexically between sweet and bitter manioc (see below).

The other example of lumping concerns the term for domesticated and nondomesticated cacao: *ako'o'ĩ*. The domesticated treelet cacao may soon be lexically distinguished from its quite common congener of old fallows, *Theobroma speciosum*. The lexical differentiation of these two species in Ka'apor has been established by two folk generics: *kaka'ĩ* for the domesticate and *kakaran'ĩ* ("cacao-false") for certain nondomesticated congeners including *T. speciosum* (Balée 1994, appendix 9) (see Chapter 9).

The linguistic evidence for and against a horticultural past for the Guajá is most economically understood in terms of the passage from one type of society to another. Some elements of the lexicon suggest the survival of horticulture. Others would indicate a lifestyle based on trekking or foraging. It is this mixed heritage, early horticulture and later foraging, that current Guajá plant terminology best reflects.

Lexical and Classificatory Implications of Foraging

Classificatory and nomenclatural consequences of foraging lead to differences between a foraging people's ethnobotany and that of a horticultural people, such as the Ka'apor. Although the Ka'apor and Guajá occupy the same habitat, and have done so for more than one hundred years, their classifications of this habitat are somewhat different. It is clear that the Ka'apor language more finely differentiates the botanical world of pre-Amazonia than the Guajá language (all data from the Guajá are available in Appendix I; comparable data from the Ka'apor are found in Balée 1994, appendix 9). Table 5.3 shows that the total number of Ka'apor generic plant names is 40 percent higher than that of the Guajá (at 483 vs. 345). Both languages exhibit three basic life forms, and these terms are roughly equal in their semantic range: "tree" (*iwira* [Guajá], *mira* [Ka'apor]), "vine" (*iwipo* [Guajá], *sipo* [Ka'apor]), and "herb" (*ka'a* [both]). In fact, these three life form terms ("tree," "vine," and "herb") seem to be practically universal in the Tupí-Guaraní family of languages, as was originally and most astutely noted by the Brazilian naturalist João Barbosa Rodrigues (1905, 7; also see Grenand 1980).

In both the Guajá and Ka'apor ethnobotanical systems, one also encounters unnamed categories of nondomesticates unclassified as to life form as well as domesticates (although the latter category is not in binary contrast to nondomesticates in the Guajá system, as it seems to be in the Ka'apor system; Balée 1989b, 1994, 179–181). The number of Ka'apor generic names for trees is only about 20 percent higher than that of the Guajá (257 vs. 214). The number of Ka'apor names for vines is actually 62 percent higher than the corresponding number in Guajá (60 to 37). Interestingly, the number of generic names for herbs in Guajá exceeds the Ka'apor number by nearly 9 percent (63 to 58).

If only the generic names in the traditional Tupí-Guaraní life forms for "tree," "vine," and "herb" are compared, the total Ka'apor number is only 16 percent greater than that of the Guajá (375 to 314). The greatest differences in numbers of folk generic names within Table 5.3 concern domesticates and unclassified nondomesticates. As for generic names for domesticates, Ka'apor has 179 percent more than does Guajá (67 to 24); regarding generic names for unclassified domesticates, Ka'apor has 486 percent more than Guajá (41 to 7). A large part of the reason there are many more folk generic names for plants in Guajá is clearly related to the numeric differences in these two categories. While it is obvious why the Ka'apor language should have a substantially higher number of generic names for domesticates than Guajá, it is perhaps less clear why Guajá should have also many fewer names for un-

Table 5.3. Comparison of types and number of generic plant names in Guajá and Ka'apor

Life Forms	No. of Generic Names in Guajá	No. of Generic Names in Ka'apor
Trees (*iwɨra* [G], *mira* [K])	214	257
Vines (*iwɨpo* [G], *sipo* [K])	37	60
Herbs (*ka'a*)	63	58
Domesticates	24	67
Unclassified domesticates	7	41

classified nondomesticates. I submit that both differences are comprehensible by reference to the distinction between foraging and horticulture. Because of horticulture, there are more "anomalous" plants, to borrow Lévi-Strauss' terminology, in the language of the people who practice it.

The differences of classification and nomenclature at the folk generic level of discrimination are substantial, but not as striking as those of lower ranks of ethnobotanical classification (Berlin 1992), especially the folk specific and varietal ranks. At this point, it is fair to say that the Guajá language does not lexically distinguish folk species of domesticated plant generics. Only in a few locales, moreover, is there a distinction between sweet and bitter manioc plants; at the time I carried out research in the Pindaré basin in 1990, both were called *tarəmə* plants (there is now the additional term borrowed from Portuguese—but the etymon having been from Proto-Tupí-Guaraní—*makači* for "sweet manioc"; Ricardo Nassif 1993). In contrast, Ka'apor has three nonsynonymous, natural folk generic names (*mani'ɨ, makaser,* and *maniaka*) and a total of twenty-four nonsynonymous folk specific names designating *Manihot esculenta* alone (Balée 1994).

Underdifferentiation of generic domesticates may be expected from a foraging society that only quite recently has begun to adopt (or readopt) a horticultural way of life (e.g., see Berlin 1992, 285–288; Brown 1985).[8] Berlin and Brown independently suggested that languages associated with foraging peoples tend to have fewer folk specifics and far fewer (if any) folk varietals than languages associated with some kind of horticultural mode of production (with the possible exception of the Seri of the Gulf Coast of Sonora; Berlin 1992, 288–290). Brown (1985) suggested that this was related to the fact that foraging peoples tend to depend on (nondomesticated) resources that are less subject to failure than are domesticated crops; therefore, horticultural peoples must be more broadly familiar with a given flora, a familiarity that presumably is encoded in the classification and nomenclature of plants, than foraging peoples. Moreover, I would argue, once horticulture has become adopted and fully a part of a society's infrastructure, domesticates, perhaps more so even than "trees" and other nondomesticates, assume a semantically primitive character (cf. Friedrich 1970, 8). This does not mean ipso facto that horticulturalists' and foragers' potentials to classify and encode biological reality are any different; indeed, I would argue categorically that they are not.

Whatever the exact reasons, it is clear that the Guajá language exhibits substantially fewer folk specifics than does Ka'apor for the flora as a whole and it has no folk varietals (Balée 1994; cf. Appendix I). I now recognize 28 folk specific names (excluding monospecific folk generic names) in Guajá and 252 folk specific names (excluding monospecific folk generic names) in Ka'apor. In other words, Ka'apor has exactly nine times more folk specific plant names than Guajá, a finding that strongly bears out the original formulations of Brown and Berlin. This difference is not merely due to the abundance of folk specific names for domesticates in Ka'apor versus the paucity (in fact, complete lack) of such names in Guajá (see Appendix I). Although it is true that the Ka'apor have 104 folk specific names for domestics (or 41 percent of all their folk specific names for plants), more than half the folk specific names in Ka'apor refer to nondomesticates. And this number of folk specific names in Ka'apor referring to nondomesticated plants, 141, is still five times greater than the corresponding number in Guajá, 28. The rank of folk varietal is completely unrepresented in Guajá; it is present in Ka'apor, referring to certain cultigens of banana and sweet potato (Balée 1994, 183–184).

Conclusions

The frequently made statement that hunter-gatherers must have more specific knowledge of all the organisms of their habitat than other peoples, as in the comment "most hunter-gatherers have an exhaustive knowledge of the resources of their habitat" (Meggers et al. 1988, 281), is not borne out in the lexical evidence for nomenclature and classification of a given flora by two different ethnobotanical systems, one that of a foraging people and the other that of a horticultural people. Elsewhere (Balée 1994) I have shown that the Guajá use and exploit far fewer plants from the same environment than do the neighboring horticultural Ka'apor. They have traditionally exerted less influence on the botanical environment than do horticultural peoples, such as the Ka'apor, because they do not create light gaps of appreciable size (such as swiddens and settlements), which seem to be correlated with increasing botanical diversity (Balée 1994, 135–138). It seems probable that in the process of agricultural regression, the Guajá lost more than just domesticated plant species themselves and the "arts" of Amazonian horticulture; they also lost knowledge about the uses to the Proto-Tupí-Guaraní people of some wild and semidomesticated plants as well. Since they did not have to fell trees for horticulture, they would not have consistently been visually exposed to the differences between the freshly cut white and red heartwoods of the two types of crabwood (*Carapa guianensis*), a difference that has been noted and linguistically encoded by loggers in the Guianas (Pennington 1981, 414) and Ka'apor swidden-clearers (Balée 1994, 183). Even "trees," therefore, can be expected to be more semantically primitive (see Friedrich 1970, 8) in the language of a horticultural people than in that of a foraging people who exploit the same, otherwise arable, environment, at least since any and all trees of the *terra firme* may be felled for making swiddens whereas not all trees are suitable for other utilitarian purposes, including firewood, to any one group.

A comparatively less finely differentiated vocabulary for these botanical domains supports such a hypothesis. It is certainly clear that the Guajá (as with the foraging Yuquí and the trekking Sirionó; Stearman 1989) also lost "useful arts" other than cultivation, such as fire-making ability from a drill and hearth (which John Cooper [1949, 283] originally did not think had occurred anywhere in lowland South America, despite scattered reports to the contrary), so this hypothesis has lost some support from other domains of the culture.

These losses of specific kinds of agroecological knowledge seem to be associated with decreased population density; increased nomadism; fewer interlocutors with whom to discuss plant names, categories, lore, and knowledge; and an increasing lexical underdifferentiation of the universe of the local flora due to its desuetude. In more general terms, the gradual transformation of a people's mode of production seems to affect botanical vocabulary in fairly distinctive and predictable ways.

6

Conquest and Migration

The differences in the culture of the various tribes cannot be explained by their living under different natural conditions, but only by their history.
—Baron Erland von Nordenskiöld (1924, 3)

Amazonian environments present human languages with a formidable job: symbolic representations of a vast domain of visible, organismic minutiae. If languages adapt to people and their environments—if languages resemble viruses in their relationship to people through time (Deacon 1997, 112)—the biological richness of the environment in which they occur should be partly evident in vocabulary. Eskimo words for snow, Australian Aboriginal words for sand, Hanunóo words for rice, and American English words for automobiles seem indexical of customary features of the respective environments of these languages (cf. Wierzbicka 1997, 10–11). In the Amazon, words for plants should be correspondingly numerous and plant classifications complex. This is, indeed, the case, but for reasons of historical ecology there are differences of degree among Amazonian languages in terms of nomenclature and classification of plants, which to the naked eye are among the most diverse of organisms in any one habitat. Historical ecology conceives of languages and landscapes as forming rational, irreducible entities that cannot be productively analyzed in isolation from each other (Balée 1998). Language changes environments through the instrument of culture; environments change languages because languages must adapt to them in the course of migration and contact of the people who speak them. Together, these interactions constitute landscapes that can only be grasped simultaneously as rule-governed products of ecology and history acting in tandem.

This chapter concerns similarities and differences between Sirionó on the one hand and other languages of the Tupí-Guaraní language family on the other in terms of the encoding of plant names and their meanings over time. I examine how Sirionó diverges from other Tupí-Guaraní languages by degree; the divergence is understood in reference to the historical interactions, as discussed below, with landscapes of the eastern Bolivian Amazon by antecedents of the Sirionó. I propose that Sirionó ethnobotany mainly supports certain general patterns of plant nomenclature and classification already observed for the Tupí-Guaraní family, putting to rest old questions about whether Sirionó qualifies as a full member or not of the Tupí-Guaraní family, from the perspective offered by the ethnobotanical data. These data suggest that Sirionó as a language and as a culture experienced a unique but not incomparable relationship with Amazonian landscapes over time.

A History of Trekking in Lowland Bolivia

The Sirionó, who speak a Tupí-Guaraní language, have lived in the biogeographic region known as the Llanos de Mojos since before the end of the seventeenth century. Ancestral Sirionó were described in 1693 as threatening to shoot arrows at a group of Jesuits and their neophytes (Anonymous 1781, 104–105; Block 1994, 38–39; Distel 1984/1985, 159; cf. Rydén 1941, 23). The missionaries were seeking to make contact with the Guarayo, who were believed to be cannibals (Cháves Suárez 1986, 236–237). The Guarayo have been historically located to the southeast of the Sirionó, and they also speak a Tupí-Guaraní language, perhaps most closely related to the Guaraçug'wé (Pauserna), whose speakers are located to the northeast of the Sirionó along the Guaporé/Iténez River.[1] In fact, the Sirionó are similar but not identical to the Guarayo and the Guaraçug'wé in language and culture, and by 1693 the Sirionó were probably already a distinct people (Anonymous 1781, 105; Holmberg 1948, 455).

The Guarayos had been known to the Spanish by name at least since 1674 (Block 1980, 35) and evidently became a separate people from the Guaraçug'wé sometime during the period 1525–1741 (Riester 1977, 31, 33). They were largely settled in a mission town by 1794 (René-Moreno 1888, 208–209). In 1696 the Sirionó were recorded as one of thirty-seven distinct Indian "nations" in the lowlands of eastern Bolivia (d'Orbigny 1946, 181, cited in Chávez Suárez 1986, 20–21). These Tupí-Guaraní–speaking peoples were distinguished from the Moxos Indians (Anonymous 1781), of an Arawakan tongue, who were the principal denizens of the Jesuit missions in the Llanos de Mojos (Block 1994; Nordenskiöld 1924, 1–2). The historical descendants of the Moxos and the mission culture that enveloped them during the 1700s became the *mestizos* of the Llanos de Mojos, known today as the Benianos (Block 1994).

But the Sirionó were not mentioned as having visited a mission until 1765 (Holmberg 1948, 455), and there is no record of the Sirionó having been missionized in the vast expanse of lands between the mission town of Trinidad on the Mamoré River and the Brazilian border during the Jesuit mission period from 1674 to 1767 and the period of mission culture, which continued into the mid-1800s (Block 1994, 143). Whereas for the Moxos dwellers of the missions before 1767 the sedentary lifestyle, profusion of Western trade goods, and agricultural surpluses caused escape into the forest to be deemed "unthinkable" (Block 1994, 143), the Sirionó were evidently throughout this period and until the early twentieth century quite independent of such missions as settlements. As late as 1888, the Sirionó (called then *los Sirionós bárbaros,* "the barbaric Sirionó") had yet to be missionized or otherwise settled in a contact situation (René-Moreno 1888, 121). There were further attempts to missionize them early in the twentieth century, and these were partially successful then (Lunardi 1938; Rydén 1941); today most Sirionó live in a settlement founded by Protestant missionaries and most of the adults seem to classify themselves as *creyentes* ("believers"). But there is no evidence that the Sirionó were brought into contact with a mission culture speaking Língua Geral, a Tupí-Guaraní creole spoken elsewhere in Amazonia such as in the northwest Amazon, since the Jesuit missionaries of the region would have spoken Mojos or Spanish in order to gain converts.

No substrate influence has been demonstrated for Sirionó, though doubts have persisted among some scholars that the language and people might have been Guaraní-ized (Cardús 1886, 280; Nordenskiöld 1924, 233; Rydén 1941, 37; Susnik 1994, 86), that is, that Sirionó had been an earlier language transformed by contact with the historic migrations of Guaraní peoples from the south, in what is now Paraguay, during the fifteenth and sixteenth centuries. I argue that what happened to the Sirionó people historically, however, as with other Tupí-Guaraní peoples who lost the practice of agriculture partially or entirely, such as the Guajá, could have influenced components of the language, especially the lexicon, quite apart from any substrate influence, real or imagined. Even though it is likely that Sirionó has had foreign influences because of its highly distinctive phonology in the Tupí-Guaraní language family (Dietrich 1990, 58, 97), it is, nevertheless, a full member of that family. And within that language family there are certain patterns of plant nomenclature (Balée 1989b) that seem to be replicated in Sirionó, to be explored below.

The opposite point of view is derived from incomplete observations on and inferences about Sirionó culture, not language. The pioneering ethnologist Nordenskiöld stated that argument about Sirionó as follows: "To judge from the few words we know of their language, they speak Guarani. Presumably this was not their original language, and they must have been Guaraniized. They are the only Indians of this area who do not raise any crops, and live exclusively on the chase, fishing, and the gathering of wild fruits. They have not settled down.... As far as we can judge, the Siriono are the tribe in this area that promises to keep its originality longest" (Nordenskiöld 1924, 233). However original and unique the Sirionó culture, it is a historical product of the gradual transformation of a more robust and diverse agricultural economy associated with primordial Tupí-Guaraní peoples. And it is now fairly clear, though perhaps it was not to Nordenskiöld, who evidently never visited a Sirionó encampment per se and only saw some who were in a mission, that the Sirionó were not entirely lacking in domesticated crops and agricultural techniques. Even the language suggests a continuous association with agriculture since Proto-Tupí-Guaraníˊan times, although some crops vanished over time.

What seems more likely than Nordenskiöld's suppositions about the origins of the language is that Sirionó is an intrusive Tupí-Guaraní language whose location in eastern Bolivia is a product only of one or more Guaraníˊan migrations, as with Chiriguano, Guarayo, Pauserna, and the very closely related Yuquí (Holmberg 1985, 11; Stearman 1989, 22; Townsend 1995, 27).[2] In other words, Sirionó is as fully a member of the Tupí-Guaraní family of languages as are those mentioned above and as are Tupí-Guaraní languages of eastern Amazonia, such as Araweté, Assurini, Guajá, Ka'apor, Tembé, and Wayãpi, and it is suitable for the purpose of cross-linguistic comparison with those languages. My primary argument is that Sirionó shares features of plant nomenclature and classification previously identified for these eastern Tupí-Guaraní languages (Balée 1989b; Balée and Moore 1991, 1994) that were not borrowed but are the historical-ecological product of the long-term encounter between human languages and Amazonian landscapes.

The argument that Sirionó is a language absorbed, and only imperfectly learned, by an ancient non-Tupian people conquered by the Guaraní or missionized by the Jesuits has an analogue, if not its source, in the view that Sirionó culture was depauperate in knowledge of religion, mythology and cosmology, construction and crafts, and technology related to exploitation of their habitat (see Isaac 1977, 138, and Stearman 1984, 630–631 for reviews; also Nordenskiöld 1924, 203; Rydén 1941, 8; Steward 1948, 899). The Sirionó were a trekking people who did not remain in a village year round. Trekking as an environmental adaptation involves some dependence on fast-producing domesticated crops combined with high mobility. It is a curious fact that it seems to be confined to lowland South America. Other groups that exhibit trekking as a mode of adaptation include some of the Maku, Yanomamö, Araweté, and Waorani. Trekking seems to be transitional between hunting-and-gathering and settled horticulture.

So the Sirionó were not in the technical sense hunter-gatherers, a technology that by the mid-twentieth century had not yet been studied in Amazonia, though it has been since for the Guajá, Yuquí, Aché-Guayaki, and Xetá.[3] Perhaps most striking in terms of technology is that the Sirionó were one of the first peoples to be described in Amazonia as lacking fire-making technology (Holmberg 1985, 11). Absence of the fire-making technology of wooden drill and hearth (from just a few select tree species) is not characteristic of any known language family in lowland South America, nor is the absence of fire-making technology more generally the case with any other human language family probably since the origin of language itself (Chapter 5). The Sirionó share this negative feature, however, with a few other groups in the Tupí-Guaraní language family, and no doubt in other families, who are either hunters and gatherers or, like themselves, trekkers, among them being the Guajá (Balée 1999), the Yuquí (Stearman 1989), and some of the Parakanã (Fausto 1997, 86). In no case is absence of fire-making technology seen with groups practicing a more sedentary horticultural mode of production in the Amazon, which has tended to include major dependence on bitter manioc among many other cultivars. Fire making is clearly a feature that they lost (Holmberg 1985, 11), not one they never had. The Sirionó have possessed domesticated crops at least since any mention was made of their agricultural technology, though these have been few and of a smaller inventory than that of Proto-Tupí-Guaraníanian peoples, which likewise suggests a depression over time in number of domesticates, because of vagaries of migration and contact with other peoples.

The environment alone cannot account for these prehistoric phenomena. The well-drained lands of the habitat, which occur exclusively on prehistoric earthworks not built by ancestors of the Sirionó, receive about 1,800 mm of rainfall per year (Townsend 1995, 18) and are rich in organic soils analogous to *terra preta* of Brazil (Langstroth 1996, 105, 250), factors that clearly favored dense populations dependent on agriculture in the past (Denevan 1966; Erickson 1995). The agricultural potential of this landscape was probably enhanced, not reduced, by prehistoric settlement and utilization, so the fact that the Sirionó planted few crops and had no means of fire making cannot be attributed to it. Rather, the source of regressive changes in Sirionó agricultural technology can only be comprehended historically, through the course of migrations and contacts to which the people and their language would have been exposed. The influences of such migrations may in general be seen in the Sirionó plant lexicon, though mechanisms have yet to be identified in order to account for the specific lexical changes.

Ant Trees, Calabashes, and Bottle Gourds

Françoise Grenand (1995) raised the point that by restricting the comparison of Tupí-Guaraní plant names to the referential level of botanical species, Balée and Moore (1991) missed showing possibly cognate forms that had wider, though related, botanical ranges of meaning. Although she did not demonstrate exactly how the referents of native terms could be restricted systemically, she did present useful objections that offer clues to language change in the tropical forest.

In the specific case of the spiny palm *Astrocaryum vulgare* Mart., Balée and Moore (1991, 232) reported *tukumã'ɨ* (K), *tukumä* (T), and *awala* (W).[4] The term *awala* is a recent borrowing from Carib (Grenand 1995, 31) and accounts for noncognacy with K and T here. The K and T names have their origins in a Proto-Tupí-Guaraní tongue. But a closely related term has not been lost in W. That term is *tukuma,* and it is cited as a "ritual name" of the closely related, and in the Wayãpi habitat far more common, *Astrocaryum paramaca* Mart. (Grenand 1995, 31–32). Grenand (1995, 23–24) argues correctly that in the course of migrations, Tupí-Guaraní words for plants have changed either internally or in their semantic range. In addition to existing plant names being adapted to new but related environmental conditions, as with the name for *Astrocaryum paramaca,* plant names may be extended to new and unrelated environmental conditions where the uses of the species are very similar or the same (Grenand 1995, 25–26). Some words do not need to adapt in the course of migration because their referents are universally present and highly distinctive, like the genus *Inga* with its extrafloral nectaries. The numerous species of *Inga* are typically grouped under a cognate generic label such as *sínga* (S) and *ínga* (K) (of which two terms the S one is probably closer to the proto-form than the K term) in Tupí-Guaraní languages throughout their range (Grenand 1995, 25). Similar processes of linguistic change and conservatism have been reported recently for British names of American birds (Brown 1992).

While a model of adaptations of plant nomenclature to new environmental conditions encountered because of the migration of Tupí-Guaraní people remains to be constructed, S also supplied evidence for such adaptations, though not all of the sort anticipated by Grenand (1995)—meaning, of course, that any such model if genuinely predictive will be necessarily complex. In part, the problem can be approached by determining how reflex forms for an original referent(s) do or do not cover the same semantic range as the original word. The terms for species of ant trees, *Tachigali* of the Fabaceae family and *Triplaris* of the Polygonaceae, are such an example. Balée and Moore (1991, 226) showed the reflexes for *Tachigali myrmecophila* Ducke or for *Tachigali paniculata* Aubl. in four languages, *táci'i* (Ar), *tači'ɨwa* (As), *taši'ɨ* (K), and *tači'ɨw* (T). In W, members of the same genus are called *tasi* (F. Grenand 1989, 102; P. Grenand 1980, 253). The ant commensals of the tree genus *Pseudomyrmex* spp. attack and sting intruders when any part of the tree, their home, is touched.[5] As for *Triplaris,* it harbors the same ants, but its distribution is not found in the habitats of Ar, As, K, and T, though it is seen in that of W, where it is named with marked forms: *móyu-tasi,* "anaconda-*tasi*," *móyu-tasi-pilã,* "anaconda-*tasi*-reddish," and *móyu-tasi-sɨ,* "anaconda-*tasi*-white" (Grenand 1989, 305). It is unlikely that a marking reversal would have occurred, since these species are not domesticated or cultivated. Therefore, it can be suggested that the proto-form **tači'ɨß* (my reconstruction) denoted one or more species of *Tachigali* first. In S, *tási* refers to *Triplaris americana* L. alone among plants and to the stinging pseudomyrmecine ants that inhabit its petioles and stems. The genus *Tachigali* is absent from the habitat. It can be hypothesized, therefore, that the aboriginal generic word adapted to changed environmental circumstances in the Llanos de Mojos and came to refer to a species whose name was derived from a more widespread genus of trees.

The disappearance of a name for a plant is also associated with this phenomenon. This is the case in S, in which the term for calabash tree (*Crescentia cujete* L. of the Bignoniaceae) is a reflex of the term for bottle gourd (*Lagenaria siceraria* Mol. of the Polygonaceae). In other words, the reflex for calabash is missing in S, though the plant is present in the habitat. Cognate terms for calabash occur in Ar, As, K, T, and W (Balée and Moore 1991, 239) and are close to *kwi* (K), whereas the term for calabash in S is *ia-í* (the fruit is *ía;* cf. Schermair 1958, 113).[6] Yet *ia-í* is in a broader sense related to *iá* (T), *ia* (Ch), *ía* (Gy) (Dietrich 1986, 354), and probably *anái* (Gw) (Horn Fitz Gibbon 1955, 25), which denote the bottle gourd. Here it can be argued that somehow S *lost* the bottle gourd while keeping its name and retained calabash while losing its name, as a result of migration and new environmental circumstances. The mechanisms that could account for such specific kinds of change remain obscure but will need to be elucidated in order to construct a systematic model for lexical change over time.

Sirionó Plant Nomenclature

Sirionó nevertheless exhibits supporting evidence for definitive patterns of plant nomenclature identified earlier in the Tupí-Guaraní family (Balée 1989b). First,

S shares with Ar, As, Gj, K, T, and W (Balée 1989b, 6) the same three botanical life form labels: *íra* (roughly, "tree"), *kiáta* (roughly, "herbs and grasses"), and *isío* (roughly, "vine"). As with the other Tupí-Guaraní languages studied, there is no single term for "plant" in the taxonomic sense and there is no separate life form term for "palms" in S. Two life form labels are also polysemous, as with the other languages: *íra* also refers to "wood" and "objects made from wood" (Schermair 1958, 133–134); *kiáta* also refers to "forest" as well as "bananas" (cf. Schermair 1958, 209), which are recent introductions to the Sirionó.

Second, plant names for nondomesticated plants are sometimes modeled on names for traditionally domesticated plants, but the reverse marking procedure does not seem to occur. In S, nondomesticated congenerics of the sweet potato (*Ipomoea batatas* Lam.) are called *ñiti-rémo,* "sweet potato-vine stem," a reference to their classification as nondomesticated vines, not the domesticated sweet potato. It is in this sense cognate with numerous terms that refer to nondomesticated *Ipomoea* as "sweet potato-similar (or false)" (Ar *yiti-rī*, As *yɨti-rána*, K *yɨtɨk-ran*, T *zɨtɨk-ran,* and W *yetɨ-lā;* Balée and Moore 1991, 233), since life form labels are semantically equivalent as markers of nondomesticated or "false" status of a marked term modeled on analogy with a domesticated plant name. Likewise, the S term *urúku-mbwé,* "*Bixa orellana*-false," which refers to nondomesticated species and varieties of annatto (*Bixa orellana* L.), is cognate with K and T *uruku-ran* (Balée 1994, 277).

Third, numerous introduced domesticated treelets (including mango, coffee, tamarind, cacao, guava, and catappa nut) are named simply *íra,* "tree," there being no other S name for them; life form labels in other Tupí-Guaraní languages are applied to domesticated plants evidently only when these are recent introductions (Balée 1989b, 12; Balée and Moore 1991, 218). Likewise, bananas and plantains, as noted above, are referred to generically as "herbs."

Finally, many plant names have been lost in S through neologisms, borrowings, and ultimately peculiarities of Sirionó history and changes in environment. But names for five traditionally domesticated species have been retained in S, as shown in Table 6.1.[7] And S has retained terms for several (though not all or even many) semidomesticated plants, such as *akyačái* (Ar *akāya'í,* As *kayuwa'ɨwa*) for edible hog plum (*Spondias mombin* L.); *imbéi* (K *ama'ɨ*) for the fibrous ambaibo (*Cecropia concolor* Willd.); *kói* (Gj *kapowa'ɨ,* K *kupa'ɨ*) for the medicinal copaiba oil tree (*Copaifera* sp.); *ñikisɨaí* (Ar *arakači'í,* Gj *arakači'a,* Gw *akasía;* T *zarakači'a'ɨw;* Horn Fitz Gibbon 1955, 25) for the edible wild papaya or *gargatea* (*Jacaratia spinosa* [Aublet] A. DC.); *urúre* (As *murure-ete,* Gj *merere'ɨ,* and K *murure'ɨ*) for the edible Amazonian mulberry tree *Brosimum acutifolium* Huber; *sínga* (K *ɨnga*) for the edible *Inga* spp.; and *tákwa* (Ar *ta'akɨ,* Gj *takwar,* K *takwar*) for a bamboo used in making arrow points (*Guadua* sp.). This proclivity to retain names for domesticates and semidomesticates seems to be generalized in the Tupí-Guaraní family, and S is divergent but not an exception, even if its botanical environment is unique.

The Sirionó language has been the subject of controversies over its origins and classification. The most extreme position has questioned the legitimacy of includ-

Table 6.1. Sirionó reflexes for traditional plant domesticates

Sirionó Name	Names in Related Languages	Botanical Referent
ibási	Ar *awači*, As *awači*, Gj *wači*, K *awaši*, Gw *avásiki*	*Zea mays* L. (maize)
níñu	Ar *minɨyu*, As *aminiyu*, K *maneyu*, T *manizu*, W *minɨyu*, Gw *maníyu*	*Gossypium barbadense* L. (cotton)
ñíti	Ar *yiti*, As *yɨtika*, K *yɨtɨk*, T *zɨtɨk*, Gw *déti* or *deti*	*Ipomoea batatas* Lam. (sweet potato)
urúku	Ar *iriko'i*, K *uruku*, W *uluku*, Gw *urúku*	*Bixa orellana* L. (domesticated annatto)
uúba	K *u'ɨwa*, T *u'ɨwa-a*	*Gynerium sagittatum* Beauvois (arrow cane)

ing Sirionó in the Tupí-Guaraní family because of a presumed substrate influence, but no empirical, linguistic evidence has been supplied for that. That assumption was associated with the mistaken axiom that every Tupí-Guaraní society known had a strong agricultural basis and therefore there could not have been exceptions. The Sirionó was one of the first of a series of trekking and foraging societies studied by professional ethnographers in Amazonia; several of those studied by participant observation since have proven to be of the Tupí-Guaraní language family also, such as Guajá, Yuquí, and Araweté. In terms of the linguistic evidence alone, Sirionó is a full member of the Tupí-Guaraní language, albeit in a subgroup of its own (Dietrich 1990; also see Priest 1987) or in questionable groupings of languages that are close to it geographically but apparently distant from it in terms of vocabulary, such as Guaraçug'wé (Pauserna) and Guarayo, which are themselves close to Chiriguano. Sirionó is, nevertheless, intrusive in lowland Bolivia and it may be that in the course of migrations, probably from the south and associated with the Guaraníán invasions of lowland Bolivia in the fifteenth and sixteenth centuries, the Sirionó agricultural economy underwent changes. And so did the Sirionó language in reflecting those changes. I suggest that the early Sirionó inhabitants had lost or were in the process of losing dependence on a larger inventory of domesticated crops than they possessed in the early twentieth century and that they were becoming less sedentary. These migrations and probably contact with other cultures, perhaps of a hostile sort, led to losses of domesticates and in some cases of words for those plants. In other cases, as with the present Sirionó word for certain ant trees, the ranges of meaning of existing vocabulary changed to accommodate new conditions encountered in the new environment. Even without specifying the exact mechanisms involved in these transformations of Sirionó vocabulary, which seem intangible at present, it can be argued that Sirionó is fully a member of the Tupí-Guaraní language family in terms of plant nomenclature and classification. The Sirionó language exhibits a nomenclatural dichotomy between traditional domesticated and nondomesticated species; modeling and marking for nondomesticates based on names of traditional domesticates; and cognates for plant life forms and traditional domesticates like many other Tupí-Guaraní languages.

These similarities together with the difference that Sirionó plant names show when compared to plant names in other Tupí-Guaraní languages can fruitfully be analyzed from the perspective of historical ecology. Historical ecology considers Sirionó language change as well as language retention to be products of a long-term entanglement between the speakers themselves and Amazonian landscapes. Efforts to document Sirionó and other Tupí-Guaraní languages should reflect that entanglement.

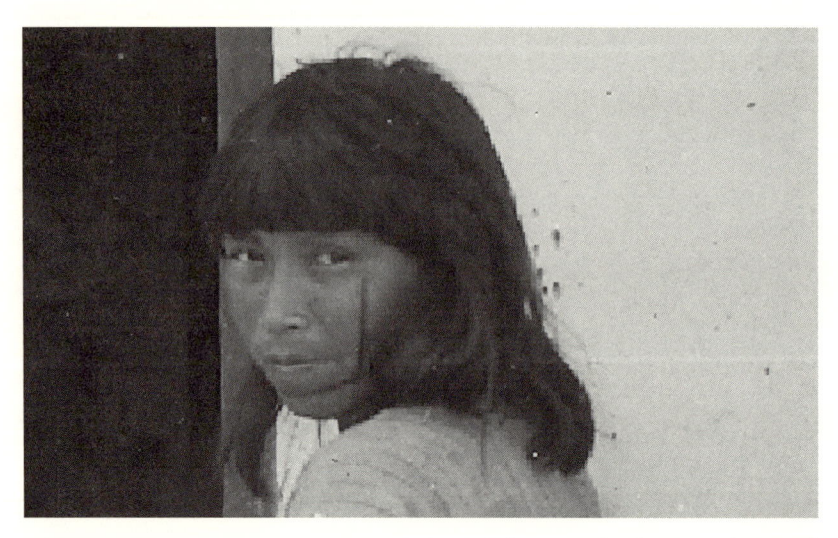

Ka'apor woman at a government outpost in 1942, with distinctive facial painting design still in use by women and men today (photo by Eduardo Galvão; original photo given to the author by Charles Wagley).

Araweté man, Igarapé Ipixuna, basin of the Xingu River, eastern Brazilian Amazon, 1985.

FUNAI worker securing bow and arrows of Aurê and Aurá, in front of the first of their last two temporary shelters, before they became wards of the Brazilian state; basin of the Tapirapé River, Brazil, October 1987.

The last temporary shelter of Aurê and Aurá; thatching is from babaçu palm (*Attalea speciosa* Mart. ex Spreng.), called *waí* in the language spoken by Aurê and Aurá.

Babaçu palm (*Attalea speciosa* Mart. ex Spreng.), a massive, solitary palm frequently found in anthropogenic forests of eastern Amazonia.

Terra preta (Amazonian Dark Earth) site on a ridge near Santarém; palm is *Acrocomia aculeata* (Jacq.) Lodd. ex Mart. (*mucajá* in the Portuguese of Brazilian Amazonia; *macaúba* in southern Brazil); other cultivated species in the photograph are young papaya trees and banana or plantain suckers.

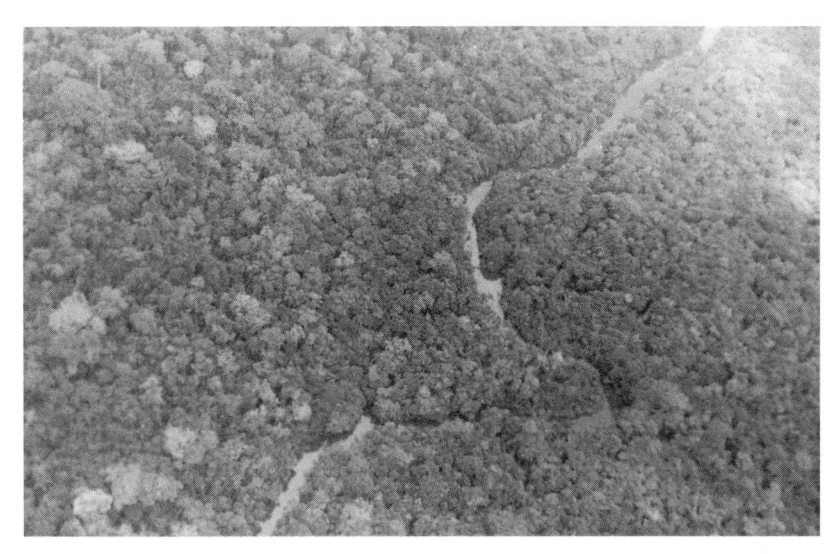

High forest along the edge of the Igarapé Gurupiuna (Gurupiuna stream), basin of the Gurupi River, near Ka'apor forest inventory sites (hectares 6 and 7 in Chapter 3 text).

Posto Indígena (P.I.) Canindé, a FUNAI field station (or *posto*) and indigenous Tembé village; it is within the Terra Indígena Alto Guamá, mostly a Tembé reserve. P.I. Canindé is on the left bank of the Gurupi River, near a high forest site discussed in the text (hectare 8 in Chapter 3 text).

An individual tree in the genus *Cecropia* (Urticaceae), reminiscent of a candelabra; generic words for *Cecropia* (*embaúba* or *imbaúba* in Portuguese) in diverse Tupí-Guaraní tongues, such as the Ka'apor and Guajá word *ama'i* (Appendix I), tend to be cognate. Photo from the collection of Meghan Kirkwood.

Plantain in an eastern Amazonian dooryard garden. The plantain is called *pako-te* ("authentic or original plantain").

Ethnozoologist Helder Lima de Queiroz records data concerning *Geochelone* spp. tortoises with a group of young Ka'apor men, village of Gurupiuna, 1989.

Yupaú (Parintintin) woman in front of a longhouse, central Rondônia, 1992; the Parintintin word for cacao (*ñumi-*) is different from that in other Tupí-Guaraní languages.

The Ka'apor name *irikiwa'i* refers to this gargantuan hardwood, which is typically found in ancient anthropogenic forests. Photo by Osmar Ka'apor; from the collection of Meghan Kirkwood.

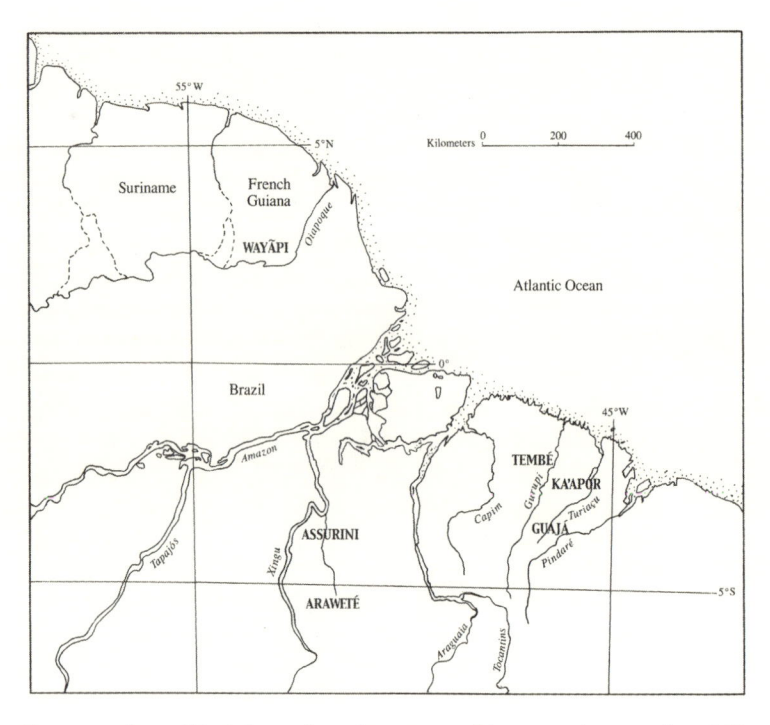

Locations of several Tupí-Guaraní–speaking groups of the eastern Amazon where field research discussed in this volume took place.

PART III

Indigenous Savoir Faire

Overview

Traditional knowledge is a fragile phenomenon, as we have seen, in which loss of agriculture, attrition in knowledge about plants and animals, disappearance of terms for biota and resources, and even the ability to make fire have occurred among native peoples over time as a result of external contacts and forces. It therefore makes sense to begin to try to understand traditional knowledge in its own living context, while it is still fresh and surviving inside indigenous languages and cultures.

The forests people made in the past, we have in fact learned, are not only useful; they have been sustainable—that means, people have continued to use them, without degrading or impoverishing them in terms of species losses. For that reason, the first philosophical voyagers, natural historians, and tropical biologists and ecologists to study them thought they were pristine forests, just like the natural forests, and they did not have a separate term for or systematic understanding of them.

In Chapter 7, I first examine the main aspects of traditional knowledge of Amazonian environments. The discussion is housed in terms of four working principles, which are (1) Amazonian languages are rich indexes of biotic diversity; (2) traditional knowledge about Amazonian environments is essentially local; (3) languages and cultures encode a tremendous amount of knowledge that is accurate and independently verifiable; and (4) Amazonian peoples have transformed landscapes for centuries if not longer, and they know they can do this. That does not mean they always recognize contingent diversity in anthropogenic landscapes, or that they always have a word for cultural versus natural forests. In some areas, such as the Llanos de Mojos of the Bolivian Amazon, it may be the case that no "natural" forests per se exist, a point originally stated by Allyn Stearman (1989) and supported in the pioneering work of William Denevan (1966). In such cases, people might be expected not to recognize anthropic versus natural causality. In other cases, indicator species might be growing in old village sites, and in forests that are anthropogenic, not because anyone planned it that way, but because there are numerous unintentional ways of modifying a landscape.

In Chapter 8, I take up the issue of the extent to which genuinely ancient knowledge of Amazonian biota has been conserved, comparing groups' native vocabularies across the Amazon Basin, as an index of how knowledge persists, and how it decays, and what kinds of knowledge are more resistant to such decay. I am looking at the Tupí-Guaraní language family specifically, which is some twenty-five hundred years old. Some things that might not be vitally important have clearly been retained over that vastly long time. The question is, why? In this chapter's discussion of the ant ordeal among the Wayãpi and Ka'apor, as part of a pubescent girl's rite of initiation, I would revise one interpretation that I had not stated clearly enough the first time. Whereas much of the rite of initiation of both groups has roots that probably go far back into the Tupí-Guaraní past, the fact that among Tupí-Guaraní groups, so far as is known, only the Wayãpi and the Ka'apor use cobra ants (*Pachycondyla commutata*) in this ordeal is more likely a shared innovation, rather than specifically something that pertained to Proto-Tupí-Guaraní society. The observation of a reasonably recent common past for the Ka'apor and Wayãpi is explored more broadly in Chapter 9 in relation to terms for the trees that chocolate is derived from.

In Chapter 9, I show how economic forces from the outside world can change even the names and classification of things that tend to be the most resistant, in an examination of the origins of the word for cacao (chocolate plant) in the Ka'apor language. The nobility of Europe quickly got a taste for chocolate; it did not take long for that to spread to the general population. That led to European demand for a tropical American product on a level comparable to that of the initial demand for gold and silver from South and Middle America in the sixteenth and seventeenth centuries, and in an earlier time, for spices and silk from Asia. Chocolate can be very hard to resist, as many people know.

I do not mean to add to debate on the origins of the domesticated cacao tree, whether both in Mesoamerica and South America independently, or in one or the other region alone; the debates over this matter of origins, distribution, and preparation of the first chocolate are well summarized in various chapters of McNeil (2006). In this context, I am specifically in agreement with Nathaniel Bletter and Douglas Daly (2006, 38) when they point out that "*Theobroma cacao* and its cultivars—and perhaps chocolate—may not have been new to South America, but cacao as a trade product of overwhelming importance certainly was." That was in the eighteenth century, a century definitive of much of Amazonian cultures on which we have ethnographic information today.

Chapter 9 is actually part of a more extended study of Amazonian historical ecology, showing how changes in technology and commerce can affect people's methods of living in forests and their means of interacting with natural resources. It wasn't globalization, but it certainly was evidence of the early modern world system. Societal and economic changes alter use and valuation of forest resources, which in turn impact the "traditional" means of classification and naming of these resources. Chapter 9 focuses on mutual engagement of nature and culture, through the medium of the interface of chocolate and linguistics, over a known time period in the eventful chronology of past Amazonia.

7
From Their Point of View

Amazonia is that region drained by the Amazon River and its tributaries together with adjacent lowlands. It represents about 4 percent of the earth's land surface. It is roughly the size of the contiguous forty-eight U.S. states or the island continent of Australia. But Amazonia far exceeds both those comparable regions in terms of the biotic and linguistic diversity of its varied landscapes. Landscapes refer not to nature paintings or manicured lawns and gardens in Euro-American tradition, but rather to distinctive juxtapositions of living species in spatial contexts that are verifiable empirically by soil, light, temperature, and historical conditions. In Amazonia, swidden fields produced by slash-and-burn horticulture, anthropogenic old-growth fallow forests of the *terra firme,* and dooryard gardens found in the context of Native American villages and camps, among other definitive zones, constitute landscapes in this sense. The environment to native Amazonians is virtually never a monolithic entity, encompassing ecosystems and ecospheres; rather, almost everywhere it represents a heterogeneous association of landscapes, each having a different character and different index of its essences. Ample diversity of native languages also exists; Amazonia has about 300 languages in 170 different family groupings.

The examination of Amazonian environmental knowledge may be facilitated by recognition of four working principles:

1. Amazonian languages and the extensive vocabularies that these comprise indicate that people recognize biotic diversity that is intrinsic to the region.
2. Environmental knowledge by Amazonian peoples tends to be restricted to local settings.
3. Amazonian languages and cultures probably encode a tremendous amount of accurate, empirically and independently verifiable knowledge about the environment.
4. Amazonian peoples since prehistory have transformed landscapes by redistributing suites of species across the region, enriching the biota in local and regional contexts, and reordering the slope and contents of the surface of the land in diverse locales.

Concomitantly, Amazonian peoples of today know that they and their antecedents have impacted and diversified Amazonian landscapes in such ways.

The Recognition of Diversity

The first principle to be apprehended in examining how native Amazonian peoples view the environment seems to be that their languages encode and reflect, to a greater or lesser extent, the high diversity of landscapes together with the living and nonliving material contents of these that surround their effective living spaces (also see Chapter 10). Amazonia contains a greater diversity of species of plants and animals than any other land region of comparable size in the world, including at least thirty thousand plant species. Excluding South America, no single, entire continent boasts more than thirty thousand plant species (Lizarralde 2001, 266). When insects are added to the list, millions of animal species may be found in Amazonia, and biological systematists are far from being able to identify and classify them all. Marked richness in fish, avian, and mammalian species has been noted as well (Lizarralde 2001). The biological diversity of Amazonia is sometimes contrasted with the seeming monotony and homogeneity—in terms of high heat and humidity, low saturation of sunlight on the forest floor, architecture of massive trees, and so on—of the forests that harbor such diversity. Native Amazonian groups as well as traditional peasant societies (*caboclos* in Lusophone Amazonia, *ribereños* in Hispanic Amazonia), however, recognize and name distinctively numerous types of forest habitats (Fleck and Harder 2000; Frechione et al. 1989; Shepard et al. 2001). And the native classifications of plant and animal species thus far described indicate a critical and deep knowledge of diversity.

In some cases, native Amazonians classify a living species known to science into more than one folk species, a phenomenon known as "overdifferentiation" (Berlin 1992, 18–19). Overdifferentiation most often occurs with domesticated plants. Manioc (cassava) (*Manihot esculenta*), a tuber crop that is a staple source of carbohydrates in many native Amazonian societies, seems to have had its origins in Amazonia in the early to middle Holocene (i.e., seven to nine thousand years ago [Piperno et al. 2000]). Although manioc is recognized as constituting but one species scientifically, it is typically subdivided into between 15 and 137 folk species in diverse Amazonian cultures, with the average number of named folk species per language being 22 (Balée 1994; Chernela 1986; Descola 1994; Grenand 1980). Overdifferentiation occasionally occurs also with a nondomesticated species. In the Ka'apor language of eastern Amazonia, the crabwood tree (*Carapa guianensis*), a forest hardwood in the mahogany family, is subdivided into "red" and "white" crabwood folk species, depending on the color of the heartwood and whether the individual is found growing in seasonally flooded forest or on high, well-drained ground—the ones with red heartwoods grow in the *terra firme,* whereas the ones with white heartwoods are found in lower elevations (Balée 1994). The Achuar people of the Peruvian Amazon recognize twelve folk species of felines, but fewer than half these folk species are distinguished as separate taxa in systematic zoology (Descola 1994, 83). The converse process of underdifferentiation, in which a biological species is grouped under one term with other biological species, seems to be less common cross-culturally, especially with regard to higher plants and vertebrates. The numbers of folk taxa in Amazonian languages for the plant and animal kingdoms, while high both when compared to classification of environments with less biotic diversity and when juxtaposed to folk Euro-American terminologies for their own natural biotas, are nevertheless not in the aggregate higher than the millions of species worldwide that are potentially recordable by science.

Limits on Systems of Classification

A second principle in understanding native Amazonian views of the environment is that while there are multiple environments to inhabit, and individual groupings of people have over time traveled across many of them in migrations and wanderings (Vidal 2000), the classification of landscapes and biotic forms is limited to localities that are familiar and more or less immediate. Folk genera of plants or animals (in folk English, monomial labels for taxa like "oak," "rabbit," and "wren") normally do not exceed five hundred terms in any language, other than the language of scientific taxonomy (Berlin 1992, 96–101). But sometimes overdifferentiated taxa are, in scientific retrospect, worth the validating of new species. The Kayapó people of central Amazonian Brazil recognize fifty-six folk species of bees of which eleven were found to be either unknown or new to science and not yet described systematically (Posey 1983b). Likewise, a new species of capuchin (organ-grinder) monkey was recognized by science after the Ka'apor of eastern Amazonia (unrelated linguistically to the Kayapó) had already encoded it as a separate folk species (see Queiroz 1992). In virtually every Amazonian language, a large percentage of vocabulary is devoted to plants and animals, and this reflects both the biotic richness of the region as well as empirical knowledge of that richness, ingrained in speech and cognition over long periods of historical time.

To some extent, throughout the Amazon, real creatures may be imbued with human qualities. Other human groups may be conceptualized as having animal attributes, and some humans are believed to be able to take on certain animal forms, while some animals may assume human forms ("shape-shifting") (Descola 1996; Slater 1994; Viveiros de Castro 1998a). To peasants who dwell along the Amazon River in Brazil, known as *caboclos,* a common shape-shifter is the gray dolphin (*Sotalea fluviatilis*). The dolphin is believed to assume the form of a strange but handsome youth at ceremonial occasions such as the feasts of Saint John and Saint Peter in June. The dolphin, known as the *boto tucuxi,* traditionally is assigned the paternity of infants born to young, unwed mothers who may have ventured too far from the party, and too close to the river, on such occasions (Wagley 1976). Much of the environmental knowledge and lore possessed by Amazonian peasants is probably syncretic, that is, a product of the collision between Native American and Iberian systems of knowledge (Pace 1998, 80). Certain surviving native systems of knowledge ascribe earthly life in most respects to some creatures that are otherwise otherworldly, as with a blue-boned, kinkajou-like animal known in the folklore of the Ka'apor people. The same people recognize an anaconda-like snake of a kilometer in length, known as the *Mayu,* to be at once both a fearful spiritual and a physical entity, capable of shifting form from a snake into a seductive woman who is imbued with dark and potent magical abilities, though the *Mayu* like the kinkajou-like creature cannot be shown to have any independent empirical existence at all. In Amazonia, one notes that some humanlike forms with humanlike personalities (though lacking sometimes in speech) are believed to be "mothers of the game animals," that is, protectors of supplies of edible meat from the forest, as seen among the Munduruku of north-central Amazonia (Murphy 1960), the Achuar (Descola 1994), and many other groups. Certain food taboos and food avoidances among the diverse Amazonian native groups, as on raptors, capybaras, and sloths, are very widespread and seem to constitute a common, ancient foundation of religious and spiritual knowledge that affects the distribution of the local biota. Religious knowledge is integrated with environmental knowledge into a single epistemological frame of reference in Amazonia. Some mythic heroes in one form or another are believed to have created various features of landscapes known to people in the present as their homeland (Vidal 2000), much like the Dreaming Beings of Australian Aboriginal lore.

Native Knowledge of the Environment

For the most part, species and landscapes are not mere products of mind; they are real entities with finite boundaries that can be described cross-culturally. In other words, native Amazonians "know" a lot about the environment, which is a third principle in understanding their views of it. The empirical landscapes of Amazonia provide the observer with a wealth of biotic and abiotic minutiae, a wealth that is premised on the staggering diversity of rainforests. The rainforest as a habitat only seems monotonous to naive, not native, observers. Similarly, the semideserts and deserts of southern Africa and Western Australia often present to Western visitors as bleak and uninhabitable landscapes (see commentaries in Tonkinson 1991, 32–34, and Lee 1993). But for the hunter-gatherers, those regions contained much natural wealth and beauty together with resources for survival. Native Amazonians tend to consider forests and diversity home. In fact, native Amazonians typically divide the "forest" into more than one type in their local milieus. Remarkably, the Matses people of Amazonian Peru recognize 178 habitats in their area alone (Fleck and Harder 2000), a number that greatly surpasses in diversity the previously existing scientific classification of their rainforest. But in retrospect, each Matses rainforest habitat is different in terms of the empirically verifiable distribution of indicator plant and animal species (Fleck and Harder 2000). In a similar fashion, and with some overlap among the various categories, the Matsigenka people, also of the Peruvian Amazon, identify sixty-nine habitats defined in terms of vegetation, twenty-nine habitats defined in terms of abiotic factors, and seven habitats understood in terms of faunal indicator species (Shepard et al. 2001).

No single Amazonian classification of plants, animals, and landscapes may be taken as a model to represent the others. For the three hundred or so native languages spoken in Amazonia, there are at least three hundred or so ethnobiological and ethnoecological systems of classification. One of the main differences concerns relations with the environment. The few hunter-gatherers (nonagricultural people) seem to have fewer folk species names (in folk English, binomial names for taxa like "live oak," "jack rabbit," and "house wren") than horticultural peoples, especially concerning the domain of plants. This is partly because horticulturalists interact with and manage more plant species than hunting-and-gathering peoples (Balée 1994). The hunting-and-gathering Guajá people of eastern Amazonia, however, seem more keenly aware of plants that are edible to particular game species than do the neighboring, horticultural Ka'apor people (Cormier 2000). Apart from this difference in mode of production (i.e., food production vs. food collecting), native Amazonian societies tend to exhibit broad similarities in terms of how they name the biota. In general, wild plants are often named for domesticates, but not vice versa, unless the plant being named is recently introduced. In the Achuar language, the wild star apple tree of the forest (*Pouteria caimito*) is called *yaas-numi*, "star apple-tree," a name that is clearly derived from the name of the domesticated and closely related star apple tree proper (*Chrysophyllum caimito*), which is simply called *yaas*, "star apple" (Descola 1994, 79). "False yam" (*kara-ran*) in Ka'apor, a name for a wild yam (*Dioscorea* sp.), is derived from *kara*, "yam," which is the name of the domesticated yam (*Dioscorea trifida*). These derivations and others like them in many languages suggest a recognition by native peoples of a conceptual dichotomy between domesticated and nondomesticated plants, though statistical patterns in the nomenclature related to the morphology of terms permit recognition of a third category—semidomesticated biota (Balée and Moore 1994). These naming patterns mean that agriculture, artificial selection, and altered genotypes of organisms that have been long subjected to human management (and are therefore quite familiar) are cognitively embedded in concepts of folk classification. So, too, is the temporal and spatial diversification of landscapes.

Diversification of Landscapes

The Amazon rainforests have long been thought to be pristine and virgin. Recently, however, many scholars have questioned that assumption, based on their growing awareness that the population of Amazonia before the arrival of the Europeans was much larger and that agriculture was much older and more widespread than had been earlier thought. These conditions led to the formation of anthropogenic forests and landscapes in the present (Balée 1989a; Dean 1995; Denevan 1992). Agriculture, which in Amazonia traditionally involves the swidden (slash-and-burn) method, has evidently resulted in an abundance of forest types in the regrowths that follow after a few years of burning. It seems that beyond about forty years and up to about two hundred years after the initial planting of a swidden field, if forest regrowth occurs, that forest tends to be high in diversity yet different in terms of the species present from surrounding, seemingly untouched pristine forest (Balée 1989a, 1994). These old forests, called fallows, have traditionally been classified as high forest (pristine forest on well-drained ground) by researchers. The new evidence suggests that these forests are of a different type, and that they would not exist had it not been because human agricultural activities changed edaphic (soil) and biotic conditions to favor them. These fallow forests are called *taper* in the Ka'apor language, whereas high forests are called *ka'a-te,* and the Ka'apor distinguish them by numerous indicator species present in one but not in the other type (Balée 1994).

Clearly Amazonian people since prehistory have impacted regional and local flora and fauna through the processes associated with domestication. Such processes have also influenced landscapes. In just a few cases, massive reconstitution of Amazonian landscapes took place in prehistory. Perhaps the most dramatic example of such manipulation is in the Baurés area of northern Bolivia, where prehistoric peoples built hydraulic earthworks stretching across more than five hundred square kilometers. These earthworks, which are etched on the land surface by linear causeways and canals, seem to have functioned as enormous fish weirs, permitting ancient people to manage potentially huge populations of fish for their sustenance. In a sense, they were ancient fish farms, to date unknown from elsewhere in the Amazonian region (Erickson 2000a). In the rainy season today, areas of artificial ponds in Baurés yield up fish including armored catfish, cichlids, and characins (such as piranhas), as well as snails. In the past, according to Clark Erickson, "Artificial ponds provided a way to store live fish and snails until needed" (Erickson 2000a, 191). The ancient earthworks of Baurés, built some five hundred to two thousand years ago, have been called the "Nazca Lines of the Amazon."[1] Probably such manipulations of the existing landscape were enabled by relatively complex sociopolitical entities, though not as complex as a civilization, which in South America is known only from the Andes.

Early Amazonian peoples domesticated several species seen pantropically today, including manioc, papaya, cashew, peanut, pineapple, arrow cane (in the grass family), tobacco, annatto (a dye tree), guava, rope plant (in the bromeliad family), cocoyam (an edible aroid, like taro), and certain chili peppers. In addition, they cultivated and managed many species of trees, such as Brazil nut, peach palm (with large edible fruit), hog plum (which has a tart vitamin C–rich fruit), soursop, and many others. Probably ancient Amazonians changed the genotypes of some of these species, in a sense bringing them within the realm of domestication, but the loss of Amazonian peoples to disease and depopulation, following the European conquest, may have occasioned a loss in the domesticated crop diversity that apparently existed on the eve of contact (Clement 1999a, 1999b; Peters 2000). In other words, a large crop inventory could have required substantial numbers of people to manage and protect it. Early Amazonian peoples appear to have contributed to diversity by spreading domesticated crops around the basin and by creating new agricultural landscapes. These new landscapes, because of the increased availability of solar radiation and probably high nutrient content in the soil, when compared to the pristine, dense forest, would have favored the introduction and growth of tree species that formerly had been limited to much more narrowly restricted habitats, such as swamp and riverine forests. Ancient Amazonian peoples also built elevated landscapes that favored the expansion of species and enrichment of local biota.

In Amazonian Bolivia, the habitat of the Sirionó people, who speak a language in the Tupi family of languages, is about two-thirds a grassy, seasonally flooded savannah, not entirely unlike the Everglades of South Florida; the rest, in contrast, consists of patches of forest on the *terra firme*. Amazonian people typically do not build settlements and sleep in marshland. The Sirionó traditionally camped in the forest patches, on the high, well-drained forest islands that dot the savanna. These forest islands, indeed, are substantially elevated above the surrounding savanna by a few to many meters, and the highest ones (up to eighteen meters in height) never flood. They are anywhere from a few acres to several square miles in extent. If one takes a historical and ecological perspective, these islands are, in fact, artificial mounds that were deliberately constructed by native peoples, probably of Arawakan linguistic affiliation, from about the beginning of the common era to about A.D. 1200 (Erickson 1995, 2000b). The Sirionó arrived in the area, probably migrating from the south out of what is now Paraguay, during the 1600s, perhaps earlier. Although the Sirionó people did not build the mounds, they do know that the mounds of their habitat are anthropogenic phenomena. The mounds are rich in artifacts such as potsherds and charcoal, moreover, that independently supply evidence for former, permanent human occupation of some definite magnitude and sophistication. The soil contents of one of these mounds was shown to contain 13 percent pure ceramics—thus, these forests have been appropriately called "ceramic forests" (Langstroth 1996). Because of all the burning for cooking and ceramic production that must have occurred in the past, the soils of the mounds are very fertile, much more so than those of the surrounding savanna. Likewise, fallow forests elsewhere in the lowland rainforest, formerly thought to be essentially homogeneous as a forest type, often harbor a wealth of animal species attracted to the many fruit trees that occur in a higher density in these forests than in high forests that have not been subjected (in recent memory at least) to any sort of agricultural manipulation. The mound forests of the Bolivian Amazon, like the many fallows throughout Amazonia generally, are, in a sense, native orchards (Balée 2000a), and thus are they understood and characterized by native societies.

The Kayapó people of the well-drained savannas of central Amazonian Brazil also have forest islands in their environment. They refer to these generically as *apêtê*. Recent researchers argued that these forest islands were anthropogenic phenomena that had been "planted" by the Kayapó (Anderson and Posey 1989). Another view, however, argued that they were relic forests of Pleistocene climate changes that had nothing to do with human intervention or creation (Parker 1992). Africanists have since reported that forest islands found dotting the well-drained savannas of Guinea (West Africa) are anthropogenic (Fairhead and Leach 1996) and would not have existed had it not been because of plantings of fruit trees and the management of these over time by successive generations of human inhabitants. Indeed, the mechanism involved and the anthropogenic evidence adduced for the forest

Table 7.1. Concepts of "planting" in Ka'apor

English Planting Concepts	Ka'apor Correspondences
Plant a swidden	*kupiša-moú*
Plant (a seed or cutting)	*yitim*
Sow (seeds only)	*omor*
Cultivate (or "raise to maturity")	*mu-tiha*

islands in Guinea constitute findings from another continent very similar to the arguments that had been made in favor of anthropogenic *apêtê*. Much of the debate on the *apêtê* hinges on whether the Kayapó themselves consider the tree species within these forests to have been planted or to have been "plantable," based on the principal interview language of Portuguese. In fact, multiple Kayapó words for "planting" do not correspond to any English or Portuguese term (Posey 1998, 112). In the Ka'apor language (again, unrelated to Kayapó) also, a single word corresponds to English "to plant" or Portuguese "*plantar*." To be sure, the Portuguese and English terms have many meanings. In Portuguese, one can accomplish the planting, sowing (*semear*), and cultivating (*cultivar*) of a crop, such as wheat or barley, by the verb "*plantar*." One can also "plant a swidden field" (*plantar uma roça*) (Holanda Ferreira n.d., 1098). In Ka'apor, all these operations are denoted by distinct, noninterchangeable terms, as shown in Table 7.1.

The four terms in Ka'apor from Table 7.1 could be used as glosses for English "to plant" or Portuguese "*plantar*" in appropriate, specified contexts. Some of the Ka'apor terms listed in Table 7.1, while not interchangeable with any of the others, may also have many meanings. *Yitim* refers not just to digging a hole in the ground and inserting seeds or cuttings therein; it also can mean burying any item, including a human corpse. *Omor* refers not only to sowing domesticated seeds (as with *pitim-ha'i-omor*, "tobacco-seeds-sow" or, idiomatically in English, "sowing of tobacco seeds"); it is used also to refer to the casual tossing away of seeds from nondomesticated fruits that may germinate and then grow into future, fruit-producing trees. And *omor* denotes tossing in general, as in "tossing out garbage or waste" or "tossing or throwing a ball." One "cultivates" (*mu-tiha*) domesticated plants; with the same verb, one "rears" children and "raises" pets. An important step toward determining the nuances in concepts of domestication and cultivation in native Amazonia involves securing an adequate understanding of native terms related (and revelatory) to these concepts.

Human Knowledge and Landscape Diversity

The fourth and final principle underlying native Amazonian views of the environment is that people recognize that human beings, past and present, have contributed to landscape diversity. Even notions of spontaneity of volunteer plants must be hedged against native constructs of the landscape. In Ka'apor, *yeye-tu-hem,* "by itself-[particle]-it is born" (or idiomatically in English, "it germinates spontaneously"), an adjectival verb phrase used to describe volunteer species in successional growth of swidden fields, may not capture the entire cultural essence of a plant species. Sometimes, it is clear that an individual of a volunteer species was "planted" by a nonhuman animal. "Deer manioc" (*arapuha-mani'i*) is sometimes said to have been "planted by deer" (*arapuha a'e yitim*), in the sense of its seeds having been interred by the deer (perhaps through feces).[2] Deer manioc is also found in old swiddens and fallows that appear in the traditional Western classification of forests to have been high forest. This is also true of many fruit trees that grow in fallows. While people may not have intended the fallow to be a repository of edible fruits, when they placed settlements within their cleared swidden fields, they also ate fruits from other fallow forests, the seeds of which they discarded (*ha'i-omor*), or sowed, about the settlement and in the garbage pits (middens) that surrounded it, though this sowing was less deliberate perhaps than their planting of manioc cuttings and seeds of other domesticated plants. Some of the seeds casually tossed away germinated and became trees, such as nondomesticated cacao, papaya, star apple, and a host of other trees more commonly found in fallow today than in high forest. Some of these fruit trees are especially attractive to game animals. Capuchin monkeys spread the seeds of nondomesticated cacao throughout the area of the fallow as they consume the sweet, edible pulp that surrounds the seeds, and the Ka'apor as well as other Amazonian peoples know this dispersal mechanism. The Ka'apor say that the monkeys "sowed" (*omor*) the seeds of nondomesticated cacao in old fallows. Although it is perhaps too early to say whether the Kayapó peoples, far to the south of the Ka'apor, "planted" the forest islands of their central Brazilian landscape, it must be remembered that there are many ways to "plant" a forest.

Four principles based on empirical research to date can be adduced in approaching how native peoples of Amazonia view the environment. First, languages encode biotic and abiotic diversity to a greater or lesser extent, seemingly greater in the case of Amazonian languages. Second, classifications of biota and landscapes are confined to local, essentially familiar environments. Third, native Amazonians know a great deal about the environments they inhabit. And fourth, native Amazonians recognize and encode linguistically the fact that people, past and present, have affected the distribution of the biota and the formation of the landscape. Much of the knowledge that informs these views has derived from about eleven thousand years of human occupation in Amazonia (Roosevelt 1998), if not more. It remains to be seen what proportion of that knowledge, and the knowledge acquired during the subsequent development of agriculture, has survived into the postmodern world of today and what prospects such knowledge may have for persisting in the day-to-day life of Amazonian societies.

8

Retention of Traditional Knowledge

In defining traditional knowledge of any sort, one of the components is age: tradition implies antiquity. Traditional ethnobiological knowledge (TEK) in Amazonia, for the present purpose, denotes specifically pre-Columbian objects of understanding that have survived to be documented ethnographically. Borrowings of ethnobiological knowledge by any one ethnic group from non-Amazonian sources that have occurred since 1492 can therefore be excluded as TEK, assuming such borrowings can be detected. Written records are often lacking and archaeological data do not necessarily yield an unbroken record of TEK to the present. Indeed, TEK has been lost autochthonously in Amazonian prehistory. It is important therefore to be able to make comparisons based on historical linguistics and modern ethnology as a proxy window onto the Amazonian past.

It is sometimes implied that ethnobiological knowledge is thousands of years or "hundreds of [human] generations" old, simply because biological resources themselves, such as varieties of domesticated and semidomesticated species, are that old (Brush 1993, 660). It is assumed that the knowledge about those resources must be as old as the resources themselves. Yet knowledge of the uses of wild species in Amazonia may be more difficult to classify as traditional, in the sense of great age, because it tends to be more labile. Names for wild plants seem to be retained over time less than are names for domesticated and semidomesticated plants (Balée and Moore 1991, 1994). It is more problematic to reconstruct uses of wild species without linguistic evidence, because such uses, if the same between disparate groups, could have been borrowed or independently discovered, unless the cultural employment of them is richly detailed and has its source in arcane knowledge. The finer the degree of detail in the usage, together with supporting correspondence from comparative linguistics and ethnology, the more likely the knowledge of the wild species in question will be traditional.

The uses of species and varieties that have been extinct for a lengthy period of time in a cultural repertoire will be obviously even more remote from observation. If loss of biological referents involves a gradual loss of knowledge about them, much native knowledge of plants must have vanished, since (as Charles Clement [1999a, 1999b] has shown) many crop genetic resources in Amazonia have disappeared since 1492. They have been lost in the Andes in both prehistoric and contemporary times too, because of monocropping and changing technologies and market conditions (Zimmerer 1996, 84–88). In addition to species, ancient technologies have been lost as well in Amazonia. Cropping of ridged fields in the Llanos de Mojos, and the exact uses to which prehistoric features such as causeways, canals, and dikes were put, disappeared for the most part by 1200—that is, well before the conquest (Denevan 1966; Erickson 1995). Although they all used fire captured from the hearths of other groups or lightning strikes and had ways of maintaining it, the actual technology of fire making was definitively lost by the Tupí-Guaraní–speaking Guajá, Sirionó, Yuquí, and some of the Parakanã (Balée 2001), as well as by the Carib-speaking Akuriyo (Kloos 1977, 120), and no doubt by other peoples at one time or another (Balée 2001). These groups are mainly hunter-gatherers or trekking peoples who had been subjected over time to the progressive loss of other aspects of aboriginal technology, such as the use and processing of bitter manioc, in addition to other domesticated crops.

Regardless of the causes for the loss of TEK, time in general effects an erasure where there is no record keeping. Even record keeping can be destroyed, as when astronomical and calendrical systems, numeration, and writing perish with withering civilizations, even though some semblances of past esoteric knowledge (such as arithmetic) may be preserved in speech and shared (but, on artificial devices, unrecorded) knowledge (Urton 1997). The Rosetta stone for TEK in Amazonia is not made of rock; rather, it is to be found in living native languages and cultural practices. For that reason if for no other, these phenomena urgently require skillful and meticulous documentation, preservation, and protection.

Knowledge itself definitely changes, as do memories of the past. Some sorts of knowledge are more ephemeral than others. Genealogies in Amazonia seem especially subject to loss (Murphy 1979). It must be remembered that in Amazonia social standing is almost never based on who one's putative ancestors were, since anyone is a potential "somebody" and the kin who count are one's living, co-resident consanguines, affines, and affinables (Basso 1988 [1973]; Rivière 1987). Although Darcy Ribeiro (1996) reported that one Ka'apor informant possessed generationally deep genealogical information, I was unable to document similar knowledge among the contemporary Ka'apor, and it seems to be an isolated case in Amazonian native societies. Ascribed relationships to the deceased guarantee no heritage in land and resources, and even if they did, the influence on Amazonian economies would be unlikely anyway. If either land or resources were at a premium in the aboriginal past, the artificial scarcities of raw materials, craft knowledge, and finished products, as documented in the intergroup trading feasts of the Upper Xingu (Oberg 1953) and the northwest Amazon (Chernela 1993; Jackson 1983), would not be expected to occur. It comes as no surprise to find but limited evidence for corporate kinship groups, except in the northwest Amazon. By definition, such groups would have to be built on genealogical links to the past, with the more prestigious and wealthiest of them being presumably the oldest (Chernela 1993). Even in the northwest Amazon, however, the ranking sibs and their redistributive power seem quite limited. True chiefs—paramount chiefs—are not attested to in the northwest Amazon except in confederacies, induced by contact (Chernela 1993, 20–22). The ethnohistory of the ancient Atlantic and Amazonian chiefdoms is ambiguous on questions of lineal authority and genealogical memory; in any event, however incipiently stratified such societies may have been (Balée 1984b; cf. Carneiro 1995; Clastres 1989), their political structures did not survive conquest and colonialism, so we should not expect to be able to study these phenomena with comparative ethnographic and linguistic methods anyway. If an Amazonian custom has been extinct for more than four hundred years, its local, linguistic vestiges will have vanished mostly as well.

TEK may have survived better than genealogy and political structure because the web of biotic communities, Amazonian landscapes, and native technologies has been less prone to shredding and extinction. In the here and now, of course, that historical-ecological web faces unprecedented destruction by expansion and growth of the global economy (UNEP 1997). Yet some native customs that are entangled with Amazonian landscapes have remained basically intact for thousands of years, and their antiquity can be proven through comparative and historical methods.

Methods of Isolating TEK

Languages may not be the most ideal data source for adducing TEK, because they are subject to fairly rapid change, in evolutionary terms anyway (see Deacon 1997). In Amazonia, languages have shown a striking tendency to diverge, unlike in Australia, where nearly all of the 260 aboriginal tongues fit fairly neatly into a single stock (Crystal 1987; Dixon 1997, 1980; Ruhlen 1991). The diversification of language per unit area in Amazonia is less dramatic than in New Guinea, but there are only two stocks (Austronesian and Indo-Pacific) found there (Crystal 1987, 317), whereas Amazonia has many stocks (Kaufman 1990). Equatorial Africa is also diverse at the language level (and the continent as a whole is the most diverse of all at this level), but not as diverse per unit area as Amazonia and with fewer stocks and families (Crystal 1987; also see Dixon 1997, 95). Even if Joseph Greenberg (1987) is right that there is a single genetic grouping, which he calls Amerind, in South America, it would have to be much older than any conventional stock or family and composed of more disparate elements than the bona fide stock of Australian aboriginal languages.

Regardless of a sharp inclination to linguistic divergence in Amazonia, in contrast to a probable tendency to converge in aboriginal Australia, some words and things have stayed the same for hundreds and even thousands of years in Amazonia. Regarding aboriginal Australia, since natural features of language drift should have brought about greater diversification, some intense and prolonged interactions among the diverse languages must have occurred throughout prehistory (Dixon 1997, 89–94). As for Amazonia, some of these retained words and things pertain to TEK, as has already been demonstrated for equatorial Africa regarding agricultural words and things among Bantu-speaking peoples (Vansina 1990). One way of exhuming the evidence is to compare at the family level, since the family has fairly knowable time depth and that time depth is circa two thousand years old

(old enough for TEK).[1] Borrowings and substratum/superstratum influences can be usually identified and excluded, and what is left that is held in common constitutes cognacy and continuity over time. One does not have to adopt a tree or wave model as the predominant explanatory device for linguistic change; rather, one only has to admit that lexical items for some kinds of things are more resistant to change than are lexical items for other kinds of things (Dixon 1997, 24–27). Because TEK covers the content of biota and relationships among them, the focus of reconstructive effort tends to be more on the lexicon than on the grammar of language.

The argument concerns conservation of culture and language over time. In Amazonia, some features widely separated in space seem to resemble one another so closely that borrowing seems to be out of the question. This can be seen best within a single genetic unit, such as a language family. Eduardo Viveiros de Castro (1992, 1) noted that "there exists something in common among the different Tupí-Guaraní societies beyond their linguistic identity and behind their apparent morphosociological diversity." While to some extent it is still true that "what to be Tupi means is an open question" (Vidal 1984–1985, 4), it is nevertheless discernible that Tupí-Guaraní peoples do have "something in common" that is unique to their language family. In part, this something in common is cosmological, mythological, and sociological, as Viveiros de Castro (1984–1985, 1992) has well described; it also refers to ethnobiological knowledge and practice that would have been distinguished from the other Amazonian language families millennia ago. But the proof of shared, unique ethnobiological knowledge in the Tupí-Guaraní family does not mean that the shared "linguistic identity" will be evident in reference to all individual plants and animals, and it does not mean that "morphosociological diversity" (in a language family that includes both the Tupinambá and Guaraní chiefdoms of centuries ago as well as the hunting-and-gathering Guajá and Yuquí of modern times) will obscure such shared knowledge. Even in closely related languages, one can find differences in the use and management of organisms (e.g., between the agriculturalist Tembé and the hunting-and-gathering Guajá, who are linguistically closely related); and in highly similar cultures (in certain respects) great language distance can be discerned (as between the Sirionó and the Araweté).

Ellen Basso (1977, 17) listed eight features that characterized a "typical Carib complex." Perhaps it is less involved to specify such features, since the Caribs are generally more concentrated geographically (along the northern South American coast east of the Negro River and north of the Amazon, with only a few outliers, such as the southerly Kalapalo and Kuikuru in the Upper Xingu). Yet several such features are general enough to apply to other language families and to be part of any description of their typicality, such as bitter manioc cultivation, bilateral descent, kindreds, shamanism, and female puberty seclusion and belief in menstrual pollution. Defining unique or nearly unique features per language family seems more challenging than characterizing the typical ones, but for the purpose of reconstruction of ethnobiological knowledge, it is methodologically necessary to do

Table 8.1. Reflexes for some traditional plant domesticates in Tupí-Guaraní languages

English	Scientific	Reflexes in Tupí-Guaraní Languages
Maize	*Zea mays* L.	Ar *awaci*, As *awači*, K *awaši*, S *ibásɨ*, W *awasi*
Cotton	*Gossypium barbadense* L.	Ar *miniyu*, As *aminiyu*, K *maneyu*, S *níñu*, T *manezu*, W *minɨyu*
Sweet potato	*Ipomoea batatas* Lam.	Ar *yiti*, As *yitíka*, K *yitɨk*, S *ñiti*, T *zitɨk*, W *yeti*
Annatto	*Bixa orellana* L.	Ar *iriko'i*, K *uruku*, S *urúku*, W *uluku*
Arrow cane	*Gynerium sagittatum* Beauvois	K *u'ɨwa*, S *uúba*, T *u'ɨwa-a*, W *wɨwa*

just that. As for Tupí-Guaraní peoples, some generalized (but not universal) and nearly unique features seem to be clear: (1) an association among exocannibalism (or only execution of foreign enemies), name bestowal, and affinity (Viveiros de Castro 1992); (2) a Dravidianate kinship terminology with elements that accommodate oblique marriage, which is more or less similar to several Carib societies (Viveiros de Castro 1998b); (3) the existence of agriculture and a strong tendency to sedentarism (Noelli 1996, 34), a minimal repertoire of domesticates (Table 8.1),[2] and associated patterns of plant nomenclature and classification (Balée and Moore 1991, 1994); and (4) certain shared ritual and mythological complexes that have relevance to ethnozoology and that exhibit shared vocabulary (Table 8.2).

What to date has seemed less certain is shared features related to exploitation of wild resources. The following sections examine how hunting behavior may be influenced by ancient ritual food prescriptions and how the medicinal uses of certain ants may be primordial in the Tupí-Guaraní family.

Women and the Yellow-Footed Tortoise

One fairly common feature in the Tupí-Guaraní family concerns activities associated with menarche and menstruation generally, which at face value should not be astonishing. The ritual restrictions placed on menstruating women may be so generalized that their origin is related to something pan-human and would therefore be necessarily very old, at least as old as the Upper Paleolithic and probably older than that (Knight 1991). Ritual restrictions on sexual intercourse, the handling of food, bathing in communal waters, and so on may have a psychoanalytic basis, such as castration anxiety (Bettelheim 1955), or may be associated with a method of female solidarity and empowerment in relation to males, who are otherwise a burden on society's system of production and reproduction (Knight 1991). The importance of female puberty rites is that they are related to reproduction in an

Table 8.2. Some ethnozoological items of ritual or medicinal importance in Amazonia

English	Portuguese/Spanish	Scientific	Tupí-Guaraní Reflexes
Yellow-footed tortoise	jabuti/peta	*Geochelone denticulata*	Ar *yaaci*, K *yaši*
Red-footed tortoise	jabuti/peta	*Geochelone carbonaria*	K *karume*
Lesser giant hunter ant	tocandira	*Paraponera clavata*	K *takangɨr*
Amazonian cobra ant	—	*Pachycondyla commutata*	K *tapiña'ĩ*, W *tapia'i*
Fever ant	tachi	*Pseudomyrmex* sp.	K *taši*, S *tási*

economy of people, not an economy of raw materials in Amazonia; people and labor—specifically the female labor that is needed in the production of flour from bitter manioc—are scarce, not raw materials (Rivière 1987).

One would expect a cultural complex that was at least thirty-five thousand years old to have undergone permutations so that whereas the kernel of the idea remains, much has changed. Indeed, the specific manifestations of such restrictions on girls at menarche and menstruating women are variable among cultures, and this variability constitutes grounds for comparison and contrast with an eye to determining original forms in proto-societies of the past, forms as old as the protolanguages they spoke. Some cultural forms can be studied, in other words, as cognate (hence as cultural reflexes of earlier ones), regardless of the differential utility (if any) they may exhibit in these modern reflexes. In the Tupí-Guaraní family, some of these cultural reflexes may be tangled up with ethnobiology.

The Ka'apor people of southeastern Amazonia have a well-documented set of ritual restrictions on menstruating women. In particular, menstruating women and girls at menarche cannot partake of any land meat other than the yellow-footed tortoise (*Geochelone denticulata*). Hunters are prescribed to fetch this tortoise for their wives and daughters, and the women are expected to eat it. The only other species of tortoise in the region, its congeneric the red-footed tortoise (*Geochelone carbonaria*), is positively tabooed as food for menstruating women, perhaps because of a folk association between it and blood (Balée 1984a, 227–228; Huxley 1956, 146, 155; Ribeiro 1976, 55).[3] There are other rules and regulations concerning menstruation, but this specific one seems to enjoin, to a degree, sustainable hunting (Balée 1984a, 1985).

Tortoises, which are slow of gait and easy to dispatch, are the first major game species to be hunted out near a village. In general, tortoise populations mark the outer boundary of the catchment area as a village ages. Other game (but usually not tortoises) are still taken near the village, especially in swiddens and their ecotones with the forest. The potential hunting pressure near the village that could be brought to bear is never realized, partly because the tortoise injunction forces hunters to the margins of the catchment area.[4] The penalties men face for failing to hunt tortoises for their menstruating women include ostracism, cuckoldry, ridicule, and divorce—none being desirable conditions for a man in Ka'apor society. The sustainable hunting of game near the village has been documented even in the very old village of Gurupiuna, in the northern part of Ka'aporland (Queiroz 1989), where the rules of providing tortoise meat to menstruating women and girls at menarche remained in effect as of 1989, when I last visited that village.

It is difficult to find a historical precedent using documentary sources alone for this Ka'apor link of menstruation and the yellow-footed tortoise. The coastal Tupinambá of Bahia may have tabooed tortoise meat for the entire society (Métraux 1979 [1950], 292–293), but perhaps it was only the red-footed tortoise that is tabooed by the Ka'apor for people in ritual states, or perhaps it was an anomalous situation for only one group of Tupinambá (who were in fact many peoples) known to the chronicler André Thevet. But another chronicler of sixteenth-century Bahia indicated that tortoises, called by their Tupinambá name of *jabutis,* were edible to the people and were considered to be especially useful as food for the sick (Soares de Sousa 1987 [1587], 255). The Tupinambá may not be ancestral in a direct historical sense to the Ka'apor, though they might have been to the Tapirapé (Wagley 1977). There were already Amazonian Tupí-Guaraní societies that predated the Tupinambá of the coast (such as ancestral groups in southwestern Amazonia) and those (like the Omágua) that simultaneously existed with them during the sixteenth century. What the Tupinambá had in common with contemporaneous groups, and with modern-day Amazonian Tupí-Guaraní groups, may be better viewed as descended from an earlier protoculture and protolanguage, a methodological usage already well established by linguists (e.g., Dietrich 1990; Rodrigues 1986, 29–39; but see Urban 1996). With the exception of Língua Geral, none of the contemporary Tupí-Guaraní languages are seen as descendants of Tupinambá. During the approximately two thousand years that have passed since the first Tupí-Guaraní protocultures, we can anticipate their heritage to have survived unevenly among the descendant groups, including the Tupinambá, for the analytic purpose at hand.

Yet it is still possible that the menstruation–tortoise complex of the Ka'apor postdates 1492, and even that it was introduced no less by European missionaries, who also had an impact on the Ka'apor language, which exhibits lexical and perhaps even syntactic influences from Língua Geral (Corrêa da Silva 1997). In the 1620s, Franciscan missionaries in the lower Tocantins River basin, which was familiar to the ancestral Ka'apor as the Smoke River (*i-takāšĩ*), had debated the sanctity of letting their native flock eat tortoise meat on holy days of obligation, such as Fridays. Some missionaries had taken the position that tortoises were not meat and could therefore be eaten, as though they were fish, on such days (Lisboa 1904 [1626–1627], 397). According to the letter from Brother Christóvão Lisboa of 2 October 1626: "They told him that we had the Letter commanding me that one abstain from the tortoises on fish days for no evidence showed them to be not meat. . . . They declared publicly from the pulpit that they [tortoises] were fish, and as such one could eat them in front of the Pope . . . and that the saintly French clerics had given that [tortoise meat] for fish" (Lisboa 1626–1627, 397).[5] If native practice had prescribed tortoise meat on ritual days of observance (such as during menstruation and the couvade), the missionaries might have adopted the tortoise as a sort of ritualized fish, but with conflicting dialogue about its propriety. Indeed, though land animals other than yellow-footed tortoises are tabooed, some scaly fish species are allowed as food (but catfish are definitely tabooed) for the menstruating Ka'apor female as well as for *caboclo* (peasant) women in the Lower Amazon. Alternatively, did the missionaries impose the idea of ritual foods in the first place on a society that primordially had none? Probably not, because the ritualization of tortoise flesh is spread more widely than the influence of the missionaries, and the vocabulary associated with it from more than one related tongue seems to be older than the conquest.

The question leads to the astute observation by Viveiros de Castro (1992, 361n6): "There appears to be a Tupí-Guaraní system of 'tortoise transformations' at work." Although originally I lacked evidence for a menstruation–tortoise complex for any group other than the Ka'apor—and I even went so far as to state, "This ritual complex seems to be uniquely Ka'apor" (Balée 1984a, 228)—subsequent research has proven that the complex is indeed shared, albeit unevenly, among diverse Tupí-Guaraní societies of the Amazon Basin. For example, the Araweté prohibit the red-footed tortoise and allow the yellow-footed tortoise as food to people in ritual states, such as the couvade and menarche, similar to the Ka'apor (Viveiros de Castro 1992, 361n6). The Parakanã, who live to the south of the Araweté, also allow yellow-footed tortoise to be eaten by people in certain ritually restricted states, such as that of a warrior who has recently killed an enemy (Fausto 1997, 200). Among the Sirionó, at the time of menarche, girls in a group had to sit for several days on ritual wooden planks that had been elevated from the ground and separated from the camp. One Sirionó woman recounted that during her puberty rite, "*Conombe mɨ'ɨ ureu*" ("The only meat we ate was from the tortoise") (Priest and Priest 1980, 124, 131). Other sources indicate that a few other meats could be eaten by Sirionó girls during their puberty rite, but tortoises are included on these lists (e.g., Holmberg 1985 [1950], 212–213). Both red- and yellow-footed tortoises occur in the Sirionó habitat (Townsend 1995). It is unclear, however, which tortoise is regarded as being edible during menarche. The Araweté term for the yellow-footed tortoise (*yaaci katɨ*) is cognate with the Ka'apor term *yaši*. Curiously, the Sirionó term *konombe* is cognate with the Ka'apor term *karume* (at the referential level of biological genus, *Geochelone*), which denotes the red-footed tortoise. *Konombe* is a folk generic name denoting both *Geochelone* species in the habitat (cf. Schermair 1962, 390). But because tortoises are among the most extensively collected of game (Holmberg 1985 [1950], 66; also see Townsend 1955, 55–56), the gross ecological effect on overall game availability induced by an injunction to capture tortoises regardless of species could be regarded as similar among the Sirionó, the Araweté, the Ka'apor, and the Parakanã. Even if the system of "tortoise transformations" (Viveiros de Castro 1992, 361n6) did not promote sustainable hunting among them all,[6] it is clear that an association between menstruation and the tortoise is autochthonous to Amazonia, predates the conquest, and may be a distinctive feature of the Tupí-Guaraní family.[7]

In other words, it is very old, and its modern reflexes have been transmitted from generation to generation over many hundreds, if not thousands (specifically, two thousand) of years.

Ant Medicine

Another feature of the Ka'apor puberty rite that is not entirely unique concerns "the ordeal." At the end of about a ten-day period of seclusion in a ritual enclosure (*kapi*), during which time a young female's feet cannot touch the ground, the Ka'apor initiate's hair is shaved off (traditionally with a bamboo knife, now with scissors). Seclusion of girls at first menses and preventing their feet from touching the ground is extremely common worldwide (Knight 1991, 383–389), and it is found in numerous Tupí-Guaraní societies. The head shaving is found also with the Sirionó (Holmberg 1985 [1950], 212), the Wayãpi (Campbell 1989, 85–96; 1995, 56), and the Tupinambá (Métraux 1979 [1950], 100; Staden 1974 [1557], 171). Perhaps the practice is not specifically Tupí-Guaraní in origin, but it may be distinctive of Amazonia. It serves a pseudocalendrical purpose, for when the girl's hair reaches her shoulders in these societies, she may be married, but not before then. After the Ka'apor initiate's hair is shaven, adult members of her uterine kin group tie long strings made from *kirawa* fiber (*Neoglaziovia variegata*) snugly about her forehead and chest. Carefully knotted about their thoraxes, on each of these strings, are about six *tapiña'i* ants (*Pachycondyla commutata* [= *Neoponera commutata* Reg. = *Termitopone commutata*]). They sting the girl repeatedly; having been stung myself by *tapiña'i* ants, I can confirm that the venomous sting is like that from a wasp—sharp, hot, and painful but still a distant second in discomfort to the sting of the dreaded lesser giant hunter ants, *Paraponera clavata*.

Remarkably, the Wayãpi of French Guiana also apply the same ants to the body in ordeals (F. Grenand 1989, 417), as do their Carib-speaking neighbors, the Wayana (Devillers 1983, 82). The Wayana may have borrowed the use of the particular ant species used in this practice, called *marake,* from the Wayãpi, since it does not seem to be used elsewhere, though ant ordeals involving other ants are common in northern South America, throughout the Cariban and Arawakan homelands. The Wayapí of northern Brazil specifically use these ants to sting female initiates in the puberty rite. Ten to fifteen of the ants, called *tapia'i,* are tied onto a belt made from palm leaf fiber (probably *Astrocaryum paramaca*) (F. Grenand 1989, 248), and then are applied to the girl's arms, legs, abdomen, back, and brow (Campbell 1995, 56). The swordleaf fiber of a closely related species of palm (*Astrocaryum vulgare* Mart.) is used by the Tupí-Guaraní–speaking Guajá of southeastern Amazonia, who do not cultivate *Neoglaziovia variegata,* to make rope and thread (Balée 1992a) (Chapters 4 and 5); perhaps there are grounds for a cultural homology here between *Astrocaryum* spp. and *Neoglaziovia variegata.* In other ways as well, including mortification of the flesh (also done by the Tupinambá) (Staden 1974 [1557], 171), the ordeals among the Ka'apor and Wayapí initiates are quite similar, despite a geographic separation of about six hundred miles. Whereas the general outlines of puberty rites and ritual avoidances of menstruants may be found throughout Amazonia and even worldwide (Knight 1991), some of the specific manifestations, such as these among contemporary Tupí-Guaraní societies, indicate something unique to them, and probably to their forebears.

The use of ants in Amazonian ordeals has been long known. The Maués of the central Amazon are perhaps most famous for letting lesser giant hunter ants sting boys in virility ordeals (Spix and von Martius 1938 [1831], 297). Pain and fever from the venomous stings last up to twenty hours or so (cf. Lenko and Papavero 1979, 297, 302–303). The use of ant stings in medicine, magic, and ritual seems especially common in northern Amazonia. Tocandira ants are used to sting men, women, and children in diverse ordeals among the Cariban Wayana (Anonymous 1982), the Apalaí (*O ritual da tocandira* 1982), and the Warao (Roth 1924, 708). Unidentified wasps are sometimes used instead of stinging ants in ordeals among the Apalaí (Farabee 1967 [1924], 223; Roth 1915, 309–310) and the Rucuyens (Roth 1915, 309–310). First-time menstruants among the Desana (Eastern Tukanoan) of the northwest Amazon actually eat toasted saúva ants (*Atta* sp.) inside a ritual enclosure as part of their initiation ceremony (Reichel-Dolmatoff 1996, 56).

Among the Kayapó, ants with the most potent stings are used in hunting magic; their bodies are crushed, mixed with *urucu* juice, and then pasted onto hunting dogs so these will "hunt with determination as the ants do" (Posey 1979a, 143). The Cariban Akawaio of Guyana also apply ant charms to their hunting dogs (Colson 1976, 454). Stinging ants were applied to hunters themselves as fetishes among the Makusis (Roth 1924, 178), the Arekunas, and the Akawaio (Im Thurn 1967 [1883], 229, cited in Roth 1915, 280). The Carib-speaking Akawaio call these ant "medicines" (i.e., catalysts or fetishes) by the generic term *murang* (Colson 1976, 454). *Murang* seems strikingly similar in phonetics and meaning to *posáõa,* a Língua Geral (Tupian) term (Balée 1994, 90, table 5.1), and suggests further evidence for borrowing between Cariban and Tupian peoples in the post-Columbian period. But it seems that only Tupí-Guaraní cultures aboriginally prescribed the stings of *tapiña'ï* ants in puberty rites.

This ritual with *tapiña'ï* ants bespeaks a common heritage of the Ka'apor and the Wayapí/Wayãpi to the far north. Since these northern groups have not lived south of the Amazon since before the early 1700s (P. Grenand 1982, 151–163), it is an old custom. The reflex terms for the ant itself (Ka'apor *tapiña'ï* and Wayãpi *tapia'i*), moreover, testify to an ancestry and familiarity with the ant that originated many hundreds of years ago, namely, from the time of the beginnings of the Tupí-Guaraní language family itself.

Another use of ants is astonishingly similar between the Ka'apor and Sirionó, who are separated by an even greater divide of some fourteen hundred miles of the Amazon Basin. Both groups employ ant species from the relatively homogeneous genus *Pseudomyrmex* (called by the reflexes *taši* in Ka'apor and *tási* in Sirionó) as fever remedies and anti-inflammatories. No evidence in Lenko and Papavero (1979) indicates any use for these ants in native or *caboclo* medicine in the vast area intervening between the Ka'apor and Sirionó (cf. Posey 1994, 58). The only other area thus far identified where their use seems likely as a febrifuge and in related contexts is among the Yanomami in northern Brazil (Milliken, personal communication, 1998). No other published medicinal references on these species are known in the Amazon. So it is likely that this usage has either been independently discovered by the Ka'apor and Sirionó, respectively, or inherited from Tupí-Guaraní protocultures of the remote past. A polysaccharide isolated from the venom of *Pseudomyrmex* sp., which is closely related to all other *Pseudomyrmex* species including those of the Ka'apor and Sirionó habitats, has activity on the human complement system and may be useful in treating rheumatoid arthritis (Schultz and Arnold 1977, cited in Hogue 1993, 454; Schultz and Arnold 1978). The retention of the term for these ants together with the common medicinal use to which they are put by the Ka'apor and the Sirionó, who could not have borrowed this knowledge from each other, suggests a parallel not a convergent process of retention. Some appeal to utilitarianism may be made regarding pseudomyrmecine ants as remedies for fever and rheumatic complaints. The venom produced is probably not simply a "counter-irritant," as has been suggested for other sorts of ant stings in the literature (Roth 1924, 708), but a biologically therapeutic treatment for these inflammatory conditions. Borrowings cannot be reconstructed, only inferred; the little the Sirionó have in common with the Ka'apor tends to be cognate. In other words, this knowledge exemplifies Amazonian TEK. It is cognate and ancient at once.

Documenting TEK in Amazonia

TEK is shared unequally in Amazonia. Even city dwellers of the Lower Amazon who abide by restrictions on catfish and other foods (generically called *comida remosa*) during ritual periods of danger (such as menstruation and sickness, generically called *resguardo*) seem to echo native constraints on behavior seen in puberty rites and the couvade, even if they do not hunt tortoises. Indeed, there are none there to hunt except the occasional backyard pet originally found in the interior forests or in the zoological park.

It cannot be argued that in all cases does TEK result in the same, similar, or any ecological effects. Even among native societies deriving their cultural personas from a common source, as among the Tupí-Guaraní societies, it can be seen that they have been exposed differentially over their historical course to contacts with other societies; perhaps this is most obvious in the Upper Xingu and the northwest Amazon. But isolated groups, such as the Tapirapé, with their central Brazilian ceremonial moieties and wrestling contests (Wagley 1977), also exhibit hybrid cultural traits, if culture is defined only in reference to language families or like genetic groupings. So we cannot assume that TEK can be traced through a language family to an origin that dates from the time of the protolanguage or protolanguages (see Dietrich [1990], who posits more than one) of that family; likewise, reconstructions of words can lead to red herrings. If one did not know the history of contact, it might be assumed that "firewater" in diverse American Indian languages could be reconstructed as an aboriginal term for whiskey (Bloomfield 1984 [1933], 455), given that "fire" and "water" are acceptable glosses for preexisting morphemes.

But we would be no further along in understanding the persistence and antiquity of TEK in Amazonia if all features of native life are ascribed to borrowings and exchanges with other groups. It would constitute a philosophical trap of particularism, and general statements concerning mechanisms of TEK would be logically impossible to make. Ethnobiological knowledge in the Tupí-Guaraní family can be and has been lost, and some of it has been borrowed through the centuries. But peculiar cultural and linguistic correspondences among diverse contemporary Tupí-Guaraní societies, such as those presented in this chapter, indicate that some knowledge of wild biological resources in Amazonia has remained essentially unchanged over thousands of years. It is the future of such knowledge that the postmodern world has cast into doubt.

9

Confection, Inflection

Historical ecology is a perspective on relations between people and the environment that, in principle, envisions how historical phenomena transform landscapes and how such transformations become conditioned and understood through local knowledge, behavior, and culture over time. The current state of landscape knowledge possessed by folk (*caboclo*) and indigenous peoples of Amazonia is, in part, a product of history. As the landscapes have changed through time, and continue to change, that knowledge, too, shows increments in some domains, losses in others. Such losses and increments of landscape knowledge are reflected in vocabulary changes, just as vocabulary can be used as an index, however crude, to knowledge of the past state of Amazonian landscapes.

Knowledge of Amazonian landscapes—in the mental control of one or more people—and based on experiential data, is at least as ancient as the Early Holocene, the presumable time of original occupation. No doubt some of that original knowledge of Amazonian landscapes has persisted, but it cannot be accessed with exactitude, since linguistic reconstruction (of words, technology, biota, and concepts) is not reliable beyond about five or six thousand years (see Kaufman 1990), given what may be considered a background rate of vocabulary loss and change similar to the more well-accepted notion of a background rate of extinction when referring to biota over long sweeps of evolutionary time.

Archaeological data alone are insufficient to probe fully ancient knowledge of Amazonian landscapes, since knowledge is more than material artifacts: it is to be sure those artifacts, but it is also behavior and cognition, which is partly reflected in real language, including written texts. Amazonia lacks written documentation, of course, before 1500, but one can utilize methods from historical linguistics in order to begin to build a model of landscape knowledge and the changes it underwent through thousands of years before the European conquest. One can demonstrate that within a five-thousand-year time period, however short from an evolutionary viewpoint, many of the landscapes and the languages associated with them in Amazonia underwent transformations, sometimes of a profound character. The landscape is that portion of the environment codified in language and subject to human intervention. A landscape represents an encounter between space and time, nature and history, biotic communities and human societies, and it is central to the conceptual apparatus of historical ecology. Landscape history is linked to environmental knowledge, and in Amazonia it is marked since the Middle Holocene by two deeply transforming phenomena: (1) the development of a system of swidden agriculture and fallow forest management by indigenous (i.e., pre-European) people and (2) the reconstitution of that system by neo-European expansionism, colonialism, and commercialization of existing landscapes in the New World, including Amazonia.

The focus of this chapter is on the second of these two historical phenomena, specifically on how eighteenth-century colonialism, Jesuit missionization, dissemination of a contact language (Língua Geral Amazônica; henceforth LGA), and the penetration of the European world system of commerce and finance changed native vocabulary and, hence, how this contact and colonialism transformed local interpretation and knowledge of Amazonian landscapes and associated biota. In particular, the emphasis here is on a single product, one of the *drogas do sertão*, cacao, and how the words and concepts for cacao underwent change in native languages given the fact that cacao and cacao beans were for some time in the eighteenth century the principal export commodity of Amazonia.

Linguistic and Cultural Background of the Ka'apor in the Eighteenth Century

The Ka'apor language (also known as Urubu, Urubú, and Urubu-Kaapor) is one of about forty languages in Tupí-Guaraní, itself one of ten branches of the Tupi family (Jensen 1999; Rodrigues 1999; Rodrigues and Cabral 2002). Ka'apor is spoken in extreme eastern Amazonia in the Brazilian state of Maranhão, specifically in the Gurupi and Turiaçu River basins, though it has recent historical origins to the west, in the present state of Pará. Eight subgroups of Tupí-Guaraní have been identified, chiefly in terms of phonological criteria (Jensen 1999; Rodrigues 1986; Rodrigues and Cabral 2002), though there is some disagreement over what languages should be included in each of the subgroups (see Mello 2002). Ka'apor has been classified in subgroup #8 in three slightly different iterations of this model; for the purpose of consistency, I will be using specifically the revised classification of Tupí-Guaraní proposed by Rodrigues and Cabral (2002), with the caveat that minor revisions in that model may become standard in the future. Subgroup #8 also includes Wayãpi, Guajá, and at least seven other living and dead languages (Jensen 1999; cf. Mello 2002).

Beatriz Corrêa da Silva (1997, 83) argued that Ka'apor is very close to Wayãpi in terms of phonological criteria. Indeed, Ka'apor informants have told me that based on their contacts with Wayãpi speakers (of the Wayampipuku dialect) in the Casa do Índio near Belém, Pará, they can understand Wayãpi better than other Tupí-Guaraní languages they have heard, in spite of unlike stress patterns in the two languages and a number of Carib borrowings in Wayãpi not occurring in Ka'apor. Detailed evidence from ritual also indicates an intimate association between Ka'apor and Wayãpi cultures that would have existed about three hundred years ago (Balée 2000b). But according to Corrêa da Silva (1997, 83), Ka'apor is unlike Wayãpi in certain morphological respects. In fact, she claims Ka'apor is more like LGA, a Tupí-Guaraní creole known also as Nhe'engatú ("the good talk," Jensen 1999, 127) and that is classified in subgroup #3, in terms of pronominal prefixes, pronominal system in general, and pronominal marking on verbs (see Corrêa da Silva 2001).

Corrêa da Silva (1997, 88–89) registered numerous, apparently borrowed, lexical items that were presumably present in Ka'apor before the Ka'apor ancestors became peaceful with Brazilian authorities and society in 1928, such as words for *caboclo* (Amazonian peasant); *padre* (Catholic priest); *camarada* (comrade, or non-Indian person); Christian; *mamãe* (mother, vocative); and *papai* (father, vocative) (also see Balée 1994, 29–30 for similar evidence). The supposition is that LGA was the donor language of these and other borrowed terms in Ka'apor. Wayãpi also underwent LGA influence (Jensen 1990, cited in Corrêa da Silva 1997, 86), but perhaps not so much as Ka'apor. Where and how did this influence originate, and what, if any, implications does it have for Ka'apor nomenclature regarding natural things in their environment?

LGA made its appearance in Amazonia sometime after the Portuguese founded a fort (Forte do Presépio) in 1616 that would become the city of Belém. LGA was based on Tupinambá, spoken in the Lower Amazonian Portuguese colony called the Province of Maranhão e Grão Pará. LGA underwent significant Portuguese lexical influence. LGA was partly the linguistic product of marriages between Tupinambá women and Portuguese soldiers and colonists (Corrêa da Silva 1997, 83–84 and passim; Rodrigues 1986, 102) and partly the influence of learned Jesuits who brought many aspects of the language with them from coastal Brazil (the uniformity of Tupinambá along the coast of Brazil has been widely noted). The Tupí-Guaraní creole of southern Brazil, Língua Geral Paulista or Tupi Austral, developed in quite parallel circumstances (Jensen 1999, 127). Jesuit missionaries arrived in the region of the estuary and Lower Amazon in 1636 (Cruz 1973), and they helped institutionalize LGA in mission settings. By 1655, there were fifty-four Jesuit missions in Amazonia, mostly along the Amazon River itself and south of it (Leonardi 1999, 56). LGA became the dominant language in the Brazilian Amazon and would be supplanted for the most part by Portuguese only about two hundred years later, during the rubber boom, beginning in the latter half of the nineteenth century, when hundreds of thousands of monolingual immigrants from northeastern Brazil arrived in the region to take up a life of rubber tapping (Leonardi 1999, 75; Moreira Neto 1988, 43–45).

By the time of the rubber cycle in Amazonian history, the Ka'apor as a people had long been isolated from and hostile to rubber tappers and Luso-Brazilian society generally (Balée 1984a). In other words, the Ka'apor were never *caboclos* per se. Rather, the close of the colonial period of Ka'apor history in the mid to late 1700s helps us to comprehend better the beginning of *caboclo* history, for it seems to be at this time that the *caboclos* emerge as a people separate from whatever indigenous roots they had, and the Ka'apor and other indigenous groups—continuing to be indigenous—on the one hand and the Lower Amazonian peasantry on the other then diverge and go their separate ways in historical time down to the present day.

At one time, however, the Ka'apor as a people apparently enjoyed relatively peaceful though probably subordinate relations with representatives of the Iberian metropole, especially Jesuit missionaries with whom they would have been in daily, face-to-face contact at least until the expulsion of the Jesuits from Brazil in 1759 (Azevedo 1930, 375, cited in Balée 1988a, 157). This is the period that in my view should be understood as being immediately at the eve of the formation of the Amazonian peasantry (see Nugent 2009, who prefers "historical peasantry" to the use of the term *caboclo* for the purpose of describing the extant, native-born, Portuguese-speaking people of Amazonia). This time of hypothesized contact between precursors of Ka'apor society and colonial Luso-Brazilian society constitutes the period shortly before the influx of African slaves into the Lower Amazon (see Guzmán 2009), who were brought to replace waning indigenous labor and populations. And this contact period occurs just before the coining of the neologism "*caboclo*," which would be used in the following years to refer to the Amazonian masses as distinct from individually named indigenous groups of Amazonia. This time frame is not coincidentally contemporary with the expulsion of the Jesuits and the demise of the Jesuit mission system in the Lower Amazon (and elsewhere) in 1759.

On the basis of ethnohistory, oral history, and linguistic data relating to toponyms, Ka'apor society originated at least four hundred kilometers to the west of their present habitat probably before 1800 in the basin of the Tocantins River (Balée 1994, 30–32) (see the map on p. 118, this volume). Before 1759, the Ka'apor probably lived even farther west, nearer to the Xingu. This is inferred because of a historical connection that seems to have existed between the Wayãpi and the Ka'apor. The Ka'apor and Wayãpi share some esoteric details of a girl's puberty rite that are most likely not due to chance (Balée 2000b, 412; see also Chapter 8). The details center on the ant ordeal, which in both societies' cultural practices involves the application of venomous stings on the initiate's skin from the same species of ant (*Pachycondyla commutata*). That ant is called by apparently cognate terms in the two languages (Ka'apor *tapiña'ĩ* and Wayãpi *tapia'ĩ*). This ant ordeal at a girl's initiation has not been described for any other pair of Tupí-Guaraní societies, which indicates evidence of shared innovation between Ka'apor and Wayãpi ancestral sociocultural and ritual systems. The ant ordeal, therefore, suggests a historical connection between the two peoples in the comparatively recent past, a connection that further supports their linguistically close pairing in subgroup #8.

Today the Wayãpi and Ka'apor live about nine hundred kilometers apart, with the estuary of the Amazon River in between them. But in the early 1700s, the precursors of the Wayãpi lived in the lower Xingu River basin and some of them became settled at one of three Jesuit missions at that time (Fisher 2000, 46; Grenand 1982, 20, cited in Corrêa da Silva 1997, 84–85). Evidence from Ka'apor mythology and concepts of species, such as the Brazil nut tree, that are not present in their habitat today, indicates a westerly origin at least as far west as the Tocantins River, called *i-takāšī* (i.e., "Smoke River") in Ka'apor (Balée 1994, 25). But because of their close linkages to Wayãpi, from the perspectives of both ritual and language, antecedents of the Ka'apor can be logically placed even farther west than the Tocantins, indeed closer to the Xingu but probably still east of that river, in the early eighteenth century.

The impact of LGA on Ka'apor language and Luso-Brazilian, Jesuit influences on Ka'apor culture seem to be related to what at one time was the principal commodity extracted by the Iberian metropole from Amazonia: cacao. The word for cacao and the origins, uses, and management of cacao by native peoples in Amazonia in prehistoric times represent an inimitable array of historical-ecological phenomena that allow us to understand Amazonian history within a capsule of a single species and its ultimate effect on the landscape. In this sense, a change in the material and economic landscape—namely, what would become the paramount importance of cacao as a commodity in the export market of the *drogas do sertão* and the acquisition of native labor to gather it—may have affected the Ka'apor language in the domain of plant nomenclature.

Origins of the Word for Cacao in Various Languages

Where did the word for cacao come from? Mesoamericanist J. Eric S. Thompson indicated that the origins of the words *cacao* and *chocolate* are not easily found. There has been a considerable amount of speculation on the subject, but it is to be doubted that any conclusions satisfactory to everyone will ever be reached (Thompson 1956, 107). Writing before an explosion of historical-linguistic and epigraphic research in Mesoamerica, Thompson was perhaps too pessimistic, though it must be granted that all linguistic reconstruction as with much of archaeological interpretation must remain speculative, however informed and enlightening.

One way of approaching Thompson's problem would be to seek the word where the plant itself originated. This exercise involves consideration of cultural factors, since the cacao of commerce (*Theobroma cacao* L.) is a domesticate. Two subspecies of cacao are recognized (Cuatrecasas 1964, 512–513), and the principal subspecies of modern commerce, *T. cacao* ssp. *cacao* (with four formae), was the only domesticated one found in Mexico and Central America at the time of the Hispanic conquest. Two commercial types are known: *criollo* (*T. cacao* ssp. *cacao*) and *forastero* (which may include other subspecies, all from South America). *Criollo* has "elongated, ridged, pointed fruits and white cotyledons" while *forastero* has "short, roundish, almost smooth fruit and purplish cotyledons" (Cuatrecasas 1964, 506; also see Coe and Coe 1996, 27; Schultes 1984). The *criollo* variety of Mexico and Central America does not grow spontaneously; in contrast, other *forastero* subspecies can be found growing spontaneously in various parts of the Amazon Basin (Cavalcante 1988, 63; Huber 1904, cited in Cuatrecasas 1964, 401) and the Guianas (Cuatrecasas 1964, 494, map). Indeed, two morphological variants are noted, an Upper Amazon *forastero* and a Lower Amazon *forastero* (Motamayor et al. 2000). Today *forastero* subspecies and varieties derived from *T. cacao* ssp. *sphaerocarpum* have become the most important in commerce (Gómez-Pompa et al. 1990, 249), accounting for about 80 percent of world production (Coe and Coe 1996, 28, 201–202). The precontact distribution of many spontaneous varieties in South America and only one, fully domesticated, variety in Mesoamerica bespeaks the possibility that cacao originated in the headwaters of the Amazon, crossed the Andes into northern Colombia, and ultimately made its way to Central America and lands farther north (Cheesman 1944, cited in Cuatrecasas 1964, 507).

The age-area (or "least-moves") hypothesis is clearly strengthened by the fact that all twenty-two known *Theobroma* species were originally found in the Amazon Basin and adjoining Guianas and only three (*T. cacao, T. angustifolium,* and *T. bicolor*) have ever grown outside that region. Cuatrecasas (1964, 507) confidently asserted, nevertheless, that the first prehistoric cultivation and selection of cacao occurred in Mexico and Central America and subsequent writers have tended to support that claim (e.g., Stone 1984, 69). Gómez-Pompa et al. (1990) presented recent evidence for a possible ancestral form to domesticated cacao, which was noted to be growing in a sinkhole in northern Yucatán. This variety is the rare *T. cacao* L. ssp. *cacao* forma *lacandonica* Cuatrecasas, which was previously only known from the Lacandon Maya area of Chiapas, Mexico (Coe and Coe 1996, 26–27). Linguistic evidence to date also seems to support an original domestication of cacao in Mesoamerica though the precise language of origin is a matter of dispute.

One account argued for a source of Mayan **kakaw* in Mixe-Zoquean (Justeson et al. 1985, 59), a putative source of borrowings in Mayan and other Mesoamerican language groups (Campbell and Kaufman 1976, 84). According to this view, many Mixe-Zoquean agricultural terms were borrowed by Mayan and other Mesoamerican language groups, reflecting perhaps the prestige of the proposed first agricultural civilization of the region, the Olmecs. The Olmec civilization might have been associated with speakers of Mixe-Zoquean (Campbell and Kaufman 1976, 84), though this inference too is debatable (Wichmann 1999). More recently, an argument has been made that *cacao* is actually a term coined by speakers of Nahuatl, perhaps the people whose capital city was Teotihuacan (Dakin and Wichmann 2000).

Regardless of which Mesoamerican linguistic group is eventually determined to be the source of the term *cacao* (i.e., Uto-Aztecan, Mayan, or Mixe-Zoquean), the use of cacao in Classic Maya culture (ca. 200 B.C.–A.D. 600) is now well established. Biochemical evidence for theobromine, one of cacao's characteristic alkaloids, has now been determined to exist on remains of spouted vessels (called "chocolate pots") in northern Belize that date from 600 B.C. to A.D. 250, that is, from the time of the Preclassic Maya culture to the beginnings of Classic Maya culture (Hurst 2001; Powis et al. 2002). By Classic times, cacao is evidently a local crop grown widely in Mesoamerica, including in peripheral areas such as the medium-sized village site of Cerén in El Salvador (Lentz and Ramírez-Sosa 2002). In other words, it was not evidently a crop only of the elite but also of the common people living on the periphery of urban civilization as well.

After the Spanish conquest of Mesoamerica, with the debut of chocolate in the European marketplace and the rapid conditioning of the Western palate by it, the term *cacao* became widely diffused to numerous languages worldwide. In Hanunóo of Mindoro Island, Philippines, two of the three words for folk species of cacao exhibit the morpheme *kakaw* (Conklin 1954, 418), no doubt borrowings from Spanish. In the Quichua language of Amazonian Ecuador, all compound names for two species of *Theobroma* (*T. cacao* and *T. subincanum*) incorporate the term *cacao* (Kohn 2002, 432). Many other Amazonian and lowland South American groups borrowed a term for cacao that entered the continent through Spanish or Portuguese. What is of most interest is why they would, and in particular why would the Ka'apor, borrow a term for a plant that they already had?

According to the historical-linguistic principle of prestige, whereby in a contact situation goods and services associated with the dominant society that were not previously present in the subordinate society tend to be borrowed by the subordinate society (see Campbell 1999, 59–60), the word for cacao would not have been borrowed by Ka'apor since it already occurred in their environment, unless cacao had acquired some prestige and economic valorization far above and beyond what it held in native Amazonia. Ethnobotanist Richard Evans Schultes (1984, 33) observed that it was difficult to explain why Amazonian Indians would have been motivated to disperse a tree the use of which lay solely in a sweet pulp on which one might suck (also see Coe and Coe 1996, 26 for a similar view). Cacao cultivation in Mesoamerica is probably as old as if not older than the Tupí-Guaraní branch of the Tupi family, dating back at least to the beginning of the common era and probably much earlier (cf. Alden 1976, 104; see Young 1994, 17), even if the word for cacao may be more recent than its original cultivation (Dakin and Wichmann 2000). Plant geneticist Charles Clement (1999b, 211) pointed out that *T. cacao* and its close relative *T. bicolor* (which may grow spontaneously in the Maya lowlands, unlike *T. cacao,* though it is of lesser quality and desirability; Thompson 1956, 107) were probably semidomesticated crops grown as stimulants in the Upper Amazon during late prehistoric times. But the use of cacao beans as stimulants is seldom found outside the Upper Amazon. The Kofán of the Ecuadorian Amazon toast and eat the beans of *T. bicolor* (which they term *mak'avɨ*) (Pinkley 1973, 69) as do the Lowland Quichua of Ecuador (Eduardo Kohn, personal communication, 2001). The practice of toasting these beans before consuming them seems fairly widespread in the Upper Amazon, despite the avowedly low quality of the beans and fruit when compared to other species of *Theobroma* (Cavalcante 1988, 66). In any case, no prehistoric Amazonian groups are known to have made chocolate (Gómez-Pompa et al. 1990, 249; Schultes 1984, 33; Stone 1984, 69).

Rather, almost everywhere outside the Upper Amazon, native Amazonians have eaten only the sweet, white pulp around the beans and then discarded the beans; in some cases, the pulp around the beans has been made into a nonfermented wine (Coe and Coe 1996, 26). Given the low aboriginal prestige of cacao in the Amazon region, the directionality of borrowing of the term is probably not, basically, Amazonia→Mesoamerica; rather, the reverse seems much more probable now. It is unlikely that Mixe-Zoquean speakers, who may have been already associated with complex, intensive agricultural society, would have borrowed an Amazonian term for a semidomesticated (or perhaps even wild) crop that had not yet developed uses as chocolate. And the possibility remains that the development of chocolate production in Mesoamerica began with *criollo* trees that had arisen from spontaneous mutations and subsequent genetic drift along the Isthmus of Panama, not far from the northernmost edge of the presumed, original distribution of cacao (Purseglove 1969, cited in Young 1994, 14–15). It is possible therefore that cacao was not dispersed into Mesoamerica by humans and was part of the original distribution of wild forms of cacao, such as the forma *lacandonica* (Gómez-Pompa et al. 1990, 249), but this remains controversial (Stone 1984; Young 1994, 14).

The first European observation of cacao occurred in 1502 along the northern coast of Honduras, on Columbus' fourth voyage (Alden 1976, 104). Rapidly, the chocolate drink made from it became highly esteemed in Europe (Alden 1976, 109), and it became well known to explorers as a valuable export crop. Cacao plantations begin in Ecuador and Venezuela by the late 1500s and early 1600s. The crop therefore may have been recognizable to Spaniard Cristoval de Acuña, who noted in 1641 that in some places groves of cacao trees along the Amazon River were so thick that the wood could serve to lodge an entire army (Acuña 1963, 76). Cacao exports from the Amazon were reported by 1678–1681, and these beans were being collected from spontaneously occurring trees, not plantation trees (Alden 1976, 114–115). By about 1725, a cacao boom started in the Amazon, and cacao became the dominant export staple of the region (Alden 1976, 118; cf. Hemming 1987, 43). By the mid-1700s, different regions of Brazil exported distinctive commodities to Lisbon: "The Rio fleet shipped gold, hides and silver; Pernambuco sent wood and sugar; and the fleets of the north [i.e., Lower Amazon], of Grão Pará and Maranhão carried cacao" (Maxwell 1973, 5). The cacao export sector of the eighteenth-century Luso-Brazilian economy was perhaps minor compared to gold in Minas Gerais and Rio de Janeiro and later coffee in São Paulo (Baer 1995, 15–19), but it seems in many ways to be the precursor of the rubber export economy of the nineteenth century as concerns the Amazon region.

The cacao export sector of the colonial Amazonian economy fell under the control of Jesuit missionaries, who induced Indians under their tutelage to collect cacao in the interior from spontaneous trees, whereas significantly less cacao came from plantations (Alden 1976, 121–122; Coe and Coe 1996, 194–195; Hemming 1987, 43). These spontaneous trees were most likely from *Theobroma cacao* and not from nondomesticated species of *Theobroma*. Although *Theobroma speciosum* Willd., a nondomesticated and very widespread cacao species known regionally as *cacauí* ("little cacao"), produces edible pulp and seeds from which chocolate can and has been made, its fruiting season is only between February and April, hardly enough time to qualify as a major export crop. Amazonian *Theobroma* cacao, in contrast, can be found for sale in all months except September to December at the market in Belém (Cavalcante 1988, 64).

Remarkably, as a percentage of the total exports from the Lower Amazon during 1730–1755, cacao ranges between 43.5 and 96.6 percent, with the highest proportion of total exports from that region occurring in the years 1730–1745 (Alden 1976, 118). The cacao trade begins to decline in the 1740s and 1750s, and this coincides with native population declines due to smallpox and measles epidemics widely reported during the period 1743–1750 (Balée 1984a, 34–35; Hemming 1987, 43; Moreira Neto 1988, 23–24). African slavery revived the trade after the 1750s, such that what is now the Brazilian state of Pará was exporting 715–850 tons of cacao per year, which constituted about 90 percent of the total exports from Brazil (Hemming 1987, 43). Even after the expulsion of the Jesuits from the Portuguese Empire in 1759–1760, most of the cacao exported from the Amazon still came from collecting expeditions rather than from cultivated trees (Alden 1976, 123–124), and cacao would not become a dominant export crop from the Brazilian state of Bahia until the late nineteenth century (Baer 1995, 19).

The impact in Amazonia of a cacao export economy combined with Jesuit control seems to have affected native languages. Indeed, the significance of the cacao export sector in the Lower Amazon cannot be overestimated in terms of its effects on local indigenous societies and their languages that were involved in it. In 1743, cacao is clearly the most important of all the *drogas do sertão* (the various forest and garden products from Amazonia that were shipped to Europe for a variety of purposes: food, spice, medicine, oil, hides, skins, timber, waxes, gums, and so on; see Cleary 2001, 83–84), for at that time cacao beans were observed to be circulating as money among the Amazonian peasantry (not dissimilar to the way cacao beans had served a monetary purpose in Aztec markets) and cacao beans owned were figured into calculations of an individual's wealth (Bruno 1966, 59). In the colonial era, cacao assumed an importance that had not been before known in aboriginal Amazonia.

Cacao was a central commodity in the "Jesuit century," as David Block (1994, 98) has so aptly described the eighteenth century in eastern Bolivia, which can apply with slight modifications also to Amazonian Brazil, coastal Brazil, and the mission zones of Paraguay and Argentina. The Jesuits introduced cacao into the Mojos Plains of eastern Bolivia (also known as the Beni), where it had not even existed in the wild before, as an export crop (Block 1994, 98). Cacao was probably not typically planted in pre-Columbian Amazonia, but the Jesuits, using native labor, cultivated it successfully in the vicinity of their Amazon missions (Aubertin 1996, 32; Bruno 1966, 61). Indeed, had it not been for the Jesuits, Amazonia could not have met European demand for chocolate (Aubertin 1996, 33). The *drogas do sertão*— a bona fide term for what today one might call TFPs (tropical forest products; see Cleary 2001, 83–86)—constitute a very long list of wild and cultivated plant and animal materials, but in terms of economic impact both in the Amazon and in the European marketplace, cacao was at the top of this list (Di Paolo 1985, 76).

The Jesuits used LGA, a creole language partially derived from Tupinambá, in their missions. Many LGA vocabulary items are borrowed from Portuguese. In cases of language contact, vocabulary items for native plants, animals, and landscape features are most often borrowed by the dominant or prestige language and vocabulary items related to politics, religion, and finance are most often borrowed by the subordinate or nonprestige language, based on contrasting principles of prestige (i.e., luxury loans) or need (Campbell 1999, 59–60). Cacao is a native Amazonian plant, so by the principle of need, it is reasonable to argue that the term came originally from some Amazonian language. *Cacao* is a Portuguese word borrowed from Spanish *cacao* that was in turn borrowed from a Mesoamerican language, where the plant first attained preeminence in terms of world commerce. Controversial evidence suggests that *cacao* can be reconstructed in Proto-Mixe-Zoquean, which dates from about 3500 B.P. (Campbell 1999, 349; also see Campbell and Kaufman 1976; Justeson et al. 1985) and was plausibly associated with the ancient Olmec civilization of the isthmus of Tehuantepec, as **kakawa* (Campbell and Kaufman 1976, 84). The possibility remains that Proto-Mixe-Zoquean borrowed the term from an Amazonian language on the basis of need, if the crop indeed originated there (though perhaps not as a fully domesticated crop) as biogeographic evidence suggests. But the prestige principle and the known time frame militate against that hypothesis.

Cacao Words and Tupí-Guaraní Languages

In several Tupí-Guaraní languages of Amazonia, *Theobroma cacao* L. ssp. *sphaerocarpum* is referred to by words that seem cognate by inspection (Table 9.1),[1] the exceptions being Ka'apor (because of an initial *k*), Parintintin, and Wayãpi. What is puzzling is that the other, seemingly cognate, words resemble the word *cacao* in their phonetic shape. At least some of these languages might be presumed to have had little if any influence from LGA, especially Guajá (a language of hunter-gatherers who have only been in contact since the 1970s) and Araweté (a language of trekkers only in contact also since the 1970s). But both these languages have a word for *comrade* (Guajá *kamarar;* Araweté *kamara;* Balée, field notes), borrowed evidently from medieval Portuguese *camarada;* corresponding borrowed terms are also known from Ka'apor (*kamarar*) and LGA (*kamarára*) (Corrêa da Silva 1997, 89), though the range of meaning among them is somewhat divergent, since at least in Guajá *kamarar* refers to the Ka'apor people, whereas in the other languages mentioned the cognate term refers to non-Indians or is even, in the case of Araweté, a personal name for a man.

These five languages are in three different subgroups (nos. 4, 5, and 8) of the eight recognized subgroups of the Tupí-Guaraní branch of the Tupi family (Jensen 1999; Rodrigues and Cabral 2002). The phonological structure of the terms apart from the word in Ka'apor in Table 9.1 does not suggest borrowing among the different languages. It is possible that Ka'apor has conserved an initial **k* in the word for cacao and that the initial consonant was deleted in Araweté, Assurini do Xingu, Guajá, and Tembé. The proto-term may have been **kaka,* and this would be far older than the cacao export economy of Lower Amazonia in the 1700s. But this hypothesis seems unlikely. The principle of prestige would tend to preclude a nonprestige language from borrowing a term for a native plant that was not of

Table 9.1. Words for cacao (*Theobroma cacao* L.) in several Tupí-Guaraní languages

Language	Subgroup #	Term	Gloss	Source
Araweté	5	aka-'i	L-tree	Balée, field notes, 1985
Assurini do Xingu	5	aka-'ɨwa	L-tree	Balée, field notes, 1986
Guajá	8	ako'o-'ɨ	L-tree	Balée, field notes, 1989
Ka'apor	8	kaka	L	Balée, field notes, 1985
Tembé	4	aka-'ɨw-ete	L-stem-true	Balée, field notes, 1986 (cf. Boudin 1978)
LGA	3	kakáu	L	Stradelli 1929
Parintintin	6	ñumi-	L	Betts 1981
Wayãpi	8	walapulu	L	Grenand 1989

commercial or agricultural importance. The nondomesticated, widely occurring cacao species, *Theobroma speciosum* Willd., is either designated by the same term (as in Assurini do Xingu and Guajá) or it is linguistically marked as though it is perceived as being a close relative of domesticated cacao (from the point of view of nomenclature, not classification per se) (see Table 9.2).[2]

In other words, in keeping with the prestige principle, one would not anticipate borrowing of terms for nondomesticated, seemingly unimportant plants (though nondomesticated cacao, especially *Theobroma speciosum* Willd., like its domesticated congener, does have a sweet, edible pulp, and people gather it for that purpose). But that evidently happened in Ka'apor. The Ka'apor words for *Protium* trees (Burseraceae), *Lacmellea* trees (Apocynaceae), and *Mabea* trees (Euphorbiaceae), all of which are found in high forest and are never cultivated per se, seem to have been borrowed also from LGA (Balée 1994). It is plausible that products from these trees were part of the *drogas do sertão* transoceanic trade; *Protium* trees, for example, exude a resin that is highly prized as boat caulking, and caulks were one of the Amazonian *drogas do sertão*.

Wayãpi, which like Ka'apor is from subgroup #8, denotes domesticated cacao as *walapulu,* clearly a borrowing from one of several Carib languages in the Guianas (Grenand 1989). Yet the Wayãpi term for nondomesticated cacao, *T. speciosum,* is *aka-'ɨ,* an apparent cognate with the terms, aside from the Ka'apor term, in Table 9.2. Françoise Grenand (1989, 121) gives the etymology as *aka* < *ãkã*, "head," and suggests also a comparison with LGA *kakao-'i*, "little cacao." Her etymology of "head" seems problematic, however, since as in Wayãpi, the vowels are also not nasalized in the cognate terms in the five other Tupí-Guaraní languages in Tables 9.1 and 9.2. It seems unlikely that nasalization for this wild cacao word would have

Table 9.2. Words for nondomesticated cacao (*Theobroma speciosum* Willd.) in several Tupí-Guaraní languages

Language	Subgroup #	Term	Gloss	Source
Araweté	5	aka-á-wi'ɨ	L-fruit-thin-stem	Balée, field notes, 1985
Assurini do Xingu	5	aka-'ɨ-wa	L-stem	Balée, field notes, 1986
Aurê and Aurá	8	aka-ú	L-large(?)	Balée, field notes, 1987
Guajá	8	ako'o-'ɨ	L-stem	Balée, field notes, 1989
Ka'apor	8	kaka-ran-'ɨ	L-false-stem	Balée, field notes, 1985
Tembé	4	aka'u-'ɨw	L-stem	Balée, field notes, 1986 (cf. Boudin 1978)
Wayãpi	8	aka-'ɨ	L-stem	Grenand 1989
Proposed reconstruction	—	*ako'o-'ɨβ	L-stem	—

been dropped in all of them, just as deletion of initial *k* in the cacao word in three different subgroups of Tupí-Guaraní also seems unlikely. Initial consonant loss is, moreover, less common than initial vowel loss (Campbell 1999, 32–33). On the basis of this mounting evidence, one can logically argue that (1) the original term in Wayãpi for nondomesticated cacao is a sequence of a literal morpheme (*aka*) and a term meaning "stem" or "tree" (*'ɨ*); (2) the Ka'apor terms for cacao and nondomesticated cacao are most likely to have been borrowed; and (3) the donor language for the cacao terms in Ka'apor was LGA.

The LGA term for cacao fruit is *kakáu* (Stradelli 1929). In LGA, diphthongs may occur in word final position. The combination of two vowels in principle represents two syllables (Taylor 1985, 11–12). But in LGA one does not canonically find the following: V [+high, +back, +vocalic]#.

This combination of phonemes is otherwise common in Portuguese, as in /páu/ ("wood, tree") and /kakáu/ ("cacao"). It can be therefore proposed that the directionality of borrowing was LGA→Ka'apor, and not the reverse. Ka'apor retained the initial *k* when it borrowed the term, and phonological substitution (in this case, by deletion of final vowel or apocope; Campbell 1999, 32, 61) accounts for the absence of the unstressed final, high back vowel in Ka'apor *kaka*. Ka'apor also extended the root lexeme's semantic range to nondomesticated cacao, analogous to the extension noted by Grenand (1989, 121) above for LGA. The reason the term for nondomesticated cacao persisted in Wayãpi is that perhaps Wayãpi was less affected by missionization influences than Ka'apor and because nondomesticated cacao was not an item of prestige, whereas domesticated cacao was a prestigious commodity thanks to Jesuit and Luso-Brazilian valorization and cultivation of it. It is striking nevertheless that the Wayãpi have a Carib loan word for domesticated cacao; it is possible from this evidence, and from the other evidence related to a strong ethnic and linguistic connection between the Ka'apor and Wayãpi, to speculate that Wayãpi once had a term like *kaka* in Ka'apor and exchanged this for *walapulu* at a later date, after they crossed the Amazon River from the south, but to date have not yet gone so far as to replace the term for nondomesticated cacao (and remodel it by analogy on *walapulu* or some other borrowed term for cacao).

This argument leaves open whether the other Tupí-Guaraní languages in the sample also borrowed the word for cacao from LGA. Although deletion of all initial *k*'s seems unlikely, the Parintintin language represents a peculiar departure from the other languages in the sample. Parintintin is from subgroup #6 of Tupí-Guaraní and it is spoken in southwestern Amazonia, close in fact to where Proto-Tupí-Guaraní is believed (using the least-moves hypothesis) to have originated, and it evidently has had little or no LGA influence, for it seems to have been beyond the distribution of the Jesuit missions. Parintintin has the focal generic name *ñumi-* for cacao and many of its relatives (Betts 1981; Waud Kracke, personal communication, 2001).[3] Parintintin is also located in the richest area of the genus *Theobroma* in the Amazon Basin. At this point, the lexeme *aka* (Araweté, Assurini do Xingu, Tembé, and Wayãpi) or *ako'o* (Guajá, which is arguably closest to the protolanguage for this term) may resemble LGA *kakáu* only by coincidence or by borrowing. It is nevertheless intriguing that whereas Ka'apor arguably borrowed *kaka* and extended it to cover nondomesticated cacao species, as discussed above Wayãpi also borrowed *walapulu* (from a Carib language) for cacao but retained *aka* for the nondomesticated cacao species. If *aka* is closer to the original Proto-Tupí-Guaraní word for cacao than *kaka,* then this word for cacao was most likely not borrowed by Mesoamerican languages from Tupí-Guaraní languages even if *aka* has cognates in Tupian language branches other than Tupí-Guaraní. That is because word-initial epenthesis of a consonant is not likely (Campbell 1999, 33). The only published data now available on another branch of Tupi are from Munduruku, of the Munduruku branch, and the word for cacao appears to be a borrowing, also from LGA, being *kakáu* (Strömer 1932, 62; but Crofts and Sheffler [1981, 18] indicate *karoba* as the Munduruku term for cacao).

Discussion

In the Ka'apor habitat of today, there are four species of *Theobroma* other than *T. cacao* and *T. speciosum.* These are *T. grandiflorum* (Willd. Ex Spreng.) Schum., called in Ka'apor *kipɨhu'ɨ*, and *T. subincanum* Mart., called *kipɨ'a'ɨ*, for which there is also a synonym, *nukipɨ'ɨ* (Balée 1994, 307). These terms do not appear to be related to the Ka'apor terms for *T. cacao* and *T. speciosum* and they are in a different folk genus. Indeed, the fruit of *kipɨhu'ɨ*, which is widely known in the Amazon region as *cupuaçu,* is apparently much more esteemed (by the significantly more time that is given to its gathering) by the Ka'apor than are its congeners, cacao and nondomesticated cacao. There is no reason to suppose that this differential appreciation was different in precontact times. The fruit of *cupuaçu* is eaten as is the fruit of cacao: it is the sweet pulp around the beans in the pod that one eats, but in the case of *cupuaçu*, this somewhat tart pulp is much more copious. *Cupuaçu* terms exhibit a tendency to cognate forms also: Guajá *kɨ-pɨ-'ɨ*, Tembé *kupi'a'ɨw*, Wayãpi *kapi-ai* (Balée 1994, 307; Grenand 1989, 112), for which one could logically propose the tentative reconstruction of **kɨpɨ'a'ɨβ* in Proto-Tupí-Guaraní. (Incidentally, Parintintin is once again the odd man out, with *ñumitahɨm,* in reference to the *cupuaçu* fruit only; Betts 1981, 268. But given the underdifferentiation of the large variety of *Theobroma* species in the Parintintin language, it is unlikely that this term is a reflex term.) These terms for *cupuaçu* in Guajá, Ka'apor, Tembé, and Wayãpi have remained phonologically similar because *cupuaçu* did not become a major export crop, as did cacao. A change in the economic landscape, in which a less-than-salient species, cacao, suddenly surged up incredibly in value and in terms of a monetary valorization system not before known in Amazonia, had the effect of influencing the language(s) most involved in its collection and exportation. Hence, whatever the original word Ka'apor had for *T. cacao* (it may have resembled, for example, *aka'ɨ* or a name containing this form), that term was replaced by a new, borrowed term from LGA, the contact language. In addition, whatever the original Ka'apor term was for wild cacao (*T. speciosum*), and this would have been very close if not identical to *aka'ɨ* (from Proto-Tupí-Guaraní **aka'ɨβ*; see Table 9.2), that term too was replaced, when the plant was modeled by analogy on *T. cacao*. In other words, Ka'apor *kakaran'ɨ,* in a broad sense, can be glossed as "that tree which resembles cacao." It is possible, indeed, that before the mercantile valorization of cacao and before Ka'apor contact with colonial Luso-Brazilian society, *T. speciosum* was more psychologically salient than *T. cacao*. That is because *T. speciosum* is much more ecologically important and common in old fallow forests (where the Ka'apor once had lived in settled villages between forty and one hundred or more years ago, but which have since seen a return of forest cover) (Balée 1994, 37). Indeed, *T. cacao* is only occasionally planted in dooryard gardens by the Ka'apor and it is not seen in the high forest or in fallow forests, which is to be expected of a domesticated species. Cacao is relatively uncommon compared to wild cacao and probably this was the case aboriginally in the Xingu and Tocantins basins also.

If *T. speciosum* were more psychologically salient than *T. cacao* before contact, it may be the case that the term for cacao was the marked form in Ka'apor, and wild cacao was unmarked linguistically. In other words, the impact of contact together with landscape modification by the Jesuit mission system, the reordering and transforming of native labor and work priorities, and the sudden high value of *T. cacao* within an imposed, alien system of exchange and valorization could have not only caused the substitution of the LGA term for the native name of cacao in Ka'apor, but that impact may also have brought about a marking reversal with regard to the Ka'apor term for *T. speciosum*. Although this assertion cannot be proven at the present moment, it is clearly a plausible scenario within the context of a historically intricate and significant contact situation.

Conclusions

In summary, it can be hypothesized that in the Ka'apor language, as in some other Amazonian languages such as Quichua, the cacao words (for *Theobroma cacao* and *Theobroma speciosum*) were borrowed and that this borrowing occurred probably because cacao, as a major export crop, had a profound impact on Indian labor of the Lower Amazon region in the eighteenth century and because that labor was to some extent controlled by Jesuit mission authorities in which LGA was the contact language. Tupí-Guaraní languages can be ruled out as sources for the Ka'apor word for cacao as well as for the English, Spanish, and Portuguese words for cacao. The evidence here presented of borrowing of the cacao term by Ka'apor further refines comprehension of the Ka'apor past and their relations to other living groups. The evidence suggests that Ka'apor culture and language were influenced by the cacao export economy on the eve of the recognition of a new Amazonian ethnic designation, that of *caboclo*. The emergence of *caboclo* culture—together with its entry into the modern world of the eighteenth century as a conceptual, real, named entity—represents the next stage of Amazonian history, after the Ka'apor peaceful experience with Luso-Brazilian society comes to a close, and the antecedents of Ka'apor society extricate themselves from the *drogas do sertão* trade and the cacao export economy, eventually to become an independent, indigenous society that until 1928 was decidedly hostile to encroachment of the state. Although the Ka'apor were therefore never incorporated into the Amazon peasantry per se, for the origin of *caboclos* as a distinctive sociocultural system postdates Ka'apor divergence from Luso-Brazilian society, this borrowing by Ka'apor of the term for cacao helps situate the antecedents of the Ka'apor historically in a setting, such as a Jesuit mission, where LGA was the contact language.

The Ka'apor borrowing of the term for cacao is most likely to have occurred farther west than the Tocantins, where the Wayãpi were also located in a Jesuit mission, along the Xingu River. This further strengthens the hypothesized close pairing of Wayãpi and Ka'apor within subgroup #8 of Tupí-Guaraní. Finally, the impact of the cacao export economy shows that a native species in the environment, even a relatively unimportant one, can be renamed in local languages when its historical-ecological setting in the world economy is completely transformed and when the people speaking those local languages are involved in the labor and technology of that transformation, as was the case with Amazonian cacao. Comprehending the history and uses of cacao and, no doubt, of other highly commercialized species of the past can be most useful for understanding the historical-ecological impact that the expansion of Luso-Brazilian society had on native Amazonian languages and associated ethnobiological vocabularies.

PART IV

Dimensions of Diversity

Overview

In these two final chapters, I engage the question of biological and ecological diversity, where it comes from in specific Amazonian contexts, and how traditional technologies of the past and present can account for it, via an initial premise of human-mediated disturbance of natural habitats. I also suggest that a new approach to understanding ecological dynamics over time in Amazonian forests is called for, one that would replace the concept of ecological succession with that of *landscape transformation,* in explicit recognition of the cultural, historical, and human mechanisms at work in altering species composition in specific locales. Such alterations have not always been toward diminishment of such diversity, though that of course has happened and is continuing to occur in the context of destructive and non-sustainable land-use strategies introduced from the outside world and still propagated in globalization today.

Most evidence points to traditional technologies, and the knowledge that underlies them, as nondestructive of local (*alpha*) species diversity; in fact, some of that diversity recorded is actually dependent, or *contingent,* upon human interference. I am convinced also that cultural diversity is an inevitable product of human history, contingency, enrichment, management, and interaction with speciose environments. Cultural diversity and species diversity constitute together a *good,* in the Platonic sense of the term. It is a good that globalization needs to begin to protect, and nurture.

The first of these concluding chapters, Chapter 10, is basically a quantitative assessment of Ka'apor knowledge of cultural forests, which they call *taper,* in light of the method of freelisting by a sample of Ka'apor adults of terms for the trees occurring in this forest type. Barlow et al. (2011) consider scientific recognition of species diversity in fallow forests (e.g., Clement and Junqueira 2010) to reflect an "anthropocentric worldview." They miss the point, however, that the diversity fostered by indigenous landscape transformation is like natural selection itself; it acts on and rearranges existing variability in the environment to begin with. If that is anthropocentric, so be it. The variability and rearrangement of species that occur in the context of indigenous forest management are recognized by indigenous cognitive systems.

The freelisting method as discussed in Chapter 10 is simple, but powerful. The most important of the fallow forest trees, by a measure called "Smith's s," appear to confirm knowledge of human effects on old growth landscapes and the cerebral intentionality and responsibility that underlie them.

The second and final chapter, Chapter 11, calls explicitly for a replacement of the terms *primary* and *secondary succession* by *primary* and *secondary landscape transformation,* in order to contextualize the relationship obtaining among people, resources, and environments over time. Some people won't like these terms, but I can't think of better ones for describing the interaction of society with biotic phenomena in the immediate environs. I believe the key to this understanding is in characterizing the essence of *human-mediated disturbance,* a term I have borrowed from European agroforestry (as in Balée 2006). Human-mediated disturbance is at the root of traditional technology. The explication of that disturbance can shed light on the origins of Amazonian forests themselves, as well as on the possible continuity of the rich and inimitable diversity, both of species and of cultures, that they have harbored.

10

Discernment of Environmental Variation

This chapter concerns how systems of traditional knowledge (TK) encode and classify the accumulated impacts of the human species on the formation and transformation of Amazonian landscapes over time. The most significant of these impacts resulted from agrarian technologies. Humans have lived in the Amazon region for thousands of years. Although debate proceeds apace as to the peopling of South America, with the focus on exactitude of dating, especially fossil versus molecular evidence and radiocarbon confidence intervals for a host of selected ancient sites, few archaeologists today dispute the findings of the past quarter century that there were pre-Clovis traditions in South America sometime before 11,400 B.P., if not probably for at least a thousand years before that (Dillehay 2008; Isbell 2008; Neves 2006). Regardless of the antiquity of the original peopling of South America, none of the credible proponents for pre-Clovis cultures in South America and, hence, Amazonia have questioned the premise that the first sapient inhabitants of the region were not only anatomically but also behaviorally modern human beings. Behavioral modernity brings with it not only certain technologies and cultural repertoires, such as probably both art and religion, but also vast social and political potential, such as the structuring of economies based on reciprocity and essentially corporate means for maintaining egalitarianism in spite of tendencies of our species to social hierarchies (Henshilwood and Marean 2003). These social abilities have been transposed to the landscape, in archaeological sites and assemblages, their conventional identifying features. On another point of agreement among archaeologists with otherwise diverse viewpoints, the first inhabitants of Amazonia would have arrived without domestication, either of plants or, with the dubious exception of the dog, of animals (Stahl 2008). In addition, it is now clear that agrarian technology arose independently in Amazonia, especially in peripheral corridors (Clement et al. 2010). The long-term human presence in Amazonia has had impacts on floristic and faunal diversity in locales and regions and these impacts are reflected in TK anthropic landscapes.

The Human Impact on Amazonian Diversity

Hunter-gatherers have arguably altered Amazonian landscapes, though clearly not as profoundly as trekking societies (Politis 2001; Rival 2002; Zent and Zent 2004) and sedentary horticultural societies (Balée 2009b; Coomes et al. 2009; Erickson and Balée 2006; Raffles and WinklerPrins 2003), with the exceptions of the *sambaqui* and shell-mound builders of the Brazilian Atlantic Coast, Lower Amazon, and Guianas, all of whom had disappeared long before the arrival of the Europeans (Gaspar et al. 2008; Roosevelt et al. 1991; Rostain 2008), and the moundbuilders of Marajó Island, the earthworks of which are surrounded by inundated lowlands, on which both significant populations of people (needed for moundbuilding) and extensive areas of domesticated food plants with an intolerance for flooding would not have been likely to coexist (Schaan 2008).

My focus on understanding diversity is at the level of species, and especially the extent to which species have been subjected to anthropic influences on the spatially limited but conceptually useful scales of alpha and beta diversity. The concern is not with the process of domestication around the Amazon Basin (e.g., Clement et al. 2010), but rather with reallocation of species distributions and turnovers in species numbers in situ and across related sites. Any emphasis on diversity, of course, could be on genetic diversity, implicating de rigueur employment of new technologies such as DNA barcoding or other molecular tools and phylogeographic methodologies (Clement et al. 2010), but that level of metamorphosis of biological material is less likely to have been observable to people involved in effecting species turnovers and domestication events in the first place. Changes on that level would not have been encoded initially in TK, because they were not directly observable, and these modifications would be perhaps ultimately immaterial when trying to discern human effects on limited yet observable spatial scales, such as those of alpha and beta.

The biomass in any one given area, as altered by human activity, clearly represents a different measure from the diversity in the species variability therein, or of the molecular differentiation within it. The global proportion of the biomass of domesticated animals and humans, for example, is probably today on the order of two-fifths (Coppinger and Smith 1983), which implies a massive reallocation of species concentrations on earth since the various Formatives from diverse continents and regions, including Amazonia (Arroyo-Kalin 2009), had taken place. Many "Holocene" environments are essentially artificial to the extent these were, in fact, transformed by cultural activity (Denevan 1992; Dickinson 2000). The ultimate emergence and expansion of agrarian technologies, domesticated organisms, and associated invasive species across the globe alone, however, do not elucidate species diversity in situ, or even across locales linked by environmental or temporal gradients (Huston 1994), which remain subordinated to the purview of still comparatively rudimentary methods of macrobiological and systematic inventories. Yet the complex local and regional phenomena documented by these inventories have been the realities acted upon and accessible to human consciousness and classification, in both folk and indigenous contexts, since the origins of agrarian technologies. Hence, standardized measurements of these possess validity when attempting to understand diversity that is associated with human-mediated disturbance. What people in indigenous contexts perceived during Amazonian prehistory, in their mundane activities on the landscape, were phenotypes and locales, and it is their variegated effects on and perceptions of these that constitute the central focus of this chapter.

Primary versus Secondary Landscape Transformation

Prehistoric societies of Amazonia engaged in human-mediated disturbance of natural biota (whether wholly, partially, or not-at-all domesticated) occurring on a multitude of landforms. That human-mediated disturbance before historical documentation became available is of considerable importance in decoding the underlying patterns of diversity on many landscapes today. Two principal types of human-mediated disturbance and the resulting change in species composition effected by humans in Amazonian indigenous contexts can be identified: primary and secondary landscape transformation (Balée 2006). Transformation in this sense is directly derived from the notion of human-mediated disturbance engendering a "fundamental change in appearance or nature" on given landscapes, with the principal methods originating in agrarian, industrial, and advanced industrial technologies (Kates et al. 1990, 5). The changes I am examining in the context of Amazonia are, for the most part, simply agrarian (Arroyo-Kalin 2009). In some cases, indigenous societies using TK created well-drained forests above seasonally inundated savanna in the form of mounds, raised fields, and causeways in eastern Bolivia (Coomes et al. 2009; Langstroth 1996). Without following any "uniform formula" (Hecht 2009, 153), certain forest islands in *cerrado* environments of central Brazil are also the result of anthropogenesis (Posey 2002). In both cases, the forest vegetation is more diverse (at least in species of trees if not other organisms, especially soil organisms, because the soils of these sites are altered and qualify as Amazonian Dark Earths [e.g., Coomes et al. 2009; Tsai et al. 2009]) than the original, presumably primeval, landscapes that surrounded them. These cases illustrate what I mean by *contingent diversity*. Specifically, diversity in these locales is, or was, contingent on human-mediated disturbance, which is another way of saying human history and agency.

This contingent diversity of forest islands, both of seasonally inundated savannas and *cerrados,* is the result of *primary landscape transformation,* which denotes a complete turnover of species due to human intervention and metamorphosis of earlier environments. That turnover in species, which can involve more or less invasive species, is often documented as resulting in a net loss of diversity, often on a large scale because of the dissemination of invasive species and their superior competitive attributes (e.g., Stahl 2009), but what we have learned from historical ecology is that primary landscape transformation can sometimes result in net increases in diversity (Balée 2006). In either case, regardless of the upward or downward trend in species abundance values, the anthropogenesis of Amazonian landscapes is increasingly documented not only in the living vegetation, but also archaeologically in the form of extensive patterns of earthworks (Arroyo-Kalin 2009; Erickson 2000a, 2000b, 2003, 2006, 2008; Heckenberger and Neves 2009;

Heckenberger et al. 2008; Roosevelt et al. 1991; Rostain 2008; Schaan 2011). In contrast, *secondary landscape transformation,* as mediated by human disturbance, results in a partial turnover in species. This does not involve significant alterations of the substrate, as in deliberate construction of earthworks seen in mounds, forest islands, causeways, raised fields, ridged fields, artificial levees, and like phenomena, but rather extensive but intermediate disturbance of the environment usually for the purpose of short-term food production or medium-term human habitation, as with swidden fields and medium-term settlements of semisedentary horticultural societies. Forests that arise in the association and aftermath of these contexts are distinct in terms of the number of species they share with adjacent nondisturbed forests on the same kinds of substrates yet quantitatively comparable in species diversity to those forests (Balée 1994).

In extreme eastern Amazonia, habitat of the Ka'apor Indians, Jaccard coefficients of alpha similarity between fallow pairs of forest ranged between 13.5 percent and 20.2 percent with an average of 17.2 percent, and coefficients of alpha similarity between high forest pairs ranged between 19.9 percent and 25.1 percent with an average of 22.8 percent. In contrast, when mixed pairs of anthropogenic fallow forest and high forest were compared, the coefficients of alpha similarity ranged between 8.7 percent and 13.0 percent with an average of only 10.9 percent, meaning that the forests produced by secondary landscape transformation are different from the native forests (Balée 1994, 134), though both exhibit similarity in terms of absolute diversity. Because there is still significant sharing of species between anthropogenic fallow forests (that is, cultural forests) and nondisturbed high forests (that is, forests on well-drained soils without evidence of human-mediated disturbance), this being an average of 10.9 percent, anthropogenesis in the turnover is partial, not total, and hence the overall sequence is one not of primary but of secondary landscape transformation. If the thesis that indigenous societies along the Amazon River and its tributaries first practiced intensive or semi-intensive agriculture (Denevan 2006; Fraser et al. 2009; Woods and McCann 1999), as suggested by the presence of *terra mulata* (dark earths without significant deposits of potsherds and other household refuse, which are evidence of human occupation in situ per se and always found with *terra preta*), withstands continued scrutiny, the referenced sites with attendant biotic diversity would have been the product of primary, not secondary, landscape transformation. The diversity of Amazonia, which on alpha and beta scales is to some extent anthropogenic, though certainly less so if seen from a gamma perspective (e.g., Notes 1999; Sombroek 2000), has been recognized in TK, that is, in local cultural contexts.

Psychological Reality of Contingent Diversity

A reasonable question concerns intention of human agents in effecting values that represent contingent diversity, and while this is always difficult to ascertain from the archaeological record, the diversity produced by human activities through primary and secondary landscape transformation may be recognized, more or less in fact, at least phenotypically, by the human agents and their predecessors who were involved in such dynamics. The question cannot be answered fully because of the retrospective character of the archaeological record, its actors having long since exited the stage of landscape anthropogenesis and, as is widely known, having bequeathed precious few symbolic artifacts, perhaps the most key of which are written documents. On the other hand, the living languages of today display landscape histories, in both their vocabularies and grammars. This is seen in ideophones that reference onomatopoeic aspects of animals as in their various vocalizations (Berlin 2006) yet also some of plants, such as the crackling sounds of trees bending in strong winds (Nuckholls 1996, 1999, 2010). Linguistics as an index of Amazonian historical ecology and hence landscape transformation in that region is noted also in marking reversals, which reference changes in the cultural and psychological familiarity and uses of organisms over time, precisely because of human-mediated disturbances (Balée 2009a; Witkowski and Brown 1983).

This recognition suggests that alpha and beta increases in diversity due to human agency probably did not go unnoticed in prehistory, because these are extant in vocabularies full of reflexes, that is, symbolic relics, in the linguistic and ethnobiological record. Such relics of landscape history are clearly most obvious inside the lexicon. Several languages explicitly distinguish between anthropogenic forests and primary forests. Such distinctions suggest recognition in TK of human-mediated disturbance and its impact on diversity. Specifically, the Ka'apor language, which is a member of the Tupí-Guaraní family of languages, denotes old fallow forest as *taper,* which is distinguished from high forest, called *ka'a-te,* on the basis of several biotic and abiotic criteria (Balée 1994), to be discussed in relation to trees, below.

Recognition of Diversity in TK

My concern here is with recognition in traditional Amazonian knowledge systems of potential noumenal diversity, which, of course, is nothing less than staggering when considering Amazonia as a whole. This would be so even if only in light of the biota of its soils (Tsai et al. 2009), or perhaps especially so in terms of that kind of diversity, let alone its better-known, elevated world proportions of marine and above-ground terrestrial organisms. In terms of global diversity, for instance, the Amazonian proportions of all species represent about one-third each for plants and fish, 30 percent for birds, and 20 percent for mammals (Lizarralde 2001). Indigenous societies tend to recognize local segments of this total diversity, and in many cases the species in certain clades, both of flora and fauna, are overdifferentiated in traditional classification systems (Balée 1994, 2003; Berlin 1992; Posey 1986, 2002; Posey and Camargo 1985).

One might ascribe such recognition to universal patterns of distinction between environments subjected to human-mediated disturbance and those not. Academic discourse on diversity begins with Plato, who prized diversity of organisms for its own sake, which he considered to be among the "brute facts" of the world

(Lovejoy 1936, 46). Fundamentally, to Plato, in what appears to be his last and in many ways most nuanced dialogue, the *Timaeus* (Cornford 1937, 1950), "it takes all kinds to make a world" (Lovejoy 1936, 51) and "the world is the better, the more [living things] it contains" (Glacken 1967, 5; Lovejoy 1936, 52). The notion has been aptly termed the principle of plenitude (Lovejoy 1936). Although it has been often argued that Plato was an essentialist who considered nature and its species diversity to be unchanging, a good in and of itself (e.g., Davies 2001), in fact, to Plato, the diversity of plants was the result of human gardening, not abstract nature in its effervescent variety; Plato indeed lamented what he considered the degradation of Attica, Greece, not because of human intervention on its landscapes, but precisely because human gardening there had ceased (Hall 2005, 218). A strong circumstantial case has been made by Charles Clement for a decrease in genetic diversity of Amazonian cultivars, if not of species, because of a cessation in the human management of these, due to human population collapse in the aftermath of European diseases and epidemics after 1500 (Clement 1999a, 1999b). It is clearly true that there are no precedents for the expansion of domesticates and invasive species across the landscapes of the world thanks to human-mediated disturbance, and regardless of global warming and other possible proximate causes of decreases in species diversity, this mega-disturbance of anthropogenic origins alone is likely to have future effects on diversity in the globe at large as well as in tropical forests such as those of Amazonia (Coppinger and Smith 1983; Gardner et al. 2009).

The focus here is not, therefore, on the potential similarity of all living things in the scope of TK, reduced to the modern *élan vital* of shape-shifting and perspectivism (Balée 2003). (My use of the term *élan vital* differs, incidentally, from that of philosopher Henri Bergson, a critic of spatialization, semiotics, and perspectivist viewpoints [Douglass 1999].) Indigenous societies of the past had altered (i.e., transformed) environments without necessarily regarding the spiritual and intellectual contents found in them. Perspectivism, with its notions of shape-shifting and living human energy looming beneath outward appearances, no doubt, exists broadly in indigenous Amazonian thought (Slater 1994; Viveiros de Castro 1998a, 2004). The concept advanced here, however, is not connected to the realization of forms in a Platonist sense, like the shadows on the wall of the cave in *The Republic*, and how these forms might transcend outward appearances or, rather, how they might exhibit something other than their shadows or "envelopes" (Uzendoski 2005).

Instead, the focus here specifically concerns the recognition of diversity among those envelopes of living things (Slater 1994; Uzendoski 2005; Viveiros de Castro 1998a, 2004), that is, their material and physical manifestations as these are perceived, named, classified, and sometimes nurtured by indigenous and folk societies, even if they do not tell the whole emic story of the forms that underlie and, in the final analysis, project them. I am also not specifically concerned, in trying to understand recognition of biotic diversity, with the fact that indigenous groups of Amazonia sometimes, if not often, exhibit more complex labeling systems of local substantive landscapes than one finds in currently dominant models of biogeography (Abraão et al. 2008; Fleck and Harder 2000; Shepard et al. 2001). The object is understanding species diversity, not landscape diversity, and this includes species diversity on contingent landscapes.

With these considerations in mind, and knowing that such phenomena are not localized to one locale, region, or linguistic or cultural grouping in Greater Amazonia, but rather found widely across it, the focus on recognition of diversity becomes an instantiation of more general principles of ethnobiology. Perhaps first among these concerns the reality of living things within the framework of spatialization in TK, which tends to be limited to the immediacy of known and historic landscapes that have molded the cultures in question that exhibit such classification (Balée 2003; Berlin 1992), and these have, in turn, been changed in historic time by cultural behaviors emanating from those very places. To continue with classical analogues, then, my object of inquiry is on the emic perception of *physis* ("what exists and grows itself") rather than on the presumed *nomos* (what human societies create, infer, and envision to underlie things) (Hughes 2001, 5).

People do tend to reduce diversity of envelopes to underlying intelligible tropes, such as tricksters illustrative of morality—or lack thereof—divinities that control game and fish supplies, and spirits that engage in all sorts of hybrid behaviors for which there is no other explanation except their otherworldliness. Those spiritual dimensions of traditional Amazonian knowledge systems, however limited in number, are always cloaked in the skins, or envelopes, of more or less familiar animals and plants. It is that familiarity with the envelopes that speaks directly to recognition of contingent diversity, as understood from the perspective of systematics, and such correspondences between the otherwise disjunctive systems of science and TK can be thereby delineated.

Reality of the Envelopes in Forests of Contingent Diversity

It was earlier determined from freelisting exercises that the Ka'apor recognize a large number of tree species spontaneously, and that the concept of tree is essentially wedded to notions of size (dominance) and hardness (density), not frequency per se (Balée and Cebolla Badie 2009). Freelisting has been used in studying people's knowledge of plants before (Nolan 2001; Quinlan 2005). In studying Ka'apor knowledge of trees in 2008 each subject was interviewed separately and each was asked in his/her language to "tell me all the tree names you know" (*"Eme'u ih? pe upa mira rer nde rekwaha nde pe"*). Respondents were not stimulated with clues or with requests to give more names, and respondents were reasonably isolated from others in order to guarantee independence of response and avoid contamination of data (Quinlan 2005). The interviews lasted about twenty-five minutes each. The data were "cleaned up" in terms of elimination of repetition, synonymy, and effects of free variation, dialect, and idiolect. There is redundancy in most freelists of domains with many items, because terms included as members of superordinate taxa are often listed together with the terms for those superordinate taxa on the same individual lists. This is widely considered to be one of the limitations of the method (Borgatti 1992, 1999). The method's utility, on the other hand, is in showing the intricacy, or not, of readily accessible vocabulary within a given semantic domain. Trees are diverse in their lexical attributes in everyday Ka'apor speech, as evidenced in freelisting studies.

In many systems of TK, it is well known that names of plants and animals will occupy a substantial portion of the lexicon (Berlin 1992). In fact, the Ka'apor distinguish 768 species of plants (Balée 1994) specifically. That knowledge would not be readily available, however, in the spontaneous exercise of freelisting, using a finite sample of Ka'apor subjects who had been given a limited time frame in which to answer a seemingly simple question. Another advantage of freelisting, apart from initially showing the intricacy of shared knowledge of a semantic domain, is it is also capable of revealing the psychological salience of items within an ethnobiological classification (Nolan 2001).

The procedure of freelisting assumes that the most psychologically salient items will be listed first. For that reason, one understands how in classes of undergraduates in the United States asked to freelist animal terms they know, dog and cat are consistently found at the top of lists that otherwise normally contain hundreds of items (Borgatti 1999). This finding also suggests reliability of the method, incidentally. Zipf's law (1949) is relevant to the underlying assumption, that the frequency of usage of a word in a natural language tends to be inversely proportional to its length, insofar as the initial terms tend to be folk generic (simple, noncompound constructions) like dog or cat, rather than folk specific constructions, like sheep dog or Maine coon cat—in other words, commonly used words, presumably of high psychological salience (that is, those readily accessible in common speech), tend to be shorter than infrequently used words. Zipf's law helps explain redundancy on freelists, since folk-specific members of a taxon will appear typically later, or further down, on the lists.

Smith's s is a means of controlling for the premise that the most psychologically important items in a domain will tend to be listed first and that if mere frequency of an item is considered as a measure of importance, the results will be blurred with numerous ties of rank order (Smith 1993). This statistic weights the order given to an item on a list as well as length of the list as an index of psychological salience for an individual on a specific item. Individual psychological salience of an item, then, is expressed as $S_j = 1 - (r_j/l_i)$, where S_j is the psychological salience of item j, r_j is the rank of item j on the individual's list, and l_i is the length of the individual's list. In determining the psychological salience of an item across a group of respondents, the s values of the item are simply averaged. In determining the content of a domain shared among a group of respondents, it is a common procedure to eliminate all items with a frequency of only 1, since presumably such items are not shared (Borgatti 1999). If the sample were substantial enough, such items either would not pertain to the domain or would have negligible psychological salience as members of the domain. The twenty-four Ka'apor informants in the earlier study listed 290 names of "trees" (*mɨra*), of which 149 had a frequency greater than 1. The psychologically most important "tree" by Smith's s in that study was *tayɨ* (*Handroanthus impetiginosus* [Mart. ex DC.] Mattos) and it was also the most frequently cited (occurring on 22 of n = 24 lists). Arguably from these data, *tayɨ* is a prototypical tree (Balée and Cebolla Badie 2009; Rosch 1978) in Ka'apor culture and Ka'apor ethnobiological classification.

In separate studies of freelists of Ka'apor and Sirionó (a Tupí-Guaraní language in eastern Bolivia), the number of valid terms in both languages for kin terms was much lower than for trees, based on unpublished research I carried out in 2007, 2008, and 2009. In the Ka'apor sample (n = 22), there were forty-eight kin terms and in the Sirionó sample (n = 20), there were twenty-six kin terms with frequencies greater than 1. But for tree names with a frequency greater than 1, as we know in the Ka'apor sample (n = 24), there were 149 terms, or about three times the number of kin terms; in the Sirionó sample (n = 18), there were fifty terms with a frequency greater than 1, or about twice the number of kin terms. The Ka'apor habitat is more diverse than the Sirionó habitat, with an average of 120 species of trees per hectare compared to about 55 species of trees per hectare, respectively (Balée 1994; Coomes et al. 2009), and this may partly explain the lower number of tree terms in Sirionó. Quantitatively speaking, in a general sense, nevertheless, both Ka'apor and Sirionó adults recognize the relatively high arboreal diversity of their respective habitats and this is reflected in tree vocabulary vis-à-vis the lexicon concerned with human kinship.

Freelisting of Trees from the Anthropogenic Forest

The inquiry here specifically involves whether Ka'apor TK encodes the anthropogenic forest, called *taper,* as distinctive in terms of biotic contents and diversity and, if so, how. Determining whether they have Platonist appreciation for the variety of beings—or whether they would subscribe, to borrow Arthur Lovejoy's phrase (Lovejoy 1936), to the principle of plenitude—including what indicates landscape transformation, is distinct from asking whether they specifically label that variety in some more or less faithful recognition of its inherent and undeniable complexity. To answer this question in a preliminary way, I will limit analysis to the domain of trees and examine whether freelists of trees from the anthropogenic forest can yield insights into shared (i.e., cultural and ethnobiological) understanding of the origins and diversity of that forest by a sample of Ka'apor adults. Determining the biota that characterize an anthropogenic landscape, such as its distinctive trees, is a related but essentially different endeavor from identifying the members of a class of organisms itself, such as birds, fish, or trees (cf. Balée and Cebolla Badie 2009). The question can be logically approached, nevertheless, also using freelisting procedures.

In August 2009, twenty-two Ka'apor adult subjects were asked to freelist "trees" (*mɨra*) of the anthropogenic forest. The question was stated as "*Ma'e mɨra ta taper rupi ha ngi nde rekwaha pe*" ("What are the trees of the anthropogenic forest?").

The same sorts of controls were in place as for the earlier study that, in essence, asked "What is a tree?" in Ka'apor (Balée and Cebolla Badie 2009). Appendix II shows the results of the present exercise, freelisting the trees of the anthropogenic forest, ranked by Smith's s and analyzed using ANTHROPAC 4.983/X software (© 1985–2002 Analytic Technologies). There were a total of 147 separate terms elicited, of which 68 were valid, that is, with a frequency higher than 1. The last column in Appendix II shows whether the term references or not species occurring either in the four-hectare inventory of anthropogenic forest or in the four-hectare inventory of high forest. If trees were distributed randomly in the two forest types, one would expect each term to reference both forests, or neither, if no difference were recognized in content, but only thirty-eight of the sixty-eight terms do so, and this difference is statistically significant ($p = .0365$, Fisher's exact test).

There is, as I noted above, a degree of sharing of species between high forests and fallow forests, on an average of about 10.9 percent (see above). The differences between these forests are most notable not so much in species that are unique (endemic), though there are such species, to one or the other forest type, but rather in quantitative measures, such as density, frequency, and dominance. These measures may be combined for an overall ecological importance value (EIV), in which the sum of the relative density, frequency, and dominance of all species in a forest inventory is 300 (Balée 1994). The twenty most ecologically important tree species from four one-hectare inventories each of anthropogenic forest and high forest, with the respective EIVs of each, are shown in Table 10.1. Only one of these thirty-nine species in total is shared.

It is at this level—ecological importance value—that one notes relevant distinctions between anthropogenic forest, the diversity of which is contingent on human history, and high forest, the diversity of which, in contrast, originates in natural drivers. The ecologically most important trees in high forest exclude palms, with one exception, the bacaba palm (*Oenocarpus distichus*); in contrast, the ecologically most important trees of the anthropogenic forest include four palms (*Attalea speciosa, Astrocaryum vulgare, Attalea maripa,* and *Astrocaryum gynacanthum*). The *Attalea* species have cryptogeal germination (Anderson et al. 1991), which helps account for their success atop once-burned substrates that hosted erstwhile swiddens and settlements of agrarian societies, which is the case with all documented secondary landscape transformations in the Ka'apor habitat. *Astrocaryum vulgare* is usually predominant in inundated forests but typically found only in old fallow, including archaeological sites (Morcote-Rios and Bernal 2001), when outside that milieu (Balée 1994). The prototypical Ka'apor tree, *tayi* (*Handroanthus impetiginosus,* known in Portuguese as *pau d'arco* or *ipê roxo*) (Balée and Cebolla Badie 2009), is essentially only viewed in the anthropogenic forest. It is an important timber species, used by the Ka'apor in making bows. Another notable contrast between these lists in Table 10.1 is that more than one-half of the trees of the anthropogenic forest are important fruit trees. These include the well-known hog plum (*Spondias mombin,* the Ka'apor name of which, incidentally, *taperiwa'i*,

Table 10.1. Comparison of twenty ecologically most important tree species in the Ka'apor habitat (cf. Table 3.3)

Rank	High Forest Species	EIV	Old Fallow Species	EIV
1	*Eschweilera coriacea*	37.83	*Jacaratia spinosa*	11.40
2	*Sagotia racemosa*	14.53	*Attalea speciosa*	9.37
3	*Tetragastris altissima*	11.60	*Astrocaryum vulgare*	7.76
4	*Protium trifoliolatum*	7.76	*Spondias mombin*	6.53
5	*Protium decandrum*	7.07	*Neea* sp. 1	6.26
6	*Protium pallidum*	6.78	*Pisonia* sp. 2	6.25
7	*Carapa guianensis*	5.69	*Pouteria macrophylla*	5.71
8	*Couepia guianensis*	5.07	*Attalea maripa*	5.40
9	*Pourouma minor*	4.54	*Platypodium elegans*	5.02
10	*Taralea oppositifolia*	4.51	*Platonia insignis*	4.32
11	*Mabea* sp.	4.06	*Simaba cedron*	4.26
12	*Pourouma guianensis*	3.28	*Hymenaea parvifolia*	4.17
13	*Dodecastigma integrifolium*	3.10	*Trichilia quadrijuga*	4.06
14	*Couratari guianensis*	2.77	*Lecythis pisonis*	3.56
15	*Oenocarpus distichus*	2.72	*Dialium guianense*	3.32
16	*Sterculia pruriens*	2.65	*Astrocaryum gynacanthum*	3.31
17	*Bagassa guianensis*	2.65	*Eschweilera coriacea*	3.19
18	*Cecropia obtusa*	2.60	*Theobroma speciosum*	3.11
19	*Newtonia psilostachya*	2.47	*Lindackeria latifolia*	3.05
20	*Chimarrhis turbinata*	2.40	*Handroanthus impetiginosus*	2.85

as given in Appendix II, literally means "fruit tree of the anthropogenic forest"); lucuma (*Pouteria macrophylla*); bacuri (*Platonia insignis*); monkey pot (*Lecythis pisonis*); and nondomesticated cacao (*Theobroma speciosum*). The only significant fruit tree on the high forest list is the aforementioned bacaba palm (*O. distichus*). The proliferation of fruit trees and other useful trees in the anthropogenic forest is in all likelihood due to initial dispersion of the seeds by human beings, whether intentional or not (Balée 2003).

Now the question becomes, how well recognized is the species composition of the anthropogenic forest? A definitive answer cannot be given from freelisting of a relatively small sample. On the other hand, it has been determined that the Ka'apor

Table 10.2. Twenty most ecologically important tree species named by twenty most psychologically important tree names in Ka'apor (n = 22)

Rank of Ka'apor Name by Smith's s	Species Name	Ecological Importance Rank Old Fallow	Ecological Importance Rank High Forest
1	*Hymenaea parvifolia*	13	—
2	*Hymenaea courbaril*	*	*
3	*Handroanthus impetiginosus*	20	—
4	*Eschweilera coriacea*	18	1
5	*Senna silvestris*	—	—
6	*Handroanthus serratifolius*	*	—
7	*Attalea maripa*	9	—
8	*Jacaranda* spp.	*	*
9	*Bagassa guianensis*	*	18
10	*Spondias mombin*	5	—
11	*Duguetia* spp.	*	*
12	*Jacaratia spinosa*	1	—
13	*Dipteryx odorata*	—	*
14	*Astrocaryum vulgare*	4	—
15	*Rinorea* sp.	*	*
16	*Helicostylis tomentosa*	—	*
17	*Pouteria macrophylla*	8	*
18	*Lecythis idatimon*	*	2
19	*Theobroma speciosum*	19	*
20	*Anacardium* sp.	*	*

distinguish lexically and psychologically between *ka'a-te* and *taper* forests. It can be also determined from Appendix II that nine of the twenty ecologically most important anthropogenic forest species are referenced by these terms, whereas only three of the high forest species are so denoted, although the difference is not statistically significant per se, in part because of small sample size.

In any event, these results are shown in Table 10.2 (in the table, an asterisk indicates that the species occurs on the inventory, but it ranks below 20). In broad terms, seventeen of the twenty psychologically most salient terms for trees of the anthropogenic forest have referents in the four-hectare inventory of anthropogenic forest; twelve have referents in the high forest. With one exception (*Eschweilera coriacea,* which is the most common tree in the habitat, probably), the seventeen terms denoting a species in the top twenty species of ecological importance of anthropogenic forest do not apply to the most important species of the high forest. The most important tree psychologically in the old fallow (*Hymenaea parvifolia*) occurs in the top twenty ecologically most important species for that forest type (at rank thirteen) and the most important tree ecologically in the old fallow (*Jacaratia spinosa*) also appears in the top twenty trees in terms of psychological salience (at rank twelve), and neither is present in inventories of high forest.

Conclusions

Plato was concerned with the loss of the gardenlike aspects of Attica, for in this degradation he perceived a reduction of diversity in living forms. Diversity was a given of the sensible world Plato inhabited, and as such, in his metaphysical paradigm it was a good in and of itself. Plato did not know of evolution by natural selection, of course, which would only first be truly understood by Darwin two thousand and some years later, but he recognized cause and effect, for he stated that "[w]e must ask the question which, it is agreed, must be asked at the outset of any inquiry concerning anything: Has it always been, without any source of becoming; or has it come to be, starting from some beginning?" (Cornford 1950, 16).

Anthropogenic forests, or *taper,* have not always been, and though it may be the case that *ka'a-te* has not always been either, its existence can be explained apart from human contingency. Rather, anthropogenic forests started from a beginning, and this beginning involved the deployment of agrarian technology enlightened and driven by TK. These landscapes are, therefore, deserving of explanation, and an adequate one can only be found in contingency, specifically that intrinsic to human history and agency, which in the initiation of landscape transformation is the same as human-mediated disturbance. The study of a system of TK, such as that of the Ka'apor of Amazonia in reference to the domain of trees, shows recognition of preexisting phenotypic diversity (Balée 1994; Balée and Cebolla Badie 2009). It also shows a clear distinction between anthropic and nonanthropic landscapes, which is in the first instance noted in the distinction of labels for the two forest types. Subsequently, the contents of the forest types are distinguished in Ka'apor TK as landscapes contingent on human history and agency and those not so contingent, as I have endeavored to show with the relatively simple yet powerful tool of freelisting. In conclusion, human societies have engendered species-rich forests, on both alpha and beta scales, and in turn, the systems of TK found in these societies reveal that the anthropic influence on the diversity of Amazonian landscapes over long periods of time has been also recognized and encoded.

11
Rethinking the Landscape

The preceding pages of this book have made a case for recognition of cultural forests in the Amazon Basin. Cultural forests exist, though they have not tended to be categorized by their particular human signature in ecological and biological science. To ignore the human factor in the formation of these forests is not only to discard history, and environmental history in particular, but it also does not help us in trying to figure out ways of seeing how people might coexist now and in the future with what remains of biological and ecological diversity on earth, especially in the Amazon Basin, where such diversity is the highest of all regions on the planet. The only biome that comes close to the species richness of Amazonia is coral reefs, and these are also on the brink of extinction, though for somewhat different reasons, mainly industrially induced changes in the chemical composition of the world's oceans. The threat to the Amazon rainforests, which include cultural forests, are principally commercial logging, ranching, and other activities that have large-scale sequelae, such as increasing the susceptibility of the region to global warming and wildfires.

Technologies of the Past

The people who made the cultural forests of Amazonia did not practice forest destruction and conversion; rather, they had technologies and ideologies about the forest that had developed over centuries, and they engaged the forest, rather than acted to merely replace it with a different biome, such as grassland, pasture, or degraded woodland, devoid of hardwood trees. To be sure, the forests they left behind were not the original landscape. At the same time, there is evidence that traditional Amazonian forest technologies had the result of enhancing local and regional diversity. In this chapter, I suggest we reconsider those enhancements and acknowledge the value of traditional knowledge in changing, even enhancing, Amazonian landscapes.

Acknowledgment of the past is a crucial first step on a new trajectory of thinking and planning for the sake of diversity, and it will be important to convey these findings not only to those who study traditional knowledge but also to proponents of conservation biology and specialists in wildlife ecology. Amazon cultural forests in the biological sciences have usually been labeled primary forests, climax forests, and mature forests (e.g., Bush et al. 2007; McMichael et al. 2012), as if they had not been altered at all by people of the past in terms of the species found in them. These terms can be misleading and when applied to ancient cultural forests they mystify the real contribution that past peoples made to the current landscapes many scientists and policy personnel in the conservation community are trying to protect today. I believe they should be protected also, but they are not incompatible with a human presence. We need to understand how people managed the forests in the past. In order to do so, it is urgent that we first revise our way of thinking about untouched nature and primeval forests in order to better understand the human impact of the past on biological and ecological diversity seen today. The time has arrived to develop methods of safeguarding the coexistence of indigenous societies and the forests, with the high diversity they harbor. The bottom line is that human impacts on nature did not necessarily always have the long-term effect of degrading the environment and causing the extinction or even extirpation of species. People as a species are not the culprit per se; rather, certain kinds of societal and economic engines, unknown in the prehistory of Amazonia, have resulted in forest transformations of a degrading, diversity-reducing sort. What is infrequently recognized is that sometimes native impacts based on traditional forest knowledge resulted in increased diversity, and this is seen in the context of archaeological sites and the forests that cover these sites. Cultural forests in the Amazon are in fact usually examples of enhanced diversity, even though for many readers that assertion, which I have made in several of the earlier chapters herein, might have seemed hard to grasp initially.

Let's begin to think about the terms that would be most appropriate in capturing the prehistoric interactions of indigenous people and Amazonian landscapes. We could even use them to describe the present, and possibly the future, if policymakers can get hold of the twin behemoth of globalization and deforestation. I would first of all point out that the reality of the concept of cultural forests is not at all adequately expressed by the term *secondary forests,* or areas of secondary succession, which since the early twentieth century and the beginnings of the discipline of ecology have been sometimes attributed to human activity, among other disturbances (Clements 1916, 293). Succession refers to species turnovers in given areas. Secondary succession in tropical forests, like those of Amazonia and elsewhere in the New World tropics, is often deemed to be the result of forest fragmentation due to deforestation, logging, and the rise of anthropogenic pastures and grasslands (Hooper 2008). The concept is incomplete, however, for ecological systems theory has essentially not incorporated the concept of humans as potential agents of primary succession. Michael Huston (1994, 234) defined primary succession as "succession on a substrate that has not been previously occupied by plants, such as volcanic ash, or lava, bare rock, or recently exposed sands or soil." The term has usually been reserved for terrestrial colonization by plants, which takes place independently of human influence or independent of, more specifically, human-mediated disturbance.

The importance of primary succession as a beginning point in the formation of a vegetational stand or association, such as a forest, has long been recognized. Frederic Clements (1916, 33) pointed to the "significance of bare areas," because here is where a vegetational association leading to a climax (maximally complex formation) commences. The phases, or "seres," of such succession "originate only in bare areas or in those in which the original population is destroyed." Although Clements has been criticized for describing climax associations as "superorganisms" (Huston 1994, 238; Raffaelli and Frid 2008, 2–3), which bespeaks a sort of metaphysical view of nature, he was, in fact, a careful observer with much scientific insight on transformations in plant communities over time. The concept of primary succession on bare substrates is clearly central to a diachronic (or time-sensitive) understanding of the origins of species turnovers. To get it going, a catalyst, or disturbance, is needed. Evidence from historical ecology on several continents suggests that in a few cases, species diversity has increased on the ground even when the disturbance was anthropogenic. This primary succession on bare areas is seen in diverse contexts, but virtually excluded in most ecological discourse is a rubric of primary succession in which the initial perturbation (what Clements called "nudation") could be characterized as human-mediated disturbance. That is why I believe we need different, and more precise, terms for understanding and classifying anthropogenic succession on earth's landscapes. These terms, reviewed in the preceding chapter, are primary and secondary landscape transformation (Balée 2006, 2009b, 2010).

Several researchers (Clement 1999a; Clement and Junqueira 2010; Erickson 2008; Heckenberger 2009; Heckenberger and Neves 2009; Junqueira et al. 2010) have proposed the alternative term of *landscape domestication* to reference the change from "bare surfaces," or the natural equivalent or near-equivalent of this, to landforms in which the array of species, including many useful species, is essentially managed, even engineered. I am in essential agreement with their findings on cultural mechanisms of change in forest composition over time, and of the often long-lasting result of such changes, sometimes including enhanced levels of diversity. The quibble I have with the term *landscape domestication* is a semantic one only. I think it has the same limitation, in fact, of Clements' term, *superorganism,* in referencing an old growth forest, when what he was talking about was really an ensemble of many organisms and species, not just one, no matter how big and complex. For the same reason, the Gaia hypothesis (that the earth itself is a living organism or system) is a metaphysical construct, though it might be a good analogy for understanding the complex interactions between global climate, the biosphere, and human agency. Landscapes, forests, and the earth are not propagated by their own individual reproduction. They consist of millions of organisms that have, individually, the genetic material (in the form of coding sequences of the DNA molecule) for such reproduction, but they alone do not have their own genomes. An organism can be domesticated, which means its sequence of base pairs can be changed by artificial selection, gene splicing, and other human-directed activities, leading ultimately to a different organism. That is a kind of transformation, but a very specific kind. It's an example of evolution. Landscapes do not evolve the way organisms do. (For similar reasons I have problems with the term *cultural evolution*, because culture is learned, not inherited biologically per se—I would prefer to talk about societal development in this case, where earlier writers used the term *evolution of culture,* or *evolution of society.*) The bottom line is landscapes do not have their own DNA, such that they could be restored or replicated by means of reproduction. Rather, the DNA of landscapes is housed in the individual biota that live, and die, in those landscapes. Those biota can undergo changes in their genetic material, sometimes even changes selected for by humans (domestication), but the landscape itself changes in a broader, even fuzzier sense.

Amazonian landscapes typically consist of both domesticated and nondomesticated species; this is the case with virtually every landscape known in the region. In the Amazon, cultural forests sometimes but not always consist of domesticated species, many semidomesticated species (species that associate well with humans, which may be preadapted to environments where fire is used in agriculture, but may not have had their DNA altered significantly), and nondomesticated species, as discussed in Chapters 2 and 3. Cultural forests are the result of both primary and secondary landscape transformation.

Primary landscape transformation exists where a complete turnover of species over a substrate characterized by no species or very different plant species has occurred. The prehistoric mounds of the large island of Marajó in the delta of the Amazon River, several of which are thirty to forty feet high and several acres each in area, appear to have been related to exploitation of fish, for in some cases the nature of earthwork construction affords perennial water sources in the dry season, and without the mounds, such areas would be completely dry (Schaan 2006, 2011). Causeways in the northern Bolivian Amazon appear to have had a similar function (Erickson 2000a, 2008). It took a significant amount of work and planning to build mounds and causeways. A feature that can be inferred from the mounds and causeways of the Bolivian Llanos de Mojos in the Upper Amazon and Marajó Island in the Amazon estuary is known as intentionality (Erickson and Balée 2006; Schaan 2006). Intentionality of human behavior actually varies in degree of purposefulness, and there are ways of measuring this. The study of the prehistoric earthworks at Ibibate (Chapter 6) and surroundings suggests that there was high purposefulness in their construction and in the resulting forested landscapes that came about; one could even use the term *landscape engineering* to describe this phenomenon. At Ibibate Mound Complex, sometime before A.D. 1500, people intentionally dug dirt out of the ground in order to raise a surface more than sixty feet high, and in so doing, they created a borrow pit that has functioned also as a perennial source of drinking water to this day. The area, except next to river courses, is otherwise completely devoid of potable water in the dry season of the Llanos de Mojos (Erickson and Balée 2006). At Marajó Island, as archaeologist Denise Schaan has shown, during the period between A.D. 480 and A.D. 700, people intentionally built large mounds, sometimes in the middle of watercourses, sometimes adjacent to them, in order to deepen or widen these watercourses as well as to form dams to control water and redirect its flow; sophisticated hydraulics is in evidence in these ancient landscape transformations (Schaan 2006, 107; 2008; 2010; 2011).

The effects of these manipulations specifically on the biota have been documented at Ibibate mound, which is the largest and highest mound yet known in the Amazon region, and which is situated near another mound, a kind of dualism that is common in prehistoric earthworks of this sort in the Bolivian Amazon (Erickson and Balée 2006). The Ibibate mound is covered in forest. Ibibate and forests like it are examples of primary landscape transformation, which as I have pointed out, involves a complete or near-complete turnover of species due to humans (Balée 2006, 2009b) (also see Chapter 10). What occurred there hundreds of years ago was very much like succession "on a bare surface," in F. E. Clements' words, except here the succession, or turnover of species, was driven by human management of resources and engineering of the landscape. The landscape began as a flooded wetland savanna, low in tree diversity, and it became, because of human intervention, a forest rich in tree species. Of course, turnovers of this primary sort could increase or decrease the total number of species present, but they always change the distributions.

The changes in abundance of species, and comparisons of abundances and distributions, are central to ecology and also to the perspective known as historical ecology, which I believe best explains the origins and continuity of cultural forests in the Amazon. Many sorts of primary landscape transformation, in theory, could reduce species diversity: these include explosions of military ordnance; construction of highways, parking lots, and apartment complexes; purposeful flooding of lands to build reservoirs, dikes, and dams; burning of large tracts of tropical forest, whether degraded by selective logging or not, to form immense cattle pastures or vast monocultures like soybean, sugar cane, rice, or what have you. The sort of primary landscape transformation that resulted from Amazon moundbuilding, however, was qualitatively different from these transformations that palpably reduce diversity. There are only about twenty species of vascular plants in the wetland savanna surrounding Ibibate mound; yet there are at least eighty-four species of trees and lianas with diameters of 10 cm or greater at breast height atop the mound (Balée 2006; Erickson and Balée 2006). Many of these species cannot tolerate flooding, seasonal or otherwise. Here we have an indigenous contribution to alpha diversity (diversity in a specific locale, defined by given environmental gradients), as a result of primary landscape transformation. In other words, biological diversity in the forest of Ibibate could never be adequately explained without referencing human and cultural activity. That is, of course, why it qualifies as a cultural forest. And the explanation for it lies in an understanding of the mechanisms involved in primary landscape transformation, specifically earthwork construction, of both an intellectually sophisticated and labor-intensive sort.

The archaeological signature of greatest significance here may be not so much the earthwork itself, including the mound's borrow pit, associated structures, and a nearby causeway, together with a vast array of mute ceramic artifacts and the skeletal materials (including human bone) that jut out of the ground, but rather the living components of the mound complex. The forest and its trees on Ibibate represent living artifacts of an anthropogenic sort; they have distinctive cultural pasts partly rooted in prehistory. The mounds of this region were mostly built between A.D. 400 and A.D. 1400 (as reviewed recently by Lombardo and Prümers [2010]). The alpha diversity, higher than the natural diversity of the surroundings, is anthropogenic. And so is the quality of biological endemism. A species is endemic if it only occurs in a very restricted area, such as about 20,000 km^2, according to the great Amazonianist field botanist and specialist of one of the plant families with many vine and liana species (Bignoniaceae), Alwyn Gentry (1986). I would argue also that a species could be considered endemic if it only occurs on a relatively restricted landform, even if such a landform is found in a landscape that may cover a much larger area than 20,000 km^2 such as with the landform occupation mounds, which are in total areal extent rather small in relation to the Llanos de Mojos, certainly less than one-third of that habitat.

Jos Barlow et al. (2011) suggest that people of the Amazonian past who might have contributed to forest diversity on an alpha scale did not enhance or protect "forest-dependent" rare and endemic species. They argue that it is not surprising that minor changes of the environment, such as by small clearings used in swidden agriculture, could have the local effect of increasing diversity: this is from the intermediate disturbance hypothesis, which holds that both no disturbance and extreme disturbance (such as deforestation of large areas) yield areas of lower species diversity than areas subjected to some kind of minor disturbance, some of the time. Yet there is a threefold problem with the view propounded by Barlow et al. (2011). Human-mediated disturbance (1) does not always result in species increases (obviously); (2) it is not always intermediate; and (3) it can indeed enhance conditions for survival of "forest-dependent" and endemic species. For example, the turumbúri tree (*Sorocea guillemiana,* mulberry family) occurs on mound sites of the Llanos de Mojos, including Ibibate. It is also a culturally important fruit tree. Its purplish fruits, which ripen in the dry season, are used in making a ceremonial beverage by the Sirionó Indians (Balée 2000a), to whom we were introduced in Chapter 6 (*turumbúri* is, incidentally, the Sirionó word for the tree—I do not know the Spanish equivalent). This species, according to Sirionó informants and my own observations, *only* occurs atop occupation mounds like Ibibate. This may be related to the fact that these mounds never flood; the dispersal agent may be a bird (Roberto Langstroth, personal communication, 2000). In other words, turumbúri is endemic to a cultural forest.

Ibibate—the mound and its forest—is a case in point of primary landscape transformation, which is, by definition, a successional process initiated by human-mediated disturbance. Here species turnover and other environmental variables had counterintuitive, positive outcomes, even including expansion of a habitat for an endemic, which is another way of saying rare, localized, and threatened, species. Barlow et al. (2011) say enhanced numbers of species due to intermediate disturbance is not "counterintuitive"; they are silent, however, about the possibility of enhanced numbers of species in a locale due to intense disturbance by humans. In other words, the intensity of anthropogenic interference at Ibibate and environs is not the same as intermediate disturbance. These transformations and others like them represent lessons from the prehistoric past that can be understood from the careful study of living artifacts that thrive on living landscapes, in which even the soil itself is partly alive, as with *terra preta* and other Amazonian Dark Earth sites (Arroyo-Kalin 2010; Navarrete et al. 2010; Woods and McCann 1999), which are always anthropogenic soil horizons. The soil of Ibibate is, in fact, highly comparable to *terra preta* (Balée 2009; see Erickson and Balée 2006, table 7.1, pp. 196, 200); 13 percent of the material is likely also from pure ceramic (Lee 1979), which is such a remarkably dense quantity of pottery in the ground that one geographer has referred to Ibibate and mounds like it as "ceramic forests" (Langstroth 1996). As to the living microbes in anthropogenic soils of the Amazon, including those of Ibibate, thousands of these species occur only in anthropogenic soils. I do not mean to exaggerate the point; soils are biologically of extreme complexity: as many as one million separate taxa can be found in a sample of only 10 g of soil and my point is that many species occurring in anthropogenic soils of the Amazon Basin are simply not found in the natural soils nearby, which as such indicates a human contribution to alpha and beta diversity, even at the microbial level of bacteria, fungi, and archaea, in cultural forests of the Amazon (Tsai et al. 2009). Such changes cannot be understood within the narrow framework of the intermediate disturbance hypothesis alone (cf. Barlow et al. 2011; McMichael et al. 2012). Human-mediated disturbance of Amazonian forests is much more nuanced and complex than that. To begin to understand these changes, I recommend adopting a historical-ecological point of view.

Historical Ecology

A considerable body of evidence suggests that cultural forests of the sort I and others have identified in Amazonia occur elsewhere in the tropical world. These observations are not based on the notion of ecological succession from a "bare surface" to a "climax" community, but rather a notion of disequilibrium, chance disturbance, and, specifically, human-mediated disturbance. That is where cultural forests, formerly misclassified as primary forests, have in fact originated. People have shaped environments for thousands of years, in the tropics and elsewhere. How should they be understood, or studied? I suggest historical ecology is the most appropriate methodological and theoretical way of comprehending these human-induced effects in nature. Historical ecology is the most accurate lens by which to grasp artificial changes in natural environments. Rebecca Dean noted that "[h]istorical ecology's focus on nonequilibrium models, and particularly their explicit recognition of the importance of contingency in landscape development, makes this the natural theoretical perspective for fully exploring the role that humans have played in shaping environments over millennia" (Dean 2010, 8). Ultimately, historical ecology gives us a window not into ecosystems and their recycling of the same species through time, but into the dynamic, unstable, contingent processes inherent in manipulations of landscapes driven by traditional knowledge. It is the landscape, not the ecosystem, that is central to this understanding. The landscape is real, and it consists of delimited space. As to the natural landscape, it exists, but not in exclusivity, as noted by Carl Sauer (1969, 333). Cultural forests are kinds of landscapes that do not reach climax states of diversity and complexity: rather, they continuously unfold, with no specific destination or endpoint. Cultural forests are a kind of landscape where, according to Sauer (1969, 333), "works of [humanity] express themselves." These forests were not necessarily made by "pristine primitives" (Wolf 1982), but rather by people continuing to use and modify traditional technologies regardless of their contact with the outside world. These are humanized landscapes made by sophisticated people, in terms of their knowledge of resources, tools, and biota, and they are useful in innumerable ways to equally knowledgeable indigenous peoples of the present.

Cultural Forests in Malesia and Africa

The region between Sumatra in the west and the Bismarck Archipelago in the east consists of numerous islands straddling the equator or located just to the north or south of the equator. In terms of its vegetation alone, the region is called Malesia: although it consists of a vast chain of islands, its scope is near-continental and comparable to large regions such as Amazonia and sub-Saharan Africa. Malesia consists of tropical landscapes, many of which were transformed in terms of their forest composition by human activity long ago. The chronology of human occupation in Malesia is, in fact, much older than that of Amazonia. The Bismarck Archipelago, which is the easternmost part of Melanesia, was colonized by thirty-five thousand years ago; Amazonia was only colonized at the end of the Pleistocene, perhaps no more than twelve or thirteen thousand years ago. Much later, in the geological time period known as the Holocene, people from Melanesia had colonized the tiny islands to the north (known as Micronesia) and east (Polynesia), where people eventually "sculpted" the landscapes seen there today (Rainbird 2004, 138). Before that, people in the equatorial zone of Southeast Asia—including Malesia—were altering forests, through disturbances such as anthropogenic fire and semidomestication and domestication of plants and animals. These alterations were principally of the sort I have been calling secondary landscape transformation. Many trees species are associated with cultural forests in this region, including trees with edible fruits or pulps, such as betel nut (*Areca* spp.) (which is an important stimulant throughout South and Southeast Asia), kenari nut and its relatives (*Canarium* spp.), sago palm (*Metroxylon* spp.), coconut palm (*Cocos nucifera*), breadfruit (*Artocarpus* spp.), sugar palm (*Arenga pinnata*), a relative of hog plum (*Spondias dulcis*), and many others (Bayliss-Smith et al. 2003; Burkill 1966; Latinis 2000; Lepofsky 1992; McClatchey et al. 2006).

These forests share some species with Amazonia, largely thanks to intercontinental exchanges of plants that took place after the era of European colonization. Indicator species of cultural forests, including domesticated or semidomesticated trees from Amazonia, which tend to persist in the mature vegetation of both regions, include cashew, guava, soursop, papaya, and cacao; indicator species originating in Malesia, found in both regions today, include breadfruit, mango, tamarind, carambola, Pacific almond, coconut, and *Citrus* spp. (Burkill 1966; Terrell et al. 2003). Often the indicator species of cultural forests in the two regions are different yet share attributes due to phylogeny (common genetic heritage). *Borassus* fan palms in Malesia are either in cultivation or in some pioneer relationship with human disturbance (Burkill 1966; Terrell et al. 2003). Counterparts of these palms might be *Attalea* palms, like babaçu (Chapter 1), all of which have cryptogeal germination, enabling them to survive fire and dominate in cultural forests where once there were agricultural villages. In addition, distinct genera of trees seem to perform similar functions that indicate past human uses (often as food), habitation, and management, such as the kenari nut trees (*Canarium* spp.) in Malesia and nondomesticated cacao trees in Amazonia (*Theobroma* spp.), both of which are important fruit tree genera most often found in the orchardlike cultural forests, and both of which have many species but only one true domesticate (as with *Canarium indicum,* the kenari nut proper, and *Theobroma cacao,* the chocolate or cacao tree) (Bayliss-Smith et al. 2003; Lepofsky 1992; McClatchey et al. 2006). Finally, both regions—Amazonia and Malesia—include indicator genera, such as bamboo (*Guadua* spp.), a point I made in Chapter 2 for the Amazon region, specifically looking at the extreme western Amazon, where bamboo forests are fairly common. In Malesia, "sometimes shifting cultivation leads to the establishment of bamboo [*buloh*], and these, when established, seem to become very stable, so that the forest takes a very long time in recovering its lost ground" (Burkill 1966, 1:293). As with old fallow and anthropogenic forest islands in Amazonia (Balée 2006; Erickson 2008) as well as tropical forest islands in West Africa (Fairhead and Leach 1996), these Melanesian old fallows were orchardlike and attracted not only people for their edible fruit but also game animals, which eat the same fruits (Bayliss-Smith et al. 2003, 347). The cultivation of these ancient forests is referred to as arboriculture throughout the region. In other islands in the New Georgia and Russell Island group, there were dense stands of *Camponesia brevipetiolata,* a disturbance indicator with a large crown. On ridgetops in some of these areas, one finds *Canarium* spp., *Prunus* spp., and *Ixora* spp., all associated with old village sites (Bayliss-Smith et al. 2003, 348).

It seems that tropical sub-Saharan Africa also contains cultural forests. In West Africa, the vast forest islands of Guinea have been shown to contain evidence of human occupation and formation. Rather than representing relics of the Pleistocene, these are forests that were formed where there were no forests before, as a result of traditional (indigenous) human occupation and activity. In the now classic account of Fairhead and Leach (1996), these primary landscape transformations involved the deliberate transplanting, planting, and cultivation of a variety of fruit trees and the establishment of firebreaks around nascent plantings to protect them. Savanna soils actually improved with human occupation and cultivation, and the development of shady orchards was favored. Trees grown in these anthropogenic forest islands include the following genera: *Canarium* (the same genus of edible nuts found on transformed summits and ridgetops in Melanesia), *Parinari* (a pantropical genus), *Ceiba* (which is found in swamps and anthropogenic forests of Amazonia; Balée 1994, 277), and *Cola* (Fairhead and Leach 1996, 208). Far to the south is further support for Fairhead and Leach's theory against what is being called the "declinist paradigm," which refers to the view that humans in African forests lead to decreases, or declines, of species and habitat diversity. In the south of tropical Africa, it has recently been shown that the advance of a fruit tree complex was deliberately caused by human propagation, the fruit trees involved being mainly marula and birdplum (Kreike 2003, 40). What these Africanist studies of cultural forests show is that forests before the Europeans were expanding, not contracting, and expanding because of people's occupation and management strategies, not in spite of the human presence.

Lasting Effects?

The evidence from Amazonia for transformations of landscapes is multifaceted, and both similar to and different from the African and Oceanic materials. Whereas one does not find terracing (for, of course, there are few if any rocklike structures), in contrast we find extensive areas of mounds, ditches, ringed villages, and other manipulations of the earth itself, such as the most recently discovered and spectacular geoglyphs of Acre, western Amazonia (Pärssinen et al. 2009; Schaan 2006, 2011). These are giant prehistoric enclosures in which ditches several meters deep were dug in the shape of circles and rectangles up to hundreds of meters in diameter; these phenomenal structures were only seen from the air when deforestation of Acre began in earnest in the late 1980s. We have found unusually high species diversity in a study of one of these geoglyphs, the only one up to now discovered completely under forest cover (Balée et al. 2013). Archaeologists and paleoecologists are working on the specific techniques people used to make these geoglyphs in the remote past.

People can still change forests in old-fashioned ways, even when using newer tools. Introduction of steel tools in Micronesia, to replace the stone tools the people had before, did accelerate swidden cultivation on some islands (Bayliss-Smith et al. 2003, 347) and presumably, therefore, secondary landscape transformation. And the Ka'apor did create old fallow, cultural forests using steel tools; indeed, they claim their ancestors never used the ancient stone axeheads occasionally found in their habitat for swidden agriculture, saying that such artifacts are *tupã-ra'i*, "thunder-seeds" (Balée 1994, 40). Yet even if intensive agriculture preceded swidden cultivation in the support of large Amazonian populations in late prehistory, which seems increasingly plausible (for a summary, see Denevan 2001, 2006), that would not preclude some form of early clearance of forest, on a limited scale, by people using stone axeheads, for limited swidden agriculture. That appears to have been the only kind of secondary landscape transformation utilized, moreover, in Atlantic Coastal forest at the time of the discovery of Brazil by the Portuguese (Dean 1995). Indeed, it seems ingenuous to assume any other reason than that for explaining the existence of stone axeheads in areas lacking *terra preta* (Chapter 2) that were nevertheless arable. In other words, while intensive agriculture seems likely to have supported vast populations, isolated pockets of swidden agriculture could have contributed to long-term forest transformations (secondary landscape transformations), in perhaps ways similar to those effected by contemporary trekking societies, such as the Hoti of Venezuela (Zent and Zent 2004) and the Nukak of Colombia (Politis 2007).

In conclusion, Amazonian anthropogenic landscapes built long ago have been the result of primary and secondary landscape transformations. These landscapes in many cases had been *intentional:* they were designed, engineered, and built in prehistory, and not just in Amazonia. Similar though not identical phenomena are found elsewhere in the equatorial regions of the globe. Landscape transformations in Amazonia often involved major alterations of the earth's surface, with movement of multitudinous tons of dirt in prearranged patterns, which in turn had effects on the subsequent distributions, diversity, and even endemism—as with *Sorocea* at Ibibate and other mounds—of the flora and fauna. Archaeological landscapes and seascapes in these regions were created by ancestors of people today who are referred to as indigenes, or indigenous people. These people typically understand and recognize the extent of the forests their ancestors created, and with that knowledge, they contribute to the richness of the planet's engagement with diversity. The Amazonian habitats that were altered by primary and secondary landscape transformation, as discussed herein, constitute prima facie evidence of an indigenous quality that still persists. It is now up to the authorities of national governments and global powers—the political powers that underwrite globalization in the aggregate—to contribute to protecting those environments. It is up to them to protect also the indigenous people who occupy them from the devastating effects of modern and postmodern land- and resource-use strategies. We need to learn some management sense from the inscriptions in the living matter of cultural forests, and the traditional people who still live in them.

APPENDIX I. GUAJÁ GENERIC PLANT NAMES

English glosses of Guajá plant names are mine. The plant terms in Guajá (left-hand column) are assigned morpheme boundaries. I did not know all these terms, so translations are not given. In many cases, I have used my knowledge of Ka'apor terms to help in glossing a Guajá word, as the plant names, while not identical, follow similar patterns in the two languages. The glosses given in this appendix are liberal. They are mostly indicative of the entire name, not the morpheme-by-morpheme breakdown, in the order of the morpheme(s) in a word. When the plant term is literal, or unanalyzable (as discussed in Balée and Moore 1991, 1994), I have not given a translation, except for names denoting domesticates that appear to be Tupí-Guaraní cognates, in which case I have given English equivalents of the term.

A. Folk Generic Names for Trees (*iwïra*)

Guajá Name	English Gloss	Latin Name	Family	Coll No.
a'ï-hu-'ï	big sloth tree	*Ampelocera edentula*	Ulmaceae	4450
ačiči-'ï		*Pterocarpus rohrii*	Fabaceae	3618
akača-'a'ï		?	?	None
akaya-rana-'ï-hu		*Simaba paraensis*	Simaroubaceae	4534
akayu		*Anacardium giganteum*	Anacardiaceae	None
		Anacardium spruceanum	Anacardiaceae	3222
ako'o-'ï		*Theobroma cacao*	Malvaceae	3487
		Theobroma speciosum		3327
akuči-akāra-'ï	agouti food tree	*Simaba cedron*	Simaroubaceae	3227
akuči-terewa-'ï	agouti fruit tree	*Pouteria macrophylla*	Sapotaceae	3753
		Pouteria trigonosperma	Sapotaceae	4452
		Pouteria venosa subsp. *amazonica*	Sapotaceae	3225
akuči-wïra	agouti tree	*Randia armata*	Rubiaceae	3690
ama'ā-wïra		*Aspidosperma album*	Apocynaceae	4227
		Aspidosperma auriculatum	Apocynaceae	3523
		Aspidosperma cf. *ateanum*	Apocynaceae	3420
ama-'ï		*Cecropia ilicifolia*	Urticaceae	4581
		Cecropia palmata	Urticaceae	3840
		Cecropia sciadophylla	Urticaceae	4582
ama-'ï-ča'ā-'ï		*Pourouma guianensis*	Urticaceae	3211

Guajá Name	English Gloss	Latin Name	Family	Coll No.
aman-'ɨ	rain tree	Senna silvestris	Fabaceae	3639
aman-turu-wi-'ɨ		Abarema cochleata	Fabaceae	4702
anita-'ɨ		Clarisia ilicifolia	Moraceae	3501
aparata-'ɨ		Micropholis acutangula	Sapotaceae	4647
aparata-'ɨ-ran		Microphilis venulosa	Sapotaceae	4609
aparayhu-'ɨ		Manilkara huberi	Sapotaceae	3474
api-'ɨ		Naucleopsis melo-barretoi	Moraceae	4787
		Pseudolmedia murure	Moraceae	4701
apiri-kowā-'ɨ	parrot calabash tree	Pouteria eugeniifolia	Sapotaceae	3166
		Sarcaulus brasiliensis	Sapotaceae	4847
ara-ka'a-'ɨ	macaw herb tree	Rauia resinosa	Rutaceae	4848
ara-kowā-'ɨ	macaw calabash tree	Poecilanthe effusa	Fabaceae	
aračiko'a-'ɨ		Annona montana	Annonaceae	3728
arakači'a		Jacaratia spinosa	Caricaceae	3379
araku-'ɨ		Bixa orellana	Bixaceae	3346
araku-rana		Bixa arborea	Bixaceae	4840
		Bixa orellana	Bixaceae	
arakwa-re-ro-'ɨ		Pouteria bilocularis	Sapotaceae	4458
arapa-pi-apo-'ɨ	deer hoofprint tree	Stryphnodendron polystachyum	Fabaceae	3238
arapari-'ɨ		Macrolobium acaciaefolium	Fabaceae	4790
arapio-'ɨ		Cochlospermum orinocense	Bixaceae	4544
arara-'ɨ-ātā-ī	little hard macaw tree	Licania sylvae	Chrysobalanaceae	3156
arara-kā-tapiwa-'ɨ		Brosimum guianense	Moraceae	3224
arati-'ɨ		Parinari rodolphii	Chrysobalanaceae	3451
atan-'ɨ	hard tree	Talisia sp. 1	Sapindaceae	3638
atarā-'ɨ		Eugenia flavescens	Myrtaceae	3874
		Eugenia lambertiana	Myrtaceae	4703
		Eugenia protracta	Myrtaceae	4730
		Eugenia sinemariensis	Myrtaceae	4553
		Eugenia schomburgkii	Myrtaceae	3386
ayā-'ɨ	ogre tree	Cordia nodosa	Boraginaceae	3530

Guajá Name	English Gloss	Latin Name	Family	Coll No.
ayá-piri-'ɨ		Eugenia modesta	Myrtaceae	4447
		Eugenia egensis	Myrtaceae	3624
		Eugenia stipitata	Myrtaceae	4668
ayá-piri-ran-'ɨ	ogre not piri tree	Campomanesia grandiflora	Myrtaceae	4726
		Myrciaria obscura	Myrtaceae	4460
ayá-wa-'ɨ	ogre fruit tree	Dulacia candida	Olacaceae	3421
čama'am-hu		Ceiba pentandra	Malvaceae	3655
čaperai-čī-'ɨ		Sacoglottis guianensis var. maior	Humiriaceae	3127
čapī-ran		Hirtella macrophylla	Chrysobalanaceae	4788
		Licania sprucei	Chrysobalanaceae	4607
čaporomo-'ɨ		?	None	
čiči-apo		Talisia acutifolia	Sapindaceae	
čičipe-'ɨ		Inga heterophylla	Fabaceae	3199
		Inga alba	Fabaceae	4528
		Inga auristellae	Fabaceae	3289
		Inga capitata	Fabaceae	3300
		Inga cecropietoreum	Fabaceae	3752
		Inga disticha		
		Inga edulis	Fabaceae	3287
		Inga falcistipula	Fabaceae	3473
		Inga marginata	Fabaceae	3719
		Inga miriantha	Fabaceae	3388
		Inga nitida	Fabaceae	3793
		Inga rubiginosa	Fabaceae	3732
		Inga thibaudiana	Fabaceae	3242
		Inga stipularis	Fabaceae	3229
čičipe-ran		Inga longiflora	Fabaceae	4753
čimiči		Newtonia suaveolens	Fabaceae	4622
		Newtonia psilostachya	Fabaceae	3780
hai-kiripi-'ɨ		Stryphnodendron barbadetiman	Fabaceae	4519
hai-kiripi-'ɨ-ran		Swartzia flaemengii	Fabaceae	
haíra-kowa-'ɨ		Eriotheca globosa	Malvaceae	3452
haíra-maka-'ɨ		Banara cf. guianensis	Salicaceae	3510
hawā-'ɨ		Simaruba amara	Simaroubaceae	3236

Guajá Name	English Gloss	Latin Name	Family	Coll No.
he-pipir-iwa-'ɨ		*Tapirira guianensis*	Anacardiaceae	3235
ɨ-tata-ĩ	not fire tree	*Senna* sp.	Fabaceae	4764
ɨnaya-'ɨ		*Attalea maripa*	Arecaceae	3377
ɨra-pihun-'ɨ	blackwood tree	*Andira retusa*	Fabaceae	3893
		Pseudoxandra cf. *polypheba*	Annonaceae	3169
ɨra-pirã-'ɨ	redwood tree	*Hirtella glabrata*	Chrysobalanceae	3725
		Hirtella racemosa var. *racemosa*	Chrysobalanceae	3259
		Hirtella tentaculata	Chrysobalanceae	3447
		Licania laevigata	Chrysobalanceae	3263
ɨra-piri-hu-'ɨ		*Thyrsodium paraense*	Anacardiaceae	3775
ɨra-põ		*Taralea oppositifolia*	Fabaceae	3288
ɨra-puku-'ɨ	tall tree	*Ouratea* cf. *coccinea*	Ochnaceae	3179
ɨra-ratã-'ɨ	hardwood tree	*Eugenia pseudopsidium*	Myrtaceae	
ɨra-rəmə-hu-'ɨ		*Guatteria dielsiana*	Annonaceae	3201
ɨra-ro-'ɨ	bitterwood tree	*Maprounea guianensis*	Euphorbiaceae	3408
		Myrciaria tenella	Myrtaceae	3343
ɨra-ru-hu-'ɨ		*Pouteria englerii*	Sapotaceae	3587
ɨra-ta-'ɨ	many trees tree	*Erisma uncinatum*	Vochysiaceae	3246
ɨra-tai	spicy tree	*Citrus* spp.	Rutaceae	None
ɨra-tata-'ĩ	fire not tree	*Cenostigma macrophyllum*	Fabaceae	4451
ɨrapə-'ɨ		*Handroanthus* sp.	Bignoniaceae	3646
		Handroanthus impetiginosus	Bignoniaceae	3635
ɨrar-'i'i ka'ĩ		*Mouriri sagotiana*	Melastomataceae	3522
ɨratɨ-'ɨ	wax tree	*Symphonia globulifera*	Clusiaceae	
irimi-riku		*Bonafousia juruana*	Apocynaceae	4797
itãĩ-wira		*Eugenia muricata*	Myrtaceae	4815
itawa-'ɨ	yellow tree	*Hymenaea courbaril*	Fabaceae	3125
iyap, raku'a		*Stigmaphyllon* cf. *martianum*	Malpighiaceae	3491

Guajá Name	English Gloss	Latin Name	Family	Coll No.
ka'i-čowa-'ɨ	monkey calabash tree	*Trichilia micrantha*	Meliaceae	3161
		Trichilia lecointei	Meliaceae	3332
		Trichilia quadrijuga	Meliaceae	3260
		Trichilia elegans ssp. *richardiana*	Meliaceae	4466
		Trichilia cipo	Meliaceae	3584
kapinia-'ɨ		?	?	None
kaiyowa'ɨ̃		*Geonoma baculifera*	Arecaceae	3347
kamiča'ɨ	shirt tree	*Ambelania acida*	Apocynaceae	3258
kamiča-apini'a-'i	striped shirt tree	*Clavija lancifolia*	Primulaceae	3362
kamiča-po-'ɨ		*Xylopia nitida*	Annonaceae	3433
kapowa-'ɨ		*Copaifera duckei* Dwyer	Fabaceae	3269
kiču-kowa-'ɨ	bearded saki calabash tree	*Apeiba echinata*	Malvaceae	3284
		Apeiba tibourbou	Malvaceae	3148
kipi-'ɨ		*Theobroma grandiflorum*	Malvaceae	3390
kipi-ran-o'o-'ɨ		*Lueheopsis duckeana*	Malvaceae	3361
kirihi-'ɨ		*Trattinickia burserifolia*	Burseraceae	3446
		Trattinickia rhoifolia	Burseraceae	3282
		Protium subserratum	Burseraceae	1391
kiripi-kowa-'ɨ		*Sapium lanceolatum*	Euphorbiaceae	4740
kiripirim-'i		*Bactris tomentosa*	Arecaceae	3495
		Bactris sp. 1	Arecaceae	3490
kɨwa-'ɨ	comb tree	*Virola michelii*	Myristicaceae	3195
kokuri-wa-hu-'ɨ		*Pouteria glomerata*	Sapotaceae	4559
kururu-wɨra	toad tree	*Chlorophora tinctoria*	Moraceae	4574
mətə-'ɨ	grey brocket deer tree	*Duguetia surinamensis*	Annonaceae	4573
mai-towa-'ɨ		*Chaetocarpus* sp.	Euphorbiaceae	4571
		Zollernia paraensis	Fabaceae	4476
mai-towa-'ɨ-rana		*Swartzia brachyrachis*	Fabaceae	4655
makari-wɨra		*Bellucia dichotoma*	Melastomataceae	4563
		Bellucia grossularioides	Melastomataceae	3373
		Miconia ciliata	Melastomataceae	4792
mani'o-'ɨ	manioc tree	*Buchenavia parvifolia*	Combretaceae	4619

Guajá Name	English Gloss	Latin Name	Family	Coll No.
marato-'i̵		Vatairea guianensis	Fabaceae	3455
maratato-wa-'i̵		Schefflera morototoni	Araliaceae	3898
mariawa; mariači		Bactris maraja	Arecaceae	3385
merere-'i̵		Batocarpus amazonicus	Moraceae	3779
		Brosimum acutifolium ssp. acutifolium	Moraceae	3450
mika-'i̵		Caryocar cf. pallidum	Caryocaraceae	3354
mirikə-'i̵	woman tree	Dodecastigma integrifolium	Euphorbiaceae	3297
		Sagotia racemosa	Euphorbiaceae	3110
mitũ-wira	turkey tree	Talisia retusa	Sapindaceae	3419
muhiri-'i̵		Byrsonima cf. amazonica	Malpighiaceae	3113
mukur-'i̵	opossum tree	Platonia insignis	Clusiaceae	3375
papara-ī		Tetragastris panamensis	Burseraceae	3293
paranã-'i̵	river tree	Markleya dahlgreniana	Arecaceae	4789
parara-'i̵		Mouriri cf. vernicosa	Melastomataceae	3593
pina-'i̵	fishhook tree	Duguetia flagelaris	Annonaceae	3521
		Duguetia sp. 1	Annonaceae	3323
		Duguetia marcgraviana	Annonaceae	3846
pini-'i̵	painted tree	Piranhea trifoliolata	Euphorbiaceae	None
pinuwa-'i̵		Oenocarpus distichus	Arecaceae	3144
		Euterpe oleracea	Arecaceae	3281
pira-čũ-'i̵	black fish tree	Ephedranthus pisocarpus	Annonaceae	3746
		Oxandra sessiliflora	Annonaceae	4693
pira-čũ-ran	looks like black fish tree	Rollinia exsucca	Annonaceae	3758
pira-mukwana-'i̵		Ouratea discophora	Ochnaceae	3445
pi̵tata-'i̵		Lecythis idatimon	Lecythidaceae	3102
pi̵ti̵m-'i̵	tobacco tree	Couratari guianensis	Lecythidaceae	3214
		Aubl. ssp. guianensis	Lecythidaceae	3693
		Couratari sp. 1	Lecythidaceae	4497
		Couratari oblongifolia		
pitōpitō-'i̵		Talisia lacerata	Sapindaceae	4557
		Cupania acutifolia	Sapindaceae	4618

Guajá Name	English Gloss	Latin Name	Family	Coll No.
puhan-wɨra; a'ɨ-rimõ	medicine tree; sloth vine	Capparis coccolobifolia	Brassicaceae	4640
təkənə haiha-ra-'ɨ		Faramea sessilifolia	Rubiaceae	3852
təkənə hə-'ɨ		Ocotea opifera	Lauraceae	3904
tikõtərə-'ɨ		Gustavia augusta	Lecythidaceae	3333
tači-'ɨ		Tachigali paniculata	Fabaceae	4643
		Tachigali myrmecophila	Fabaceae	3112
təkə-haira-'ɨ		Guarea kunthiana	Meliaceae	3404
takamã-'ɨ		Astrocaryum vulgare	Arecaceae	3481
tamamari-'ɨ	tamarind monkey tree	Brosimum lactescens	Moraceae	4478
tami-'ɨ		Bauhinia viridiflorens	Fabaceae	3339
		Bauhinia macrostachya var. obtusifolia	Fabaceae	3472
tapika-'ɨ		?	?	
tapi'ir-pami-'ɨ		Sterculia pruriens	Malvaceae	3196
tapuru-'ɨ	grasshopper tree	?	?	
taraka-ɨ		Bagassa guianensis	Moraceae	3498
		Helicostylis tomentosa	Moraceae	3109
tare-'ɨ		Trichilia areolata	Meliaceae	4454
		Trichilia cf. smithii	Meliaceae	4685
		Trichilia schomburgkii	Meliaceae	4704
tari-'ɨ		Ilex parviflora	Aquifoliaceae	3625
tata-kaya-'ɨ		Croton cajucara	Euphorbiaceae	3475
tato-wa-'ɨ		Ormosia flava	Fabaceae	3709
tawa-wa-'ɨ		Spondias mombin	Anacardiaceae	3380
tayahu-wɨra		Parkia pendula	Fabaceae	3301
timičikərə-'ɨ		Simaba amara	Simaroubaceae	4456
wənənə-'ɨ		Casearia arborea	Salicaceae	4545
		Casearia javitensis	Salicaceae	
		Laetia corymbulosa	Salicaceae	4549
wənənə-'ɨ-rana		Eupatorium sp.	Asteraceae	4817
wə'ən		Bactris acanthocarpa var. acanthocarpa	Arecaceae	3355

Guajá Name	English Gloss	Latin Name	Family	Coll No.
wa'u-wa'a-'ɨ	yellow fruit tree	*Ziziphus itacaiunensis*	Rhamnaceae	4552
wa'ĩ-'ɨ	not fruit tree	*Attalea speciosa*	Arecaceae	3376
wa-čiči-'ɨ		*Sterculia alata*	Malvaceae	4836
wa-hawa-'ɨ		*Pilocarpus macrophyllus*	Rutaceae	4867
wari-rana-ka'aw	not howler monkey bowl	*Lindackeria latifolia*	Salicaceae	3338
wari-wa-'ɨ	howler monkey fruit tree	*Brosimum acutifolium* subsp. *interjectum*	Moraceae	3450
wari-rana-ka'aw	not howler monkey bowl	*Sloanea grandis*	Elaeocarpaceae	4486
wari-wa-'ɨ	howler monkey fruit tree	*Brosimum acutifolium*	Moraceae	4471
wa-we-pire-'ɨ		?	?	None
wa-pe-piri-hu-'ɨ	big flat striped fruit tree	*Pouteria* sp.	Sapotaceae	4498
wa-pini-hu-'ɨ	big striped fruit tree	*Couepia guianensis* ssp. *divaricata*	Chrysobalanaceae	3413
			Chrysobalanaceae	3209
		Licania canescens	Chrysobalanaceae	3700
		Licania kunthiana	Chrysobalanaceae	3403
		Licania membranacea	Chrysobalanaceae	3257
wa-pipiri-hu-'ɨ		*Pouteria bangii*	Sapotaceae	4476
wa-pirã-'ɨ	red fruit tree	*Protium robustum*	Burseraceae	3243
		Tetragastris altissima	Burseraceae	3133
wa-tərə-hə-'ɨ		*Lecythis jarana*	Lecythidaceae	4468
wa-yu-'ɨ	yellow fruit tree	*Chrysophyllum sparsiflorum*	Sapotaceae	3833
		Pouteria caimito	Sapotaceae	
		Pouteria sp. 86	Sapotaceae	3240
		Pouteria cf. *glomerata*	Sapotaceae	3740
		Pouteria sp. 15	Sapotaceae	3226
		Pouteria reticulata ssp. *reticulata*	Sapotaceae	3858
			Sapotaceae	
		Pouteria hispida		4449
		Pouteria sagotiana	Sapotaceae	3213
		Pouteria gongrijpii	Sapotaceae	3245
		Pouteria jariensis	Sapotaceae	

Guajá Name	English Gloss	Latin Name	Family	Coll No.
wahara-pɨ'a-ran		*Faramea multiflora*	Rubiaceae	4859
wai-ha-'ɨ	small fuzzy fruit tree	*Eschweilera pedicellata*	Lecythidaceae	4469
		Lecythis chartacea	Lecythidaceae	3137
wəkəwa-'ɨ		*Mabea occidentalis*	Euphorbiaceae	3143
wapu-wa-hu-'ɨ		*Ficus paraensis*	Moraceae	4799
warara-'ɨ		*Calliandra* sp. 1	Fabaceae	
		Hymenolobium excelsum	Fabaceae	4765
		Pithecellobium racemosum	Fabaceae	3723
		Parkia ulei	Fabaceae	3492
		Pithecellobium corymbosum	Fabaceae	3596
		Pithecellobium gongrijpii	Fabaceae	3184
wari-makwa-'ɨ		*Abuta grandifolia*	Menispermaceae	4861
waya-ran-'ɨ-piči'a		*Myrciaria floribunda*	Myrtaceae	3497
		Myrciaria disticha	Myrtaceae	4490
wayuwa-'ɨ		*Acroclidium aureum*	Lauraceae	3436
		Aiouea cf. *multiflora*	Lauraceae	3616
		Aniba burchelii	Lauraceae	3206
		Endlicheria sp. 1	Lauraceae	3188
		Ocotea caudata	Lauraceae	4649
		Ocotea cf. *caudata*	Lauraceae	3409
		Ocotea fasciculata	Lauraceae	3696
		Ocotea laxiflora	Lauraceae	4763
		Ocotea marmellensis	Lauraceae	4493
		Ocotea sp. 1	Lauraceae	3406
		Systemonodaphne mezii	Lauraceae	4474
wayuwa-'ɨ-ran		*Mezilaurus lindaviana*	Lauraceae	4769
		Sloanea parviflora	Elaeocarpaceae	4555
wɨra-him-hu-'ɨ	big slippery tree	*Drypetes variabilis*	Euphorbiaceae	3132
wɨra-i-ku		*Calyptranthes* sp. 1	Myrtaceae	3469
wɨra-kači	spiny tree	*Anaxagorea acuminata*	Annonaceae	3397
		Pseudoxandra polyphleba	Annonaceae	3408
wɨra-ke'e		*Diospyros duckei*	Ebenaceae	3247
wɨra-ki'i	pepper tree	*Pogonophora schomburkiana*	Euphorbiaceae	3524
		Dulacia candida	Euphorbiaceae	3421
wɨra-kurum-hu-'ɨ		*Croton matouriensis*	Euphorbiaceae	3175

Guajá Name	English Gloss	Latin Name	Family	Coll No.
wɨra-pirã	redwood	Eugenia exalta	Myrtaceae	4605
wɨra-piri-hu-'I	big striped tree	Schistostemon macrophyllum	Humiriaceae	4522
wɨra-ra'i	little tree	Vismia guianensis	Clusiaceae	3471
wɨra-ra-o'o-'ɨ; wɨra-iči-'i račiči		Sloanea guianensis	Elaeocarpaceae	3294
wɨra-rači	spiny tree	Solanum crinitum	Solanaceae	3366
wɨra-račio		Astronium lecointei	Anacardiaceae	3591
		Casearia commersoniana	Salicaceae	4692
wɨra-him-hu	big slippery tree	Siparuna amazonica	Monimiaceae	3337
wɨra-ro-'ɨ	bitter wood tree	Parkia nitida	Fabaceae	3442
		Abarema jupunba var. jupunba	Fabaceae	3204
		Inga gracifolia	Fabaceae	3197
wɨra-tata-'ɨ	fire tree	Neea oppositifolia	Nyctaginaceae	4540
		Pisonia sp. 2	Nyctaginaceae	4517
		Neea cf. ovalifolia	Nyctaginaceae	3458
		Pisonia sp. 1	Nyctaginaceae	3241
		Neea sp. 1	Nyctaginaceae	3290
wɨra-ratã	hardwood	Amanoa guianensis	Euphorbiaceae	3231
		Minquartia guianensis	Olacaceae	3454
wɨra-te	true tree	Pera ferruginea	Euphorbiaceae	3610
		Pseudima frutescens	Sapindaceae	3914
wɨra-uhu (?)	big tree	Conceveiba guianensis	Euphorbiaceae	3597
wɨra-yoki'a		Maytenus sp. 29	Celastraceae	3668
wɨrayuwahu'i		Himatanthus sucuuba	Apocynaceae	3513
wɨra-yu	yellow tree	Trema micrantha	Ulmaceae	3370
wɨra-riku		Trichilia cf. septentrionalis	Meliaceae	3358
wɨri-'ɨ	lashing material tree	Eschweilera amazonica	Lecythidaceae	4615
		Eschweilera apiculata	Lecythidaceae	4633
		Eschweilera bracteosa	Lecythidaceae	4648
		Eschweilera coriacea	Lecythidaceae	3104
		Eschweilera micrantha	Lecythidaceae	4485
		Eschweilera ovata	Lecythidaceae	4455
		Eschweilera parviflora	Lecythidaceae	4612
wɨri-pətai-'ɨ		Ormosia paraensis	Fabaceae	3777

Guajá Name	English Gloss	Latin Name	Family	Coll No.
wiriri-'ɨ	swallow tree	*Cordia* indt.	Boraginaceae	3641
		Cordia sp. 1	Boraginaceae	3437
		Cordia scabrifolia	Boraginaceae	3181
		Cordia scabrida	Boraginaceae	3589
wɨra-ho-hu	big leaf tree	*Swartzia* sp. 15	Fabaceae	3623
wɨta-'ɨ		*Hymenaea parvifolia*	Fabaceae	4487
wopo-'ɨ		*Ficus* sp. 1	Moraceae	3878
ya'pi'amõ		*Combretum* sp. 1	Combretaceae	3508
ya-məkai-'ɨ		*Lecythis pisonis*	Lecythidaceae	3382
yači-purum-'ɨ	tortoise tree	*Jacaranda heterophylla*	Bignoniaceae	4793
		Jacaranda copaia	Bignoniaceae	3372
yai-'ɨ		*Rinorea publiflora*	Violaceae	3315
		Amphirrox longifolia	Violaceae	3321
		Rudgia cornifolia	Rubiaceae	3314
yakəmĩ-wɨra	pheasant tree	*Hirtella fasciculata*	Chyrsobalanceae	3702
		Hirtella triandra	Chrysobalanceae	633
		Exellondendron barbatum	Chyrsobalanceae	3149
yakamĩ-təməkərə-'ɨ	pheasant food tree	*Matayba arborescens*	Sapindaceae	4742
		Cupania scrobiculata	Sapindaceae	3598
yakarata'a-'ɨ		*Fusaea longifolia*	Annonaceae	3120
yakare-wɨra	alligator tree	*Zanthoxylum rhoifolium*	Rutaceae	3900
		Zanthoxylum sp. 1	Rutaceae	3141
yaku-wi-hu-'ɨ	long thin pheasant tree	*Goupia glabra*	Goupiaceae	3142
yami-'ɨ		*Platypodium elegans*	Fabaceae	3865
		Acacia multipinnata	Fabaceae	3422
yare-ro-'ɨ		*Carapa guianensis*	Meliaceae	3212
yare-ro-ran-'ɨ		*Cedrela fissilis*	Meliaceae	4673
		Guarea macrophylla ssp. *pendulispica*	Meliaceae	4510
yanu-'ɨ	spider tree	*Campomanesa aromatica*	Myrtaceae	4467
		Eugenia sp. 36	Myrtaceae	4479
		Eugenia patrisii	Myrtaceae	4750
		Eugenia omissa	Myrtaceae	3312
		Eugenia punicifolia	Myrtaceae	
yanupa-'ɨ		*Genipa* sp. indet.	Rubiaceae	3488

Guajá Name	English Gloss	Latin Name	Family	Coll No.
yapi-'iwɨr		*Bauhinia acreana*	Fabaceae	4687
		Bauhinia corniculata	Fabaceae	4491
yapu-'ɨ	oropendola tree	*Apuleia leiocarpa* var. *molaris*	Fabaceae	3774
		Qualea dinizii	Vochysiaceae	4823
yapu-čū-ran-'ɨ	not white oropendola tree	*Myrcia paivae*	Myrtaceae	4733
yawa-'ɨ	dog tree	*Inga crassiflora*	Fabaceae	3516
yapu-čū-wɨra	black oropendola tree	*Coccoloba latifolia*	Polygonaceae	3850
yawa-pəkə	dog banana	*Castilla ulei*	Moraceae	3745
yawa-tara-'iran		*Coccoloba lehmanii*	Polygonaceae	4586
yawar-'a-'ɨ	jaguar fruit tree	*Protium altsonii*	Burseraceae	3177
		Protium aracouchini	Burseraceae	3189
		Protium polybotryum	Burseraceae	3418
		Protium trifoliolatum	Burseraceae	3159
		Protium decandrum	Burseraceae	3191
		Protium heptaphyllum	Burseraceae	3688
		Protium krukoffii	Burseraceae	3121
		Protium spruceanum	Burseraceae	3428
		Protium tenuifolium	Burseraceae	4614
yawar-'a-čū-'ɨ	black jaguar fruit tree	*Protium sagotianum*	Burseraceae	3108
		Protium pallidum	Burseraceae	3117
		Protium nodulosum	Burseraceae	3138
yeta-'i-pipiru-'ɨ		*Dialium guianense*	Fabaceae	3580
yeyu-'ɨ	gold wolf fish (*Hoplerythrinus sp.*) tree	*Amphirrhox surinamensis*	Violaceae	4577

B. Folk Generic Names for Vines (*iwɨpo*)

Guajá Name	English Gloss	Latin Name	Family	Coll. No.
a'i-mo-wɨpo	sloth vine	*Tontelea laxiflora*	Celastraceae	
		Derris amazonica	Celastraceae	3906
ana-wɨpo	frog vine	*Omphalea diandra*	Euphorbiaceae	4827
		Clytostoma sp. 2	Bignoniaceae	3509(?)
arapa-puhã	red brocket deer remedy	*Fridericia chica*	Bignoniaceae	
čimo-ate	authentic fish poison	*Serjania paucidentata*	Sapindaceae	3506
ɨpo-pi-rana		*Merremia macrocalyx*	Convolvulaceae	4846
ɨpo-piri-hu		*Melothria* sp. 1	Cucurbitaceae	3925
ɨra-pərəhəmə-'ɨ		*Guarea guidonia*	Meliaceae	3328
ka'i-ka-mo	*Cebus* monkey vine	*Memora schomburgkii*	Bignoniaceae	4864
karuwa-'ɨ-rana	ogre not tree	*Vigna lasiocarpus*	Fabaceae	4796
karuwa-'irimõ	ogre vine	*Dioclea reflexa*	Fabaceae	3344
karuru-čimo		*Derris utilis*	Fabaceae	3462
kururu-wɨpo	toad vine	*Marsdenia* indt. sp.	Apocynaceae	4835
		Pereskia grandifolia	Cactaceae	4853
mai-rimõ	god vine	*Machaerium quinata*	Fabaceae	4744
mainimi-kwe-'a-rana	hummingbird not herb	*Melothria trilobata*	Cucurbitaceae	4841
mainumi-kwe-'a	hummingbird herb	*Passiflora glandulosa*	Passifloraceae	4818
matə-puhã	grey brocket deer remedy	*Bignonia sordida*	Bignoniaceae	3517
mayu-ɨpo	snake vine	?	?	—
parakwa-čiči		*Heteropsis jenmanii*	Araceae	3503
parakwana-miti		*Cissus erosa*	Vitaceae	3533
tərakoa-rimõ		*Philodendron distantilobium*	Araceae	3336
		Philodendron sp. 25	Araceae	3335
		Anthurium sinuatum	Araceae	4851
tərakoa-ama'ã-kiri		*Monstera subpinata*	Araceae	3912
tatu-puhã	armadillo remedy	*Fridericia oligantha*	Bignoniaceae	3520

Guajá Name	English Gloss	Latin Name	Family	Coll. No.
tukur-iwɨpo	grasshopper vine	*Cissus cissyoides*	Vitaceae	4800
tukur-uhu-rimõ	big grasshopper vine	*Ipomoea* sp. 25	Convolvulaceae	4830
wɨpo-kərəm		*Mussatia* sp. 1	Bignoniaceae	4505
		Hippocratea sp. 1	Celastraceae	3219
		Masagnia macrodisca	Malpighiaceae	3671
		Anemopaegma aff.	Bignoniaceae	3186
		Salacia macrantha	Celastraceae	3221
		Cheliloclinium hippocrateoides	Celastraceae	3219
wɨpo-pe	flat vine	*Bauhinia* sp. 26	Fabaceae	3105
		Bauhinia coronata	Fabaceae	3689
		Bauhinia guianensis	Fabaceae	3239
		Bauhinia splendens	Fabaceae	3365
		Senna chrysocarpa	Fabaceae	3809
wɨpo-piuna	black vine	*Adenocalymma* cf.	Bignoniaceae	
		Clystoma sp. 1	Bignoniaceae	3251
		Hippocratea sp. 2 *purpurescens*	Celastraceae	3296
wɨpo-pi		*Mascagnia anisepetala*	Malpighiaceae	3325
wɨpo-rači	spiny vine	*Uncaria guianensis*	Rubiaceae	3423
wɨpo-rətə	hard vine	*Capparis lineata*	Brassicaceae	3907
		Heteropterys multiflora	Malpighiaceae	3129
		Memora allamandiflora		T392
wɨpo-ya'a		?	?	3444
wɨpo-'aru-i		*Cheiloclinium cognatum*	Celastraceae	3106
wɨpo-čə'a	black vine	*Guatteria scandens*	Annonaceae	3507
wɨpo-hu	big vine	*Combretum fruticosum*	Combretaceae	3918
		Connarus favsus	Connaraceae	3920
		Schwannia sp. (?)	Malpighiaceae	3923
yači-mõ-rana	tortoise not vine	*Entada polystachya*	Fabaceae	4805
yaiči-amõ	tortoise vine	*Coccoloba* cf. *excelsa*	Polygonaceae	3278
yamakara-mõ		*Doliocarpus* sp. 1	Dilleniaceae	3135
		Davilla sp. 25	Dilleniaceae	3480
yami-'i-mõ		*Machaerium augustifolium*	Fabaceae	4767

C. Folk Generic Names for Herbs (*ka'a*)

Guajá Name	English Gloss	Latin Name	Family	Coll. No.
akana-hu		*Calathea* sp. 1	Marantaceae	3364
akaru-hu-ruwi		*Calathea ornate*	Marantaceae	4858
añi		*Dieffenbachia* sp. 1	Araceae	3320
ara-pia-'i	little macaw egg	*Bidens bipinnatus*	Asteraceae	4850
arapa-mani'ɨ-'ɨ	red brocket deer manioc plant	*Manihot leptophylla*	Euphorbiaceae	3909
arira-ki		*Diplasia karataefolia*	Cyperaceae	
arira-popo		*Polypodium ciliatum*	Polypodiaceae	3329
ariwi		*Ischnosiphon obliquus*	Marantaceae	3494
ariwi-kəmakərə		*Mimosa pigra*	Fabaceae	4802
čo-ka'a-ran	black not herb	*Heliconia densiflora* ssp. *densiflora*	Heliconiaceae	3511
i-yar	its owner	*Pistia stratiodes*	Araceae	4810
hairi-rawai		*Adiantun cayennense*	Adiantaceae	3341
ipayu-rána		*Ludwigia leptocarpa*	Onagraceae	4806
ira-parakwa		*Gouania* cf. *cornifolia*	Rhamnaceae	3371
iwir	lashing material	*Urena lobata*	Malvaceae	3349
ka'a-tamki anama-uhu		*Costus scaber*	Zingiberaceae	3913
ka'a-'i	little herb	*Cyathula prostrata* aff.	Amaranthaceae	3340
ka'a-čũ	black herb	*Rhizomorpha* sp. 1	Mycota	3464
ka'a-hu	big herb	*Calathea grandis*	Marantaceae	3331
ka'a-ka-čĩ	black white herb	*Renealmia breviscapa*	Zingiberaceae	4825
		Renealmia floribunda	Zingiberaceae1	3318
ka'a-mururu		*Geophila repens*	Rubiaceae	4844
		Salvinia auriculata	Salviniaceae	4812
ka'a-pipi'a		*Lantana moritziana*	Verbenaceae	3529
ka'a-rače	spiny herb	*Solanum subinerme*	Solanaceae	3351
ka'a-ro	bitter herb	*Calathea sellowii*	Marantaceae	3384
ka'a-ta'i		*Calathea capitata*	Marantaceae	3345
ka'i- kəmə-'i		*Costus scaber*	Zingiberaceae	3319

Guajá Name	English Gloss	Latin Name	Family	Coll. No.
kamiča-tu-kənə	multiple shirts sugar cane	*Arthrostylidium*	Poaceae	3496
kwači-ka'a	coati herb	*Heliconia psittacorum*	Heliconiaceae	3342
kwarape-mino-rána		*Ludwigia helminthorriza*	Onagraceae	4813
paininə		*Manihot baccata*	Euphorbiaceae	3528
		Lomariopsis prieuriana	Lomariopsidaceae	
pariri		*Calathea elliptica*	Marantaceae	3363
tamakai-ča		*Cyperus luzulae*	Cyperaceae	3369
tamata-ka'a		*Monotagma* sp.	Marantaceae	3359
urupe		?	Mycota	3470
warara-ha-ka'a		*Dichorisandra affinis*	Commelinaceae	3915
		Miconia cf. *tomentosa*	Melastomataceae	3470
		Miconia nervosa	Melastomataceae	3317
warara-piya-ran		*Psychotria racemosa*	Rubiaceae	3477
warara-wɨra		*Psychotria hoffmannseggiana*	Rubiaceae	3911
		Psychotria racemosa	Rubiaceae	3326
		Psychotria rosea var. *trichophora*	Rubiaceae	3493
wari-rawai	howler monkey tail	*Polypodium* sp. 1	Polypodiaceae	3334
wari-rawai-ran	not howler monkey tail	*Adiantum latifolium*	Adiantaceae	3479
		Trichomanes vittoria	Hymenophyllaceae	3193
wɨra	tree	*Ludwigia hyssopifolia*	Onagraceae	3368
wɨra-ha	hair tree	*Neckeropsis* sp. 1	Neckeraceae	3398
wɨra-kerem-ti		*Phytolacca rivinoides*	Phytolaccaceae	3489
wira-ki (=*ka'aki*)	bird pepper	*Conyza bonariensis*	Asteraceae	3374
wiripipi		*Eichornia crassipes*	Pontederiaceae	4811
		Reussia rotundifolia	Pontederiaceae	4809
yakamĩ-təməkənə	pheasant food	*Piper divaricatum*	Piperaceae	3461
		Ischnosiphon puberulus var. *scaber*	Marantaceae	4866
		Piper carniconnectivum	Piperaceae	3478
		Pothomorphe peltata	Piperaceae	3352
yaməkara-ma'a-i		*Coccoloba* cf. *scandens*	Polygonaceae	3220

Guajá Name	English Gloss	Latin Name	Family	Coll. No.
yapura-ka'a	weasel herb	*Stigmaphyllon sinuatum*		3491
yawar-ka'a-hu	big jaguar herb	*Phenakospermum guyannense*	Strelitziaceae	3350
yawewi-račī	spiny stingray	*Desmoncus polyacanthus*	Arecaceae	3500
yawi-ka'a		?	?	—
yawpe		*Bactris hirta* var. *pectinata*	Arecaceae	3499
yu-aran-uhu	big not spiny (plant)	*Cyphomandra* sp. 1	Solanaceae	4833
yu-'i	little spiny (plant)	*Astrocaryum gynacanthum*	Arecaceae	3484
yapu-ruwai-ran	looks like oropendola tail feather	*Lindsaea lancea*	Lindsaeaseae	3194
yawar-i-monom-'im-ti	little dog vine	*Bauhinia* sp. 98	Fabaceae	3466
yamokō		*Smilax* sp. 1	Smilacaceae	3192
yawa-tərə-waya (= *karata-ran*)	dog's guaja fruit	*Xiphidium caeruleum*	Haemodoraceae	3485

D. Folk Generic Names for Nondomesticates Unclassified as to Life Form

Guajá Name	English Gloss	Latin Name	Family	Coll. No.
ka'api'i	grass	*Urochloa humidicola*	Poaceae	3503
		Fimbristylis annua	Poaceae	3502
		Lindernia crustaceae	Scrophulariaceae	3531
		Indeterminate genus sp. 2	Poaceae	4814
maratə		*Dioscorea* indt. sp.	Dioscoreaceae	4857
nana'ĩ	not pineapple	*Aechmea bromeliifolia*	Bromeliaceae	3383
		Billbergia sp. 1	Bromeliaceae	4832
		Bromelia balansae	Bromeliaceae	4843
		Bromelia goeldiana	Bromeliaceae	3348
takwarə		*Guadua glomerata*	Poaceae	3463
takwar-i		*Olyra caudata*	Poaceae	4628
		Olyra cardifolia	Poaceae	4842
		Ichnanthus brevicrobs	Poaceae	3324
takwar-i-ran		*Olyra cordifolia*	Poaceae	3322

E. Folk Generic Names for Domesticated Plants

Guajá Name	English Gloss	Latin Name	Family	Coll. No.
akayu-rána	false cashew	*Anacardium occidentale*	Anacardiaceae	3391
arakačiahú (= *məmə*)		*Carica papaya*	Caricaceae	3467
araku-ate		*Bixa orellana*	Bixaceae	
kamana'ĩ	not bean	*Phaseolus* sp. 1	Fabaceae	None
kara	yam	*Dioscorea trifida*	Dioscoreaceae	None
kiki tai-pičika'e	small chile pepper	*Capsicum frutescens*	Solanaceae	3534
kwi		*Crescentia cujete*	Bignoniaceae	None
ma-'ɨwa		*Mangifera indica*	Anacardiaceae	3483
makači	sweet manioc (borrowed from Portuguese)	*Manihot esculenta*	Euphorbiaceae	None
matak	sweet potato (borrowed from Portuguese)	*Ipomoea batatas*	Convolvulaceae	3482
məmə-ra'ĩ'ɨ	papaya seed tree	*Ricinus communis*	Euphorbiaceae	3468
nana'ĩ	not pineapple	*Ananas comosus*	Bromeliaceae	3392
Paku	banana, plantain	*Musa* spp.	Musaceae	None
tərəmə		*Manihot esculenta*	Euphorbiaceae	3367
tai-hamãi	very spicy treelet	*Citrus* sp. 1	Rutaceae	None
tai-rana	false spicy treelet	*Averrhoa carambola*	Oxalidaceae	4837
		Lantana camara	Verbenaceae	4808
		Odontocarya duckei	Menispermaceae	4807
takwari-ake		*Oryza sativa*	Poaceae	None
takwar-rã		*Saccharum officinarum*	Poaceae	None
urumũ	squash	*Cucurbita moschata*	Cucurbitaceae	
wači; kwači	field corn	*Zea mays*	Poaceae	3526
waya	guava	*Psidium guajava*	Myrtaceae	3393
yanači	watermelon	*Citrullus lanatus*	Cucurbitaceae	None

APPENDIX II. TREES OF THE ANTHROPOGENIC FOREST (*TAPER*)

These names were elicited by freelisting with a sample of Ka'apor adults (n = 22) and are ranked in order of psychological salience by Smith's s. Presence (+) or absence (−) in anthropogenic forest (F) or high forest (H) inventories are indicated in the last column (see Chapter 10 text).

Rank	Ka'apor Name	Scientific Name	Freq.	Smith's s	F/H
1	yetai'ɨ	*Hymenaea parvifolia*	15	0.451	+/−
2	tarapai'ɨ	*Hymenaea* spp.	14	0.437	+/+
3	tayɨ	*Handroanthus impetiginosus*	12	0.405	+/−
4	parawa'ɨ	*Eschweilera coriacea*	13	0.39	+/+
5	šimoran	*Senna silvestris*	11	0.301	−/−
6	tayɨpo	*Handroanthus serratifolius*	10	0.285	+/−
7	inaya	*Attalea maripa*	10	0.211	+/−
8	para'ɨ	*Jacaranda* spp.	7	0.206	+/+
9	tareka'ɨ	*Bagassa guianensis*	7	0.203	+/+
10	taperiwa'ɨ	*Spondias mombin*	5	0.183	+/−
11	pina'ɨ	*Duguetia* spp.; *Unonopsis rufescens*	6	0.168	+/+
12	mamawiran	*Jacaratia spinosa*	5	0.158	+/−
13	kumaru'ɨ	*Dipteryx odorata*	4	0.151	−/+
14	tukumã	*Astrocaryum vulgare*	7	0.143	+/−
15	piwa'ɨ	*Rinorea* spp.	6	0.143	+/+
16	akaú'ɨ	*Helicostylis tomentosa*	6	0.143	−/+
17	akušitɨrɨwa'ɨ	*Pouteria macrophylla*	7	0.141	+/+
18	yašiamɨr	*Lecythis idatimon*	5	0.123	+/+
19	kakawiran	*Theobroma speciosum*	4	0.121	+/+
20	akayu'ɨ	*Anacardium* spp.	5	0.117	+/+
21	ama'ɨ	*Cecropia* spp.	4	0.117	+/+
22	payu'ã'ɨ	*Couepia* spp.; *Parinari* sp.; *Hirtella bicornis*	3	0.107	+/+
23	yapukwai'y	*Lecythis pisonis*	5	0.107	+/+

Rank	Ka'apor Name	Scientific Name	Freq.	Smith's s	F/H
24	irikiwa'ɨ	Manilkara huberi	4	0.104	+/+
25	ɨwise'ɨ	Simarouba amara	4	0.102	+/+
26	kupa'ɨ	Copaifera spp.	3	0.102	+/+
27	inga'ɨ	Inga spp.	5	0.093	+/+
28	paruru'ɨ	Sacoglottis spp.	3	0.091	+/+
29	merahɨtawa	Byrsonima sp.	4	0.088	+/+
30	kanei'ɨ	Protium spp.	3	0.081	+/+
31	payangi'ɨ	Vismia guianensis	2	0.08	–/–
32	pitɨminem	Couratari oblongifolia	4	0.079	+/–
33	šamato'ɨ	?	3	0.078	?/?
34	mɨratā	Erythroxylum citrifolium	2	0.072	–/–
35	yurupepe'ɨ	Dialium guianense	3	0.071	+/+
36	kipɨhu'ɨ	Theobroma grandiflorum	3	0.066	–/+
37	yiniro'ɨ	Genipa americana	2	0.065	–/–
38	taši'ɨ	Tachigali spp.	2	0.062	+/+
39	tekweripihun	Rollinia exsucca	3	0.061	+/–
40	kururu'ɨ	Taralea oppositifolia	3	0.059	–/+
41	karatu'ā'ɨ	Fusaea longifolia	3	0.059	+/+
42	pu'ɨpirang'ɨ	Ormosia coccinea	2	0.055	–/–
43	kupapa'ɨ	Pouteria spp.	2	0.053	+/+
44	ayuwa'y	Lauraceae spp.	2	0.053	+/+
45	ašiwa'ɨ	?	2	0.052	?/?
46	irarɨ	Cedrela fissilis	4	0.052	+/–
47	marato'ɨ	Schefflera morototoni	2	0.051	+/+
48	tayɨtawa	Handroanthus sp.	2	0.05	+/–
49	kuyeri'ɨ	Lacmellea aculeata; Ambelania acida	2	0.05	–/+
50	ama'ɨatā	Cecropia sp.	2	0.049	+/+
51	yeyu'ɨ	Astronium lecointei	2	0.047	+/–
52	āyākɨwa'ɨ	Apeiba spp.	2	0.045	+/+
53	apa'ɨ	Parahancornia spp.	2	0.043	–/+
54	šiširupe'ɨ	Inga spp.	3	0.04	+/+

Rank	Ka'apor Name	Scientific Name	Freq.	Smith's s	F/H
55	inayaɨvɨ	?	2	0.038	?/?
56	kɨkɨ'ɨ	*Newtonia* spp.; *Pithecellobium* sp.	2	0.036	+/+
57	pytymyte	*Couratari guianensis*	3	0.034	+/+
58	amangaputɨr'ɨ	*Senna pendula; Cassia fastuosa*	2	0.034	−/−
59	pakuri'ɨ	*Platonia insignis*	2	0.034	+/+
60	ɨraki'i'ɨ	*Myrciaria tenella*	3	0.033	+/−
61	paraku'ɨ	*Chimarrhis turbinata*	2	0.031	+/+
62	ingahu'ɨ	*Inga* spp.	2	0.031	+/+
63	tamaran'ɨ	*Zollernia paraensis*	4	0.028	+/−
64	akušitɨrɨwahu'ɨ	*Pouteria macrocarpa; Franchetella* sp. 1	2	0.027	+/+
65	pinuwa'ɨ	*Oenocarpus distichus*	2	0.027	+/+
66	tekwerɨ	*Cordia* spp.	2	0.021	+/+
67	pɨkɨ'a'ɨ	*Caryocar villosum*	2	0.02	−/−
68	pakurisōsō'ɨ	*Rheedia* spp.	2	0.005	+/+

NOTES

Chapter 2

1. Roskoski (1982; cited in Ewel 1986, 251) found that at least one species of *Inga* fixes no nitrogen.

2. All collection vouchers cited herein are in the series of the author and are deposited at the New York Botanical Garden with duplicates at the Museu Paraense Emílio Goeldi.

3. Although they did not report the presence of *caiauê,* Simões and Lopes (1987) did note deep deposits of *terra preta* on all thirty-one archaeological sites that they excavated along the lower and middle courses of the Madeira River.

4. The pottery of this culture is notable for being coiled, corrugated, tempered by quartz, feldspar, or mica as opposed to *caraipé* (*Licania* spp.), and, if painted, the style is black-and-red-on-white (Meggers and Evans 1973; Willey 1971).

5. Chemical analysis of *terra preta* from the Araweté swidden was carried out at the Faculdade de Ciências Agrárias do Pará (Belém), with the exception of carbon and nitrogen, which were analyzed at EMBRAPA/CPATU (Belém).

6. The Assurini also cultivate several varieties of maize, one of which (*awači-pinimu*) exhibits the opaque phenotype as well and is also being studied. The Assurini also prefer *terra preta* (*iwi-piúnə*) for maize cultivation.

7. Collection numbers are in the series Balée with vouchers at the New York Botanical Garden and duplicates at the Museu Paraense Emílio Goeldi.

The ecological importance value of a species is the sum of its relative frequency, relative density, and relative dominance. The formulae for calculating these values are given elsewhere: Campbell et al. (1986); Cottam and Curtis (1956); Curtis and Cottam (1962); Mori et al. (1983). The relative values for frequency, density, and dominance as well as the importance values of all species ≥10 cm dbh from the Araweté and Assurini inventories appeared in Balée and Campbell (1990).

8. In addition, the Indians believe the stone axeheads and archaic potsherds found in their habitats to be of divine, not human, origin so long forgotten are the cultures that produced them.

9. The radiocarbon date is in years before A.D. 1950, using the traditional half-life of C-14 of 5,568 years. The dating of this charcoal (sample no. RL-2034) was carried out at Radiocarbon, Ltd., at Lampasas, Texas.

10. The total extent of Brazilian Amazonian *terra firme* is 3,303,000 km^2. Therefore, 389,370 km^2 of anthropogenic forest (totaled in Table 2.4) represents 11.8 percent of Brazilian Amazonian *terra firme.*

Chapter 3

1. Perhaps the seminal statement for this point of view in anthropology is "[e]cology involves one unalterable factor, the natural environment" (Steward 1938, 261).

Chapter 5

1. In light of this and related evidence, the hypothesis that agricultural regression (from horticulture to foraging) in lowland South America may have occurred also in prehistoric times—for reasons perhaps unrelated to epidemiological factors and massive depopulation—may also be supported. If agricultural regression in lowland South America did occur in prehistoric times (Chapter 4), however, empirical instances are still wanting and the prehistoric ecological conditions that might have been associated with such changes in society remain speculative. More paleoethnobotanical research is needed for probing this intriguing possibility.

2. The terms *degraded* and *devolved* have been used by other authors, including Lathrap, in place of the term *regressed,* which I prefer because of its greater precision (see Dole 1991) (see Chapter 4).

3. Several Guajá generic terms for domesticated plants are terms (perhaps recently) borrowed from Portuguese, such as *matak* (from "*batata,*" sweet potato) and *makači* (from "*macaxeira,*" certain varieties of sweet manioc—the Portuguese words, incidentally, have their etymons in an earlier Tupí-Guaraní language). Others are terms for recently introduced domesticates based on analogy with nondomesticates, a pattern also found in Ka'apor (see Balée 1989; 1994, 196–200). Still others seem to have no natural analogues in other related languages, or if they do it is due to chance or similar interpretations of some salient characteristic of the plant.

4. The terms glossed in Table 5.2 as "manioc," namely, *mani'o* and *mani'ɨ,* no longer refer to manioc (*Manihot esculenta*). A term polysemous with babaçu mesocarp flour, *tərəmə,* rather, is the modern term for the manioc plant and food products now derived from it. I hypothesize that these terms, however, did at one time refer to manioc in the Guajá language (with *mani'ɨ* more specifically referring to the plant itself and *mani'o* to its tubers). Whereas the referential meaning of the unmarked term was replaced (not lost) by another term (*tərəmə*) during agricultural regression, the marked terms were somehow retained. In particular, the term "brocket deer-manioc-stem" seems to be cognate with terms in other Tupí-Guaraní languages.

5. The terms *pacova* and *pacová* that also refer to bananas and some other related monocots in certain dialects of Brazilian Portuguese are credited by Brazilian lexicographers as having derived from Tupí-Guaraní languages (Holanda Ferreira n.d., 1015).

6. The term cited here is from the Pindaré River basin; the Guajá of the Turiaçu basin call papaya by a term borrowed from the Portuguese, *məmə* (from *mamão*).

7. This analysis may also partially apply to the term for the common bean, *kamana'ɨ* (see Table 5.1); however, this term does not seem to refer to any nondomesticated plants.

8. Headland (1983), who presented the first evidence as to few "binomial specifics" in the language of a foraging people, the Agta of the Philippines, offered somewhat different explanations from those of Berlin and Brown for this phenomenon.

Chapter 6

1. In point of fact, disagreement exists on just how genetically close Guarayo and Sirionó are. There is little doubt that Guarayo and Guaraçug'wé or Pauserna are very closely related linguistically and culturally (Riester 1977; Stearman 1984, 640), more so than either is to Sirionó. Lemle (1971, 128), Loukotka (1968, 119), and Rodrigues (1984/1985, 38) included them in the same proposed subgroup of the Tupí-Guaraní family on the basis of phonological criteria alone, but Wolf Dietrich (1990, 115) finds Sirionó to form a subgroup of its own based on phonological, morphological, and grammatical criteria. Dietrich's arguments are persuasive, but his scheme disallows a subgroup of Wayãpi and Urubú (Ka'apor) to be formed because Ka'apor is a low-rate language in terms of retention of the criteria he uses. Yet Wayãpi seems to be more closely intelligible with Ka'apor than any other known Tupí-Guaraní language. Guarayo is understood by some Sirionó informants also, but there has been some intermarriage among them and it is not clear how much of this mutual intelligibility has been recently learned. In any case, the terms for calabash and numerous other plants are divergent between them.

2. On the basis of phonological, grammatical, and morphological evidence, Dietrich (1990, 58) views Sirionó as a "peculiar case" (also see Priest 1987 on an item of phonology only); for him it is a pivotal link between Amazonian, Bolivian, and southern branches of the Tupí-Guaraní language family, forming an isolated subgroup of its own within the family. Dietrich (1986, 204) also argues, without clear supporting evidence in my view, for a northern rather than southern origin for the Sirionó. Stearman (1984, 640; 1989, 22) suggested a southerly origin for the Yuquí and Sirionó based on linguistic, ethnohistoric, and cross-cultural evidence, and Priest (1987) showed an apparently high index of cognacy between Yuquí and Sirionó. I do agree with Dietrich (1986, 204), however, that the Sirionó are intrusive in the Llanos de Mojos and not representatives of a proto-population, protoculture, or non-Tupian protolanguage that supposedly existed there before the Guaraní migrations of the 1400s and 1500s. This view does not presuppose that the region was uninhabited before they arrived.

3. Aché-Guayaki and Xetá might have been Guaraní-ized (Dietrich 1990), unlike the probable scenario with the others mentioned here.

4. The following Tupí-Guaraní language abbreviations will be used in the remainder of this chapter: Ar = Araweté, As = Assurini do Xingu, Ch = Chiriguano, Gj = Guajá, Gw = Guaraçug'wé, Gy = Guarayo, K = Ka'apor, S = Sirionó, T = Tembé, and W = Wayãpi of French Guiana.

5. The proteinaceous venom ejected by the lancets of these ants is used to reduce fever among both the Sirionó and the Ka'apor, though the same ants come from different tree species in the respective cases and though the Sirionó and Ka'apor are separated by about 2,000 km of Amazon forest (Chapter 7). The venom of a closely related species contains a polysaccharide that deactivates part of the human complement system, perhaps being useful therefore in the treatment of rheumatoid arthritis (Schultz and Arnold 1977). The similarities between the Sirionó and the Ka'apor in the use of pseudomyrmecine ants seem to be historically parallel rather than convergent or borrowed.

6. In this regard, it is interesting to note that the bottle gourd is called *anái* (Horn

Fitz Gibbon 1955, 25) in Gw. Gw speakers lived near the Sirionó along the Guaporé River, which suggests a different migratory and contact history for them, in spite of possibly a common origin with S. The term is clearly cognate if referents are restricted to botanical species (or even families, in this case).

7. In Table 6.1, the data sources for the Ar, As, Gj, and K terms are found in Balée 1994 and Balée and Moore 1991. The Gw terms are from Horn Fitz Gibbon (1955, 25). The S data are from my unpublished notes for the most part and Schermair (1958, 1962). Regarding *nĩñu,* it seems that several S reflexes for plant names (and no doubt other vocabulary) have been subjected to syncope. Languages in the Tupí-Guaraní family with a stress tendency toward the penultimate syllable as with S are more likely to experience syncope of words than those with a tendency toward stress on the final syllable (Dietrich 1990, 16), although it is true that As and W also exhibit a tendency toward penultimate stress and may show less syncope than S in the domain of plant names for unexplained reasons. In any case, the S reflex for cotton evinces syncope of the first syllable based on the examples in the table.

Chapter 7

1. The Nazca lines are on the desert plains of south-central Peru. The lines represent figures such as birds and other animals, geometric figures, and sometimes enclosures. There are many hypotheses for their existence. Some researchers believe they had to do with the water supply; some that they were purely for religious reasons. See the article "Nazca Lines" in Selin 1997, 777–780.

2. They believe that deer eat parts of wild manioc plants, and research shows that the deer definitely eat parts of domesticated ones.

Chapter 8

1. Although Johanna Nichols (1992, 24) posits an older time depth for language families, at about twenty-five hundred to four thousand years, the genetic units recognized as families in Amazonia seem usually younger than this; Nichols' figure for the family in fact better overlaps with the ranges of the stock in Amazonia (see Kaufman 1990, 51). In the schema proposed by Mary Haas (1969, 60), the protolanguage of the family (such as Proto-Tupí-Guaraní) would be of the first order, since it is based on reconstruction made from comparison of living languages; the protolanguage of a stock, such as Proto-Tupi, would be of the second order, based on the comparison of reconstructed languages. It is probably wisest to work with both orders and between them for more accurate reconstructions.

2. The abbreviations in Tables 8.1 and 8.2 refer to language names as follows: Ar = Arawaeté, As = Assurini, K = Ka'apor, S = Sirionó, T = Tembé, and W = Wayãpi. The data sources for the Ar, As, and K terms are found in Balée 1994 and Balée and Moore 1991, 1994. The S terms are from my unpublished notes and Schermair 1958 and 1962.

3. Curiously, the red-footed tortoise was traditionally prescribed as food for Ka'apor women with late onset of menstruation or menarche to induce menstruation (Ribeiro 1996, 584).

4. There might be other reasons why it is never realized, but these have not been identified. If all tortoises were actually tabooed while other game were always allowed, perhaps a greater hunting pressure would be witnessed in the vicinity of Ka'apor villages.

5. "Lhe dizerem que tinhamos nós a Carta mandando eu que se abstivessem dos Jabotins nos dias de peixe por não haver razão que mostrasse não serem Carne . . . disserão no pulpito publicamente que elles erão peixe e por tal se comião diante do Papa . . . e que os religiosos santos francezes por peixe o derão" (Lisboa 1904 [1626–1627], 397).

6. That tortoises are quickly hunted out and serve as an index on the availability of other game is supported by evidence from the Mekranoti Kayapó (Werner 1984, 193, cited in Balée 1985, 499–500), who do not use the species for the same ritual purposes as do the Tupí-Guaraní societies mentioned here. In general, the ecological implications of tortoise hunting may be similar, regardless of ritual taboos and injunctions that surround it. Yet its continuity through time may be unrelated to any functional role it may play in the ecology and economy of native society, since it has persisted only among a few of many surviving Tupí-Guaraní groups. In other words, Galton's problem may be invoked as easily as the functionalist argument to explain the survival, in distinctive versions, of this trait into modern times.

7. Ka'apor informants say that the menstruation–tortoise complex began with the origin of the moon. Moreover, the words for moon (*yahɨ*), menstruation (*yaɨ*), and the tortoise (*yaši*) are very similar phonetically. Moon (*yahɨ*) was a boy who surreptitiously had sexual intercourse with his sister; he had used genipapo juice to paint his face and thus disguise himself during these rendezvous. Upon being discovered, however, he fled from his punitive covillagers to the sky on a ladder of arrows. His sister followed. He became the moon, and she, Venus (Huxley 1956, 164–166). As part of Moon's revenge on those who forced him into his celestial exile, he imposed menstruation and concomitant meat taboos, except for yellow-footed tortoises, on women (Balée 1984a, 247). A very similar myth was recorded among the *caboclos* on the Acará River (Oliveira 1951, 29–30), where the Ka'apor people once lived in the early nineteenth century. It is intriguingly similar in certain respects to a belief of the Panoan-speaking Cashinahua of eastern Peru. According to ethnographer Kenneth Kensinger (1995, 35): "Women, unlike men, menstruate, a physiological process which according to the Cashinahua has its origins in a curse placed on women by an incestuous male who had been fatally cursed by his sister before going on a raid. His severed head, with the telltale black paint on its face, cursed all women as it climbed into the sky to become the moon, and each month both men and women are reminded of the consequences of incest" (also see Lévi-Strauss 1978, 97–99).

The association of the moon, menstruation, incest, and genipapo juice (from *Genipa americana*) is found also in the mythology of the Ecuadorian jungle Quichua (Whitten 1976, 51) and the Arawakan Kuniba (Lévi-Strauss 1978, 94). Although menstruation, the moon, and incest are common in mythology (Knight 1991), these specific associations seem peculiarly Amazonian. If an original, Amazonian, or American mythic foundation cannot be found, perhaps some borrowing occurred among diverse Amazonian language families and stocks at some remote point in the past. In any event, myths linking the moon and menstruation across fundamentally unlike societies must be more common, and more alike, than ritual practices conjoining menstruation and a preference or even prescription for tortoise meat, which suggest a more recent though still ancient origin.

Chapter 9

1. In Tables 9.1 and 9.2 the hyphens in the glosses indicate morpheme boundaries; L refers to a literal, monomorphemic, essentially nonpolysemous plant term (see Balée and Moore 1991). English plant morphemes that heuristically meet this criterion would be "oak," "maple," "pine." In Table 9.1, the LGA term refers to the fruit of the cacao tree only.

2. Aurê and Aurá, listed in Table 9.2, are the only known speakers of a newly recorded Tupí-Guaraní language originally spoken between the Xingu and Tocantins Rivers (Jensen 1999, 128; Mello 1996) (see Chapter 1).

3. The closely related Uru-eu-wau-wau language (also called Yupaú or Tupi-Kawahib [James Welch, personal communication, 2002; Rodrigues and Cabral 2002]), which is also in subgroup #6, denotes a nondomesticated cacao of the forest of central Rondônia (in the southwestern Amazon region of Brazil) as *ni-mi-tahi-ma* or *i-mi-ta-hi-ma* (Balée, field notes, 1992). These terms can arguably be glossed as "smooth cacao." The initial syllables in these terms (the differences between which may be due to free variation), therefore, are quite similar to Parintintin *ñumi*.

WORKS CITED

Abraão, M. B., B. W. Nelson, J. C. Baniwa, D. W. Yu, and G. H. Shepard Jr. 2008. Ethnobotanical ground-truthing: Indigenous knowledge, floristic inventories and satellite imagery in the upper Rio Negro, Brazil. *Journal of Biogeography* 35:2237–2248.
Acuña, Cristoval de. 1963. A new discovery of the great river of the Amazons. Trans. Clement Markham (from original of 1641). In *Expeditions into the Valley of the Amazons*. New York: B. Franklin. (Orig. pub. 1859, Hakluyt Society.)
Albert, Bruce. 1985. *Temps du sang, temps des cendres: Représentation de la maladie, système rituel et espace politique chez les Yanomami du sud-est (Amazonie brésilenne)*. Paris: Laboratoire d'Ethnologie et de la Sociologie Comparative, Université de Paris X.
Alcorn, Janice B. 1984. *Huastec Mayan ethnobotany*. Austin: University of Texas Press.
———. 1989. Process as resource: The traditional agricultural ideology of Bora and Huastec resource management and its implications for research. In *Resource management in Amazonia: Indigenous and folk strategies*, edited by Darrell A. Posey and William Balée, pp. 63–77. Bronx: New York Botanical Garden.
Alden, D. 1976. The significance of cacao production in the Amazon region during the late colonial period: An essay in comparative economic history. *Proceedings of the American Philosophical Society* 120(2):103–135.
Alland, Alexander. 1975. Adaptation. *Annual Review of Anthropology* 4:59–73.
Anderson, Anthony B. 1981. White-sand vegetation of Brazilian Amazonia. *Brittonia* 13(3): 199–210.
———. 1983. The biology of *Orbignya martiana* (Palmae): A tropical dry forest dominant in Brazil. PhD diss., University of Florida, Gainesville.
Anderson, Anthony B., and Suely Anderson. 1983. A "tree of life" grows in Brazil. *Natural History* 94(12):41–46.
Anderson, Anthony B., and W. Benson. 1980. On the number of tree species in Amazonian forests. *Biotropica* 12(3):235–237.
Anderson, Anthony B., Peter H. May, and Michael J. Balick. 1991. *The subsidy from nature. Palm forests, peasantry, and development on an Amazon frontier*. New York: Columbia University Press.
Anderson, Anthony B., and Darrell A. Posey. 1985. Manejo de cerrado pelos Índios Kayapó. *Boletim do Museu Paraense Emílio Goeldi, Botânica* 2(1):77–98.
———. 1989. Management of a tropical scrub savanna by the Gorotire Kayapó of Brazil. In *Resource management in Amazonia: Indigenous and folk strategies*, edited by Darrell A. Posey and William Balée, pp. 159–173. Bronx: New York Botanical Garden.
Andrade, E. B. 1983. Relatório da expedição para coleta de germoplasma de caiauê, *Elaeis oleifera* (HBK) Cortez, na Amazônia brasileira. Belém, Brazil: EMBRAPA/CPATU.

Anonymous. 1781 (orig. 1702). Abrégé d'une relation Espagnole.... In *Lettres édifiantes et curieuses, écrites des missions étrangères,* vol. 8. Paris: J. G. Merigot le jeune.

Anonymous. 1982. O ritual da tocandira. *Atualidade Indígena* 22:50–62.

Araújo Brusque, F. C. de. 1862. *Relatório apresentado à Assemblea Legislativa da Província do Pará na primeira sessão da XIII legislatura.* Belém, Brazil: Typographia Frederico Carlos Rhossard.

Araújo-Costa, Fernanda. 1983. Projeto Baixo Tocantins: Salvamento arqueológico na região de Tucurui (Pará). Master's thesis, Universidade de São Paulo, São Paulo, Brazil.

Arroyo-Kalin, Manuel. 2009. *Domesticação na paisagem: Os solos antropogênicos e o Formativo na Amazônia.* XV Congresso Sociedade de Arqueologia Brasileira, Belém, Brazil, 20–23 September.

———. 2010. The Amazonian Formative: Crop domestication and anthropogenic soils. *Diversity* 2(4):473–504.

Aspelin, Paulo, and Sílvio Coelho dos Santos. 1981. *Indian areas threatened by hydroelectric projects in Brazil.* Copenhagen: International Work Group for Indigenous Affairs.

Aubertin, C. 1996. Heurs et malheurs des ressources naturelles en Amazonie Brésilienne. *Cahier des Sciences Humaines* 32(1):29–50.

Azevedo, José L. 1930. *Os Jesuítas no Grã-Pará.* Coimbra, Portugal: Imprensa da Universidade.

Baer, W. 1995. *The Brazilian economy: Growth and development.* 4th ed. Westport, CT: Praeger.

Bailey, Robert C., G. Head, M. Jenike, G. Owen, R. Rechtman, and E. Zechenter. 1989. Hunting and gathering in tropical rain forest: Is it possible? *American Anthropologist* 91:59–82.

Bailey, Robert C., and Thomas N. Headland. 1992. The tropical rain forest: Is it a productive environment for human foragers? *Human Ecology* 19(2):261–285.

Baldus, Herbert. 1936. Ligeiras notas sobre duas tribos Tupi da margem paraguaya do Alto Paraná (Guayaki e Chiripá). *Revista do Museu Paulista* 20:749–756.

Balée, William. 1984a. The persistence of Ka'apor culture. PhD diss., Columbia University, New York.

———. 1984b. The ecology of ancient Tupi warfare. In *Warfare, culture, and environment,* edited by R. Brian Ferguson, pp. 241–265. Orlando, FL: Academic.

———. 1985. Ka'apor ritual hunting. *Human Ecology* 13(4):485–510.

———. 1986. Análise preliminar de inventário florestal e a etnobotânica Ka'apor (Maranhão). *Boletim do Museu Paraense Emílio Goeldi, sér Bot.* 2(2):141–167.

———. 1987a. A etnobotânica quantitativa dos índios Tembé (Rio Gurupi, Para). *Boletim do Museu Paraense Emílio Goeldi, sér. Bot.* 3(1):29–50.

———. 1987b. Relatório etnológico sobre os últimos dias da Frente de Atração do Rio Tapirapé (Município de Marabá, Estado do Pará) em 22–29 de outubro de 1987. Brasília: Fundação Nacional do Índio.

———. 1987c. Cultural forests of the Amazon. *Garden* (11)6:12–14, 32.

———. 1988a. The Ka'apor Indian wars of lower Amazonia, ca. 1825–1928. In *Dialects and gender: Anthropological approaches,* edited by Richard R. Randolph, David M. Schneider, and May N. Diaz, pp. 155–169. Boulder, CO: Westview.

———. 1988b Indigenous adaptation to Amazonian palm forests. *Principes* 32(2):47–54.

———. 1989a. The culture of Amazonian forests. In *Resource management in Amazonia: Indigenous and folk strategies,* edited by Darrell A. Posey and William Balée, pp. 1–21. Bronx: New York Botanical Garden.

———. 1989b. Nomenclatural patterns in Ka'apor ethnobotany. *Journal of Ethnobiology* 9(1):1–24.

———. 1992a. People of the fallow: A historical ecology of foraging in lowland South America. In *Conservation of neotropical forests: Building on traditional resource use,* edited by Kent H. Redford and Christine Padoch, pp. 35–57. New York: Columbia University Press.

———. 1992b. Indigenous history and Amazonian biodiversity. In *Changing tropical forests,* edited by H. K. Steen and R. P. Tucker, pp. 185–197. Durham, NC: Forest History Society.

———. 1994. *Footprints of the forest: Ka'apor ethnobotany—The historical ecology of plant utilization by an Amazonian people.* New York: Columbia University Press.

———. 1995. Historical ecology of Amazonia. In *Indigenous peoples and the future of Amazonia: An ecological anthropology of an endangered world,* edited by Leslie E. Sponsel, pp. 97–110. Tucson: University of Arizona Press.

———. 1998. Historical ecology: premises and postulates. In *Advances in historical ecology,* edited by William Balée, pp. 13–29. New York: Columbia University Press.

———. 1999. Mode of production and ethnobotanical vocabulary: A controlled comparison of Guajá and Ka'apor (Eastern Amazonian Brazil). In *Ethnoecology: Knowledge, resources, and rights,* edited by Ted L. Gragson and Ben Blount, pp. 24–40. Athens: University of Georgia Press.

———. 2000a. Elevating the Amazonian landscape. *Forum for Applied Research and Public Policy* 15(3):28–32.

———. 2000b. Antiquity of traditional ethnobiological knowledge in Amazonia: The Tupí-Guaraní family and time. *Ethnohistory* 47(2):399–422.

———. 2001. Environment, culture, and Sirionó plant names. In *On biocultural diversity: Linking language, knowledge, and the environment,* edited by Luisa Maffi, pp. 298–310. Washington, DC: Smithsonian Institution Press.

———. 2003. Native views of the environment in Amazonia. In *Nature across cultures: Views of nature and the environment in non-western cultures,* edited by Helaine Selin, pp. 277–288. Dordrecht: Kluwer Academic.

———. 2006. The research program of historical ecology. *Annual Review of Anthropology* 35:75–98.

———. 2009a. Landscape transformation and language change: A case study in Amazonian historical ecology. In *Amazon peasant societies in a changing environment: Political ecology, invisibility and modernity in the rainforest,* edited by Cristina Adams, Rui Murrieta, Walter Neves, and Mark Harris, pp. 33–53. New York: Springer.

———. 2009b. Culturas de distúrbio em substratos amazônicos. In *As terras pretas de índio da Amazônia: Sua caracterização e uso deste conhecimento na criação de novas áreas,* edited by W. G. Teixeira, D. C. Kern, B. E. Madari, H. N. Lima, and W. Woods, pp. 48–52. Manaus, Brazil: EMBRAPA Amazônia Oriental.

———. 2010. Writing South American archaeology after Steward: A review essay. *Latin American Antiquity* 21(4):451–468.

———. 2012. *Inside cultures: A new introduction to cultural anthropology.* Walnut Creek, CA: Left Coast.

Balée, William, and Marilyn Cebolla Badie. 2009. The meaning of "tree" in two different Tupí-Guaraní languages from two different Neotropical forests. *Amazônica. Revista de Antropologia* 1:96–135.

Balée, William, and David G. Campbell. 1990. Evidence for the successional status of liana forest (Xingu River basin, Amazonian Brazil). *Biotropica* 22(1):36–47.

Balée, William, and Anne Gély. 1989. Managed forest succession in Amazonia: The Ka'apor case. In *Resource management in Amazonia: Indigenous and folk strategies,* edited by Darrell A. Posey and William Balée, pp. 129–158. Bronx: New York Botanical Garden.

Balée, William, and Denny Moore. 1991. Similarity and variation in plant names in five Tupi-Guarani languages (eastern Amazonia). *Bulletin of the Florida Museum of Natural History (Biological Sciences)* 35(4):209–262.

———. 1994. Language, culture, and environment: Tupí-Guaraní plant names over time. In *Amazonian Indians from prehistory to the present: Anthropological perspectives,* edited by Anna C. Roosevelt, pp. 363–380. Tucson: University of Arizona Press.

Balée, William, Denise P. Schaan, and Rosângela Holanda. 2013. Florestas antrópicas no Acre: Inventário florestal do geoglifo Três Vertentes, Acrelândia, AC. In *Etnobotânica e botânica econômica do Acre,* edited by Amauri Siveiro. Rio Branco, Brazil: EMBRAPA. (in press).

Balick, Michael J. 1984. Ethnobotany of palms in the Neotropics. In *Ethnobotany in the Neotropics,* edited by Ghillean T. Prance and Jackie Kallunkie, pp. 9–23. Bronx: New York Botanical Garden.

Barbosa Rodrigues, João. 1898. *Palmae Mattogrossenses, novae vel minus cognitae.* Rio de Janeiro: Typographia Leuzinger.

———. 1899. *Palmae Novae Paraguayensis.* Rio de Janeiro: Typographia Leuzinger.

———. 1905. Mbaé kaá Tapyeté Enoyndaua ou Botânica da nomenclatura indígena. *3rd Congresso Científico Latinoamericano,* vol. 1. Rio de Janeiro: Impresso Nacional.

Barlow, Jos, Toby A. Gardner, Alexander C. Lees, Luke Parry, and Carlos A. Peres. 2011. How pristine are tropical forests? An ecological perspective on the pre-Columbian human footprint in Amazonia and implications for contemporary conservation. *Biological Conservation,* doi:10.1016/j.biocon.2011.10.013.

Basso, Ellen B. 1977. Introduction: The status of Carib ethnology. In *Carib-speaking Indians: Culture, society, and language,* edited by Ellen B. Basso, pp. 9–22. Anthropological Papers of the University of Arizona, 28. Tucson: University of Arizona Press.

———. 1988 (orig. 1973). *The Kalapalo Indians of Central Brazil.* Prospect Heights, IL: Waveland.
Bayliss-Smith, T., E. Hviding, and T. Whitmore. 2003. Rainforest composition and histories of human disturbance in Solomon Islands. *Ambio* 32(5):346–352.
Beckerman, Stephen. 1979. The abundance of protein in Amazonia: A reply to Gross. *American Anthropologist* 81:533–560.
———. 1980. More on Amazon cultural ecology. *Current Anthropology* 21:540–541.
———. 1983. *Carpe diem:* An optimal foraging approach to Bari fishing and hunting. In *Adaptive responses of native Amazonians,* edited by Raymond B. Hames and William T. Vickers, pp. 269–299. New York: Academic.
Benitez, J.P. 1967. *Formación social de pueblo Paraguayo.* Asunción, Paraguay: Ediciones Nizza.
Berlin, Brent. 1992. *Ethnobiological classification: Principles of categorization of plants and animals in traditional societies.* Princeton, NJ: Princeton University Press.
———. 2006. The first congress of ethnozoological nomenclature. *Journal of the Royal Anthropological Institute* 12:23–44.
Bettelheim, Bruno. 1955. *Symbolic wounds.* London: Thames and Hudson.
Betts, La Vera. 1981. *Dicionário Parintintin–Português, Português–Parintintin.* Brasília: Summer Institute of Linguistics.
Black, Francis L. 1992. Why did they die? *Science* 258:1739–1740.
Bletter, Nathaniel, and Douglas C. Daly. 2006. Cacao and its relatives in South America: An overview of taxonomy, ecology, biogeography, chemistry, and ethnobotany. In *Chocolate in Mesoamerica: A cultural history of cacao,* edited by Cameron L. McNeil, pp. 31–68. Gainesville: University Press of Florida.
Block, David. 1980. In search of El Dorado: Spanish entry into Moxos, a tropical frontier, 1550–1767. PhD diss., University of Texas, Austin.
———. 1994. *Mission culture on the Upper Amazon: Native tradition, Jesuit enterprise, and secular policy in Moxos, 1660–1880.* Lincoln: University of Nebraska Press.
Bloomfield, Leonard. 1984 (orig. 1933). *Language.* Chicago: University of Chicago Press.
Boom, Brian M. 1986. A forest inventory in Amazonian Bolivia. *Biotropica* 18:287–294.
———. 1989. Use of plant resources by the Chácobo. In *Resource management in Amazonia: Indigenous and folk strategies,* edited by Darrell A. Posey and William Balée, pp. 78–96. Bronx: New York Botanical Garden.
Borgatti, Stephen P. 1992. *ANTHROPAC 4.983/x.* Columbia, SC: Analytic Technologies.
———. 1999. Enhanced elicitation techniques for cultural domain analysis. In *Enhanced ethnographic methods. Audiovisual techniques, focused group interviews, and elicitation techniques,* edited by J. J. Schensul, M. D. LeCompte, B. K. Nastasi, and Stephen P. Borgatti, pp. 115–151. Walnut Creek, CA: AltaMira.
Boudin, Max H. 1978. *Dicionário de Tupi Moderno: Dialeto Tembé-Ténêtéhar do Alto Rio Gurupi.* 2 vols. São Paulo, Brazil: Conselho Estadual de Artes e Ciências Humanas.
Braga, P. 1979. Subdivisão fitogeográfica, tipos de vegetação, conservação e inventário florístico da floresta amazônica. *Acta Amazônica* 9(4):53–80.

Brochado, José P. 1977. *Alimentação na floresta tropical*. Caderno no. 2. Porto Alegre, Brazil: Universidade Federal do Rio Grande do Sul.

Brown, Cecil H. 1985. Mode of subsistence and folk biological taxonomy. *Current Anthropology* 26:43–62.

———. 1992. British names for American birds. *Journal of Linguistic Anthropology* 2(1):30–50.

Brown, S., and A. E. Lugo. 1990. Tropical secondary forests. *Journal of Tropical Ecology* 6:1–32.

Bruno, E. S. 1966. *Amazônia*. Vol. 1 of *História do Brasil: Geral e regional*. São Paulo, Brazil: Editora Cultrix.

Brush, Stephen B. 1993. Indigenous knowledge of biological resources and intellectual property rights: The role of anthropology. *American Anthropologist* 95:653–671.

Burkill, I. H. 1966. *A dictionary of the economic products of the Malay Peninsula*. 2 vols. Kuala Lumpur: Ministry of Agriculture and Co-operatives.

Bush, Mark B., R. P. Dolores, and P. A. Colinvaux. 1989. A 6,000 year history of Amazonian maize cultivation. *Nature* 340:303–305.

Bush, Mark B., M. R. Silman, M. B. de Toledo, C. Listopad, W. D. Gosling, C. Williams, P. E. de Oliveira, and C. Krisel. 2007. Holocene fire and occupation in Amazonia: Records from two lake districts. *Philosophical Transactions of the Royal Society, B* 362:209–218.

Campbell, Allan T. 1989. *To square with genesis: Causal statements and shamanic ideas in Wayãpi*. Iowa City: University of Iowa Press.

———. 1995. *Getting to know Wai Wai: An Amazonian ethnography*. New York: Routledge.

Campbell, David G., Douglas C. Daly, Ghillean T. Prance, and Ubirajara N. Maciel. 1986. Quantitative ecological inventory of terra firme and várzea tropical forest on the Rio Xingu, Brazilian Amazon. *Brittonia* 38(4):369–393.

Campbell, Lyle. 1999. *Historical linguistics: An introduction*. Cambridge, MA: MIT Press.

Campbell, Lyle, and Terrence Kaufman. 1976. A linguistic look at the Olmecs. *American Antiquity* 41(1):80–89.

Cardús, J. 1886. *Las misiones Franciscanas entre los infieles de Bolivia*. Barcelona.

Carneiro, Robert L. 1978. The knowledge and use of rain forest trees by the Kuikuru Indians of central Brazil. In *The nature and status of ethnobotany*, edited by Richard Ford, pp. 210–216. Ann Arbor: University of Michigan Press.

———. 1983. The cultivation of manioc among the Kuikuru of the upper Xingu. In *Adaptive responses of native Amazonians*, edited by Raymond B. Hames and William T. Vickers, pp. 65–111. New York: Academic.

———. 1985. Hunting and hunting magic among the Amahuaca of the Peruvian Montaña. In *Native South Americans: Ethnology of the least known continent*, edited by Patricia J. Lyon, pp. 122–132. Prospect Heights, IL: Waveland.

———. 1995. The history of ecological interpretations of Amazonia: Does Roosevelt have it right? In *Indigenous peoples and the future of Amazonia: An ecological anthropology of an endangered world*, edited by Leslie E. Sponsel, pp. 45–70. Tucson: University of Arizona Press.

Cavalcante, Paulo B. 1988. *Frutas comestíveis da Amazônia.* 4th ed. Belém, Brazil: MCT/CNPq, Museu Paraense Emílio Goeldi.

Chagnon, Napolen A. 1977. *Yanomamö: The fierce people.* 2nd ed. New York: Holt, Rinehart and Winston.

Chagnon, Napolen A., and Raymond B. Hames. 1979. Protein deficiency and tribal warfare in Amazonia: New data. *Science* 20(3):910–913.

Chambers, J. Q., N. Higuchi, and J. P. Schimel. 1998. Ancient trees in Amazonia. *Nature* 391:135–136.

Charnov, Eric. 1976. Optimal foraging: The marginal value theorem. *Theoretical Plant Biology* 9:129–136.

Charnov, Eric, Gordon H. Orians, and Kim Hyatt. 1976. Ecological implications of resource depression. *American Naturalist* 110:247–259.

Chávez Suárez, J. 1986. *Historia de Moxos.* 2nd ed. La Paz, Bolivia: Editorial Don Bosco.

Cheesman, E. E. 1944. Notes on the nomenclature, classification and possible relationships of cacao populations. *Tropical Agriculture* 21(8):144–159.

Chernela, Janet M. 1982. An indigenous system of forest and fisheries management in the Uaupes Basin of Brazil. *Cultural Survival Quarterly* 6(2):17–18.

———. 1986. Os cultivares de mandioca (tucano). In *Etnobiologia.* Vol. 1 of *SUMA: Etnologia Brasileira,* edited by Berta Ribeiro, pp. 151–158. Rio de Janeiro: Vozes.

———. 1989. Managing rivers of hunger: The Tukano of Brazil. In *Resource management in Amazonia: Indigenous and folk strategies,* edited by Darrell A. Posey and William Balée, pp. 238–248. Bronx: New York Botanical Garden.

———. 1993. *The Wanano Indians of the Brazilian Amazon: A sense of space.* Austin: University of Texas Press.

Childe, V. Gordon. 1956. *A short introduction to archaeology.* London: Frederick Muller.

Chmyz, I., and Z. C. Sauner. 1971. Nota prévia sobre as pesquisas arqueológicas no Vale do Rio Piquiri. *Dédalo* 7(13):7–35.

Clastres, Hélène. 1995. *The land-without-evil.* Urbana: University of Illinois Press.

Clastres, Pierre. 1968. Ethnographie des Indiens Guayaki (Paraguay-Brésil). *Journal de la Société des Americanistes* 57:9–61.

———. 1972. The Guayaki. In *Hunters and gatherers today,* edited by M. G. Bicchieri, pp. 138–174. New York: Holt, Rinehart, and Winston.

———. 1973. Eléments de démographie amérindienne. *L'Homme* 13(1–2):23–36.

———. 1989. *Society against the state.* Trans. R. Hurley and A. Stein. New York: Zone.

Cleary, David. 2001. Towards an environmental history of the Amazon: From prehistory to the nineteenth century. *Latin American Research Review* 36(2):65–96.

Clement, Charles R. 1989. A center of crop genetic diversity in western Amazonia. *Bioscience* 39(2):624–631.

———. 1999a. 1492 and the loss of Amazonian crop genetic resources: I. The relation between domestication and human population decline. *Economic Botany* 53:188–199.

———. 1999b. 1492 and the loss of Amazonian crop genetic resources: II. Crop biogeography at contact. *Economic Botany* 53:203–216.

Clement, Charles R., M. de Cristo-Araújo, G. C. d'Eeckenbrugge, A. Alves Pereira, and

D. Picanço-Rodrigues. 2010. Origin and domestication of native Amazonian crops. *Diversity* 2:72–106.

Clement, Charles R., and André B. Junqueira. 2010. Between a pristine myth and an impoverished future. *Biotropica* 42(5):534–536.

Clements, Frederic E. 1916. *Plant succession: An analysis of the development of vegetation.* Washington, DC: Carnegie Institution of Washington.

Cochrane, T., and P. Sanchez. 1982. Land resources, soil properties, and their management in the Amazon region: A state of knowledge report. In *Amazonia: Agriculture and land use research,* edited by Susanna Hecht, pp. 137–209. Cali, Colombia: CIAT.

Coe, Sophie D., and Michael D. Coe. 1996. *The true history of chocolate.* New York: Thames and Hudson.

Colson, A. B. 1976. Binary oppositions and the treatment of sickness among the Akawaio. In *Social anthropology and medicine,* edited by J. B. Louden, pp. 422–499. London: Academic.

Conklin, Harold C. 1954. The relation of Hanunóo culture to the plant world. PhD diss., Yale University, New Haven, CT.

Cook, N. D., and W. G. Lovell, eds. 1992. *Secret judgments of God: Old World disease in colonial Spanish America.* Norman: University of Oklahoma Press.

Coomes, Oliver T., C. Abizaid, and M. Lapointe. 2009. Human modification of a large meandering Amazonian river: Genesis, ecological and economic consequences of the Masiseh cutoff on the Central Ucayali, Peru. *Ambio* 38:130–134.

Cooper, John M. 1949. Fire making. In *The comparative ethnology of South American Indians.* Vol. 5 of *Handbook of South American Indians,* edited by Julian H. Steward, pp. 283–292. Bureau of American Ethnology, Bulletin 143. Washington, DC: US Government Printing Office.

Coppens, W. 1975. Contribución al estudio de las actividades de subsistencia de los Hotis del Río Kaima. *Boletin Indigenista Venezolano* 16(12):65–78.

Coppens, W., and P. Mitrani. 1974. Les Indiens Hoti. *L'Homme* 14(3–4):131–142.

Coppinger, R. P., and C. K. Smith. 1983. The domestication of evolution. *Environmental Conservation* 10:283–292.

Cormier, Loretta A. 2000. The ethnoprimatology of the Guajá Indians of Maranhão, Brazil. PhD diss., Tulane University, New Orleans.

———. 2006. Between the ship and the bulldozer: Historical ecology of Guajá subsistence, sociality, and symbolism. In *Time and complexity in historical ecology: Studies in the Neotropical lowlands,* edited by William Balée and Clark L. Erickson, pp. 341–363. New York: Columbia University Press.

Cornford, F. M. 1937. *Plato's cosmology.* New York: Harcourt, Brace.

———. 1950. *Plato's Timaeus.* Trans. F. M. Cornford. Indianapolis: Liberal Arts Press.

Corrêa, C. 1985. Fases ceramistas não-sambaqueiras do litoral do Pará. Master's thesis, Universidade Federal do Pernambuco, Recife.

Corrêa da Silva, Beatriz C. 1997. Urubú-Ka'apor: Da gramática à história: A trajetória de um povo. Master's thesis, Universidade de Brasília, Brasília.

———. 2001. A codificação dos argumentos em Ka'apor: Sincronia e diacroniá. In *Línguas indígenas Brasileiras: Fonologia, gramática e história,* edited by A. S. A. C. Cabral and Aryon D. Rodrigues, pp. 343–351. Atas do I Encontro Internacional do Grupo de Trabalho sobre Línguas Indígenas ANPOLL. Belém, Brazil: EDUFPA.

Cottam, G., and J. T. Curtis. 1956. The use of distance measurement in phytosociological sampling. *Ecology* 37:451–460.

Crofts, M., and M. Sheffler. 1981. *Dicionário bilíngue em Português e Mundurukú, Soat a'õ dup wuya'õm pariwat a'om tak.* 2nd ed. Brasília: Summer Institute of Linguistics.

Crumley, Carole L. 1994. Historical ecology: A multidimensional ecological orientation. In *Historical ecology: Cultural knowledge and changing landscapes,* edited by Carole L. Crumley, pp. 1–16. Santa Fe, NM: School of American Research Press.

———. 1998. Foreword. In *Advances in historical ecology,* edited by William Balée, pp. ix–xvi. New York: Columbia University Press.

———. 2006. Historical ecology: Integrated thinking at multiple temporal and spatial scales. In *The world system and the earth system: Global socio-environmental change and sustainability since the Neolithic,* edited by Alf Hornborg and Carole L. Crumley, pp. 15–28. Walnut Creek, CA: Left Coast.

Cruz, Ernesto. 1973. *História do Pará,* vol. 1. Belém, Brazil: Governo do Estado do Pará.

Crystal, D. 1987. *The Cambridge encyclopedia of language.* New York: Cambridge University Press.

Cuatrecasas, José. 1964. Cacao and its allies: A taxonomic revision of the genus *Theobroma. Contributions from the United States National Herbarium* 35(6):378–614. Washington, DC: Smithsonian Institution.

Culbert, T. P. 1988. The collapse of Classic Maya civilization. In *The collapse of ancient states and civilizations,* edited by N. Yoffee and G. Cowgill, pp. 60–101. Tucson: University of Arizona Press.

Curtis, J. T., and G. Cottam. 1962. *Plant ecology workbook.* Minneapolis, Burgess.

d'Évreux, Yves. 1864. *Voyage dans le nord du Brésil fait durant les années 1613 et 1614.* Leipzig, Germany: Librairie A. Franck.

d'Orbigny, A. 1946. *Descripción geográfica, histórica y estadística de Bolivia.* La Paz, Bolivia.

Dakin, K., and S. Wichmann. 2000. Cacao and chocolate (A Uto-Aztecan perspective). *Ancient Mesoamerica* 11:55–75.

Daly, Douglas C., and Ghillean T. Prance. 1989. Brazilian Amazon. In *Floristic inventory of tropical countries: The status of plant systematics, collections, and vegetation plus recommendations for the future,* edited by David G. Campbell and H. D. Hammond, pp. 401–425. Bronx: New York Botanical Garden.

Davies, K. G. 2001. What makes genetically modified organisms so distasteful? *Trends in Biotechnology* 19:424–427.

Deacon, Terrence W. 1997. *The symbolic species: The co-evolution of language and the brain.* New York: W. W. Norton.

Dean, Rebecca M. 2010. The importance of anthropogenic environments. In *The archaeology of anthropogenic environments,* edited by Rebecca M. Dean, pp. 3–14. Occasional Paper 37, Center for Archaeological Investigations. Carbondale: Southern Illinois University.

Dean, Warren. 1984. Indigenous populations of the São Paulo–Rio de Janeiro coast: Trade, aldeamento, slavery and extinction. *Revista de História,* n.s., 117:3–26.
———. 1995. *With broadax and firebrand: The destruction of the Brazilian Atlantic forest.* Berkeley: University of California Press.
Denevan, William M. 1966. *The aboriginal cultural geography of the Llanos de Mojos of Bolivia.* Berkeley: University of California Press.
———. 1976. The aboriginal population of Amazonia. In *The native population of the Americas in 1492,* edited by William M. Denevan, pp. 205–235. Madison: University of Wisconsin Press.
———. 1992. The pristine myth: The landscape of the Americas in 1492. *Annals of the Association of American Geographers* 82(3):369–385.
———. 2001. *Cultivated landscapes of native Amazonia and the Andes.* New York: Oxford University Press.
———. 2006. Pre-European forest cultivation in Amazonia. In *Time and complexity in historical ecology: Studies in the Neotropical lowlands,* edited by William Balée and Clark L. Erickson. New York: Columbia University Press.
Denevan, William M., and Christine Padoch, eds. 1988. *Swidden fallow agroforestry in the Peruvian Amazon.* Advances in Economic Botany 5. Bronx: New York Botanical Garden.
Denevan, William M., John M. Treacy, Janis B. Alcorn, C. Padoch, J. Denslow, and S. F. Paitan. 1984. Indigenous agroforestry in the Peruvian Amazon. *Interciencia* 9(6):346–357.
Denevan, William M., and Albert Zucchi (1978). Ridged-field excavations in the central Orinoco Llanos, Venezuela. In *Advances in Andean archaeology,* edited by D. L. Browman, pp. 235–245. The Hague: Mouton.
Denslow, Julie S. 1987. Tropical rainforest gaps and tree species diversity. *Annual Review of Ecology and Systematics* 18:431–451.
Descola, Philippe. 1994. *In the society of nature: A native ecology in Amazonia.* Trans. N. Scott. Cambridge: Cambridge University Press.
———. 1996. Constructing natures: Symbolic ecology and social practice. In *Nature and society: Anthropological perspectives,* edited by Philippe Descola and G. Pálsson, pp. 82–102. London: Routledge.
Devillers, C. 1983. What future for the Wayana Indians? *National Geographic Magazine* 163:66–83.
Di Paolo, P. 1985. *Cabanagem: A revolução popular da Amazônia.* Belém, Brazil: CEJUP.
Dickinson, W. R. 2000. Changing times: The Holocene legacy. *Environmental History* 5:483–502.
Dietrich, Wolf. 1986. *El idioma Chiriguano: Gramática, textos, vocabulario.* Madrid: Instituto de Cooperación Iberoamericana.
———. 1990. *More evidence for an internal classification of the Tupí-Guaraní language.* Indiana Beiheft/Suplemento/Supplement 12, pp. 1–136. Berlin: Gebr. Mann Verlag.
Dillehay, Thomas D. 2008. Profiles in Pleistocene history. In *Handbook of South American archaeology,* edited by H. Silverman and W. H. Isbell, pp. 29–43. New York: Springer.

Distel, A. A. F. 1984/1985. Hábitos funerarios de los Sirionó (Oriente de Bolivia). *Acta Praehistorica et Archaeologica* 16:159–182.

Divale, William, and Marvin Harris. 1976. Population, warfare, and the male supremacist complex. *American Anthropologist* 78:521–538.

Dixon, R. M. W. 1980. *The languages of Australia.* New York: Cambridge University Press.

———. 1997. *The rise and fall of languages.* New York: Cambridge University Press.

Dodt, Gustavo. 1939. *Descripção dos rios Paranaíba e Gurupy.* Coleção Brasiliana 138. São Paulo, Brazil: Companhia Editora Nacional.

Dole, Gertrude E. 1961/1962. A preliminary consideration of the prehistory of the upper Xingu Basin. *Revista do Museu Paulista,* n.s., 13:399–432.

———. 1991. The development of kinship in tropical South America. In *Profiles in cultural evolution,* edited by A. T. Rambo and K. Killogly, pp. 373–403. Anthropological Papers, 85, Museum of Anthropology, University of Michigan. Ann Arbor: University of Michigan.

Donkin, R. A. 1985. *The peccary—with observations on the introduction of pigs to the New World.* Philadelphia: American Philosophical Society.

Douglass, P. 1999. Bergson and cinema: Friends or foes? In *The new Bergson,* edited by J. Mullarkey, pp. 209–227. Manchester, UK: Manchester University Press.

Ducke, Adolfo. 1946. Plantas de cultura precolombiana na Amazônia brasileira: Notas sobre as espécies ou formas espontáneas que supostamente lhes teriam dado origem. *Boletim do Instituto Agronômico do Norte* 8:1–24.

Ducke, Adolfo, and G. A. Black. 1953. Phytogeographical notes on the Brazilian Amazon. *Anais da Academia Brasileira de Ciências* 25(1):1–46.

Eden, Michael J., Warwick Bray, Leonor Herrera, and Colin McEwan. 1984. *Terra preta* soils and their archaeological context in the Caquetá Basin of southeast Colombia. *American Antiquity* 49(1):124–140.

Eisenberg, J., M. A. O'Connel, and P. August. 1984. Density, productivity, and distribution of mammals in two Venezuelan habitats. In *Vertebrate ecology in the northern Neotropics,* edited by John Eisenberg, pp. 187–207. Washington, DC: Smithsonian Institution Press.

Emmons, Louise H. 1984. Geographic variation in densities and diversities of non-flying mammals in Amazonia. *Biotropica* 16(3):210–222.

Erickson, Clark L. 1995. Archaeological methods for the study of ancient landscapes of the Llanos de Mojos in the Bolivian Amazon. In *Archaeology in the lowland American tropics,* edited by Peter W. Stahl, pp. 66–95. Cambridge: Cambridge University Press.

———. 2000a. An artificial landscape-scale fishery in the Bolivian Amazon. *Nature* 408: 190–193.

———. 2000b. Lomas de ocupación en los Llanos de Moxos. In *Arqueologia de las Tierras Bajas,* edited by Alicia Durán Coirolo and Roberto Bracco Boksar, pp. 207–226. Montevideo, Uruguay: Min. de Educación y Cultura, Comisión Nacional de Arqueologia.

———. 2003. Historical ecology and future explorations. In *Amazonian dark earths: Wim*

Sombroek's vision, edited by W. I. Woods, W. G. Teixeira, J. Lehmann, C. Steiner, A. M. G. A. WinklerPrins, and L. Rebellato, pp. 455–500. New York: Springer.

———. 2006. The domesticated landscapes of the Bolivian Amazon. In *Time and complexity in historical ecology: Studies in the Neotropical lowlands,* edited by William Balée and Clark L. Erickson, pp. 235–278. New York: Columbia University Press.

———. 2008. Amazonia: The historical ecology of a domesticated landscape. In *Handbook of South American archaeology,* edited by H. Silverman and W. H. Isbell, pp. 157–183. New York: Springer.

Erickson, Clark L., and William Balée. 2006. The historical ecology of a complex landscape in Bolivia. In *Time and complexity in historical ecology: Studies in the Neotropical lowlands,* edited by William Balée and Clark L. Erickson, pp. 187–233. New York: Columbia University Press.

Eriksen, Love. 2011. *Nature and culture in prehistoric Amazonia: Using G.I.S. to reconstruct ancient ethnogenetic processes from archaeology, linguistics, geography, and ethnohistory.* Lund Studies in Human Ecology 12. Lund, Sweden: Lund University.

Evans, Clifford, and Betty J. Meggers. 1960. *Archaeological investigations in British Guiana.* Washington, DC: Smithsonian Institution.

Ewel, J. J. 1986. Designing agricultural ecosystems for the humid tropics. *Annual Review of Ecology and Systematics* 17:245–271.

Fairhead, James, and Melissa Leach. 1996. *Misreading the African landscape: Society and ecology in a forest-savanna mosaic.* Cambridge: Cambridge University Press.

Falesi, Ítalo C. 1972. *Solos da rodovia transamazônica.* Boletim Técnico, 55. Belém, Brazil: IPEAN.

Fanshawe, D. B. 1954. Forest types of British Guiana. *Caribbean Forester* 15(3):73–111.

Farabee, William C. 1967. *The Central Caribs.* Oosterhout N.B., The Netherlands: Anthropological Publications. (orig. pub. 1924, University Museum, University of Pennsylvania, Anthropological Publications 10.)

Fausto, Carlos. 1997. A dialética da predação e familiarização entre os Parakanã da Amazônia Oriental: Por uma teoria da guerra ameríndia. PhD diss., Museu Nacional, Rio de Janeiro.

Fischer, G. R. 1987. *Manejo sustentado de florestas nativas.* Joinville, Brazil: G. R. Fischer.

Fisher, William. 2000. *Rainforest exchanges: Industry and community on an Amazonian frontier.* Washington, DC: Smithsonian Institution Press.

Fittkau, G., and H. Klinge. 1973. On biomass and trophic structure of the central Amazon rain forest ecosystem. *Biotropica* 5:1–14.

Fleck, David W., and John D. Harder. 2000. Matses Indian rainforest habitat classification and mammalian diversity in Amazonian Peru. *Journal of Ethnobiology* 20(1):1–36.

Foley, Robert. 1987. *Another unique species: Patterns in human evolutionary ecology.* Hong Kong: Longman Group.

Fraser, James, T. Cardoso, André Junqueira, N. P. S. Falcão, and Charles R. Clement. 2009. Historical ecology and dark earths in whitewater and blackwater landscapes: Comparing the middle Madeira and lower Negro rivers. In *Amazonian dark earths: Wim*

Sombroek's vision, edited by W. I. Woods, W. G. Teixeira, J. Lehmann, C. Steiner, A. M. G. A. WinklerPrins, and L. Rebellato, pp. 229–264. New York: Springer.

Frechione, John, Darrell A. Posey, and Luiz Francelino da Silva. 1989. The perception of ecological zones and natural resources in the Brazilian Amazon: An ethnoecology of Lake Coari. In *Resource management in Amazonia: Indigenous and folk strategies,* edited by Darrell A. Posey and William Balée, pp. 260–282. Bronx: New York Botanical Garden.

Friedrich, Paul. 1970. *Proto-Indo-European trees: The arboreal system of a prehistoric people.* Chicago: University of Chicago Press.

Frikel, Protásio. 1959. Agricultura dos índios Munduruku. *Boletim do Museu Paraense Emílio Goeldi,* n.s., Antropologia, 4:1–35.

———. 1968. Os Xikrin. *Publicações Avulsas do Museu Goeldi* 7:3–119.

Fróis, Ricardo L. 1953. Estudo sôbre a Amazônia maranhense e seus limites florísticos. *Revista Brasileira de Geografia* 15(1):96–100.

Galvão, Eduardo. 1979. *Encontro de sociedades: Índios e brancos no Brasil.* Rio de Janeiro: Paz e Terra.

Gardner, T. A., J. Barlow, R. Chazdon, K. M. Ewers, C. A. Harvey, C. A. Perls, and N. S. Sodhi. 2009. Prospects for tropical forest biodiversity in a human-modified world. *Ecology Letters* 12:561–582.

Gaspar, M. D., P. DeBlasis, S. K. Fish, and P. R. Fish. 2008. Sambaqui (shell mound) societies of coastal Brazil. In *Handbook of South American archaeology,* edited by H. Silverman and W. H. Isbell, pp. 319–335. New York: Springer.

Gellner, Ernest. 1988. *Plough, sword, and book: The structure of human history.* London: Collins Harvill.

Gentry, Alwyn H. 1986. Endemism in tropical vs temperate communities. In *Conservation biology: The science of scarcity and diversity,* edited by M. Soule, pp. 153–181. Sunderland, MA: Sinauer.

———. 1988. Tree species richness of Upper Amazonian forests. *Proceedings of the National Academy of Sciences (Ecology)* 85:156–159.

Giddens, Anthony. 1987. *Social theory and modern sociology.* Stanford, CA: Stanford University Press.

Glacken, C. J. 1967. *Traces on the Rhodian shore: Nature and culture in Western thought from ancient times to the end of the eighteenth century.* Los Angeles: University of California Press.

Glanz, W. 1982. The terrestrial mammal fauna of Barro Colorado Island: Censuses and long-term changes. In *The ecology of a tropical forest: Seasonal rhythms and long term changes,* edited by J. E. G. Leigh, A. S. Rand, and D. M. Windsor, pp. 455–468. Washington, DC: Smithsonian Institution Press.

Glaser, Bruno, and William I. Woods, eds. 2004. *Amazonian dark earths: Explorations in space and time.* Berlin: Springer-Verlag.

Gómez-Pompa, Arturo, J. Salvadore Flores, and M. A. Fernández. 1990. The sacred cacao groves of the Maya. *Latin American Antiquity* 1(3):247–257.

Gómez-Pompa, Arturo, J. Salvadore Flores, and V. Sosa. 1987. The "pet-kot": A man-made tropical forest of the Maya. *Interciencia* 12(1):10–15.

Goodland, Robert, and Howard Irwin. 1975. *Amazon jungle: Green hell to red desert?* New York: Elsevier.

Goody, Jack. 1977. *The domestication of the savage mind.* New York: Cambridge University Press.

Gottsberger, G. 1978. Seed dispersal by fish in the inundated regions of Humaitá, Amazonia. *Biotropica* 10(3):170–183.

Goulding, Michael. 1980. *The fishes and the forest.* Berkeley: University of California Press.

Graffam, G. 1992. Beyond state collapse: Rural history, raised fields, and pastoralism in the South Andes. *American Anthropologist* 94(4):882–904.

Greenberg, Joseph H. 1987. *Language in the Americas.* Stanford, CA: Stanford University Press.

Greig-Smith, P. 1983. *Quantitative plant ecology.* Berkeley: University of California Press.

Grenand, Françoise. 1989. *Dictionnaire Wayãpi–Français.* Paris: Peeters/SELAF.

———. 1995. Le voyage des mots—Logique de la nomination des plantes: Exemples dans des langues Tupi du Brésil. *Revue d'Ethnolinguistique (Cahiers du LACITO)* 7:23–42.

Grenand, Pierre. 1980. *Introduction à l'étude de l'univers Wayãpi: Ethnoécologie des indiens du Haut-Oyapock (Guyane Française).* Langues et Civilisations a Traditional Orale 40. Paris: SELAF.

———. 1982. *Ainsi parlaient nos ancêtres: Essai d'ethnohistoire "Wayãpi."* Travaux et documents de l'ORSTOM, 148. Paris: ORSTOM.

Gross, Daniel R. 1975. Protein capture and cultural development in the Amazon Basin. *American Anthropologist* 77:526–549.

———. 1979. A new approach to central Brazilian social organization. In *Brazil: Anthropological perspectives. Essays in honor of Charles Wagley,* edited by Maxine Margolis and William Carter, pp. 321–342. New York: Columbia University Press.

———. 1982. Proteina y cultura en la Amazonia: Una segunda revisión. *Amazonia Peruana* 3(6):127–144.

Guzmán, D. de A. 2009. Mixed Indians, caboclos and curibocas: Historical analysis of a process of miscegenation; Rio Negro (Brazil), 18th and 19th centuries. In *Amazon peasant societies in a changing environment: Political ecology, invisibility and modernity in the rainforest,* edited by Cristina Adams, Rui Murrieta, Walter Neves, and Mark Harris, pp. 55–68. New York: Springer.

Haas, Mary R. 1969. *The prehistory of languages.* The Hague: Mouton.

Hall, M. 2005. *Earth repair: A transatlantic history of environmental restoration.* Charlottesville: University of Virginia Press.

Hames, Raymond B. 1980. Game depletion and hunting zone rotation among the Ye'kwana and Yanomamo of Amazonas, Venezuela. In *Working papers on South American Indians,* no. 2, edited by Raymond B. Hames, pp. 31–66. Bennington, VT: Bennington College.

———. 1983. The settlement pattern of a Yanomami population bloc. In *Adaptive responses of native Amazonians,* edited by Raymond B. Hames and William T. Vickers, pp. 393–427. New York: Academic.

Hames, Raymond B., and William T. Vickers. 1983. Introduction. In *Adaptive responses of native Amazonians* edited by Raymond B. Hames and William T. Vickers, pp. 1–26. New York: Academic.

Harlan, Jack R. 1992. *Crops and man.* 2nd ed. Madison, WI: American Society of Agronomy, Crop Science Society of America.

Harris, Marvin. 1974. *Cows, pigs, wars and witches: The riddles of culture.* New York: Random House.

———. 1984. A cultural materialist theory of band and village warfare. In *Warfare, culture, and environment,* edited by R. Brian Ferguson, pp. 111–140. Orlando, FL: Academic.

Hays, T. E. 1982. Utilitarian/adaptationist explanations of folk biological classification: Some cautionary notes. *Journal of Ethnobiology* 2(1):89–94.

Headland, Thomas N. 1983. An ethnobotanical anomaly: The dearth of binomial specifics in a folk taxonomy of a Negrito hunter-gatherer society in the Philippines. *Journal of Ethnobiology* 3(2):109–120.

———. 1987. The wild yam question: How well could independent hunter-gatherers live in a tropical rain forest ecosystem? *Human Ecology* 15(4):463–491.

Headland, Thomas N., and L. A. Reid. 1989. Hunter-gatherers and their neighbors from prehistory to the present. *Current Anthropology* 30:43–66.

Hecht, Susanna B. 2009. Kayapó savanna management: Fire, soils, and forest islands in a threatened biome. *Amazonian dark earths: Wim Sombroek's vision,* edited by W. I. Woods, W. G. Teixeira, J. Lehmann, C. Steiner, A. M. G. A. WinklerPrins, and L. Rebellato, pp. 143–162. New York: Springer.

Hecht, Susanna B., and Darrell A. Posey. 1989. Preliminary results on soil management techniques of the Kayapó Indians. In *Resource management in Amazonia: Indigenous and folk strategies,* edited by Darrell A. Posey and William Balée, pp. 174–188. Bronx: New York Botanical Garden.

Heckenberger, Michael J. 2005. *The ecology of power: Culture, place, and personhood in the southern Amazon, AD 1000–2000.* New York: Routledge.

———. 2009. Biocultural diversity in the southern Amazon. *Diversity* 2(1):1–16.

Heckenberger, Michael J., and Eduardo Góes Neves. 2009. Amazonian archaeology. *Annual Review of Anthropology* 38:251–266.

Heckenberger, Michael J., A. Kuikuro, U. T. Kuikuro, J. C. Russell, M. Schmidt, C. Fausto, C. Franchetto, and B. Franchetto. 2003. Amazonia 1492: Pristine forest or cultural parkland? *Science* 301:1710–1713.

Heckenberger, Michael J., J. C. Russell, C. Fausto, J. R. Toney, M. J. Schmidt, E. Pereira, B. Franchetto, and A. Kuikuro. 2008. A pre-Columbian urbanism, anthropogenic landscapes and the future of the Amazon. *Science* 321:1214–1217.

Heinen, H. Dieter, and Kenneth Ruddle. 1974. Ecology, ritual, and economic organization in the distribution of palm starch among the Warao of the Orinoco Delta. *Journal of Anthropological Research* 30(2):116–138.

Heinsdijk, D. 1957. *Forest inventory in the Amazon Valley (region between the Rio Tapajós and Rio Xingu)*. Rome: Food and Agriculture Organization.

Hemming, John. 1987. *Amazon frontier: The defeat of the Brazilian Indians*. London: MacMillan.

Henderson, Andrew. 1995. *The palms of the Amazon*. New York: Oxford University Press.

Henshilwood, C. S., and C. W. Marean. 2003. The origin of modern human behavior. *Current Anthropology* 44:627–651.

Hilbert, P. P. 1955. *A cerâmica arqueológica da região de Oriximiná*. Instituto de Antropologia e Etnologia do Pará, Publicação 9. Belém, Brazil: Museu Paraense Emílio Goeldi.

Hill, Jonathan, and Emilio F. Moran. 1983. Adaptive strategies of Wakuénai peoples to the oligotrophic rain forest of the Rio Negro basin. In *Adaptive responses of native Amazonians,* edited by Raymond B. Hames and William T. Vickers, pp. 113–135. New York: Academic.

Hill, Kim, and Kristen Hawkes. 1983. Neotropical hunting among the Aché of eastern Paraguay. In *Adaptive responses of native Amazonians,* edited by Raymond B. Hames and William T. Vickers, pp. 139–188. New York: Academic.

Hill, Kim, Kristen Hawkes, Magdalena Hurtado, and Hillary Kaplan. 1984. Seasonal variance in the diet of Aché hunter-gatherers of eastern Paraguay. *Human Ecology* 12(2): 101–136.

Hogue, C. L. 1993. *Latin American insects and entomology*. Berkeley: University of California Press.

Holanda Ferreira, A. B. de. n.d. *Novo dicionário da língua portuguesa*. 14th ed. Rio de Janeiro: Editora Nova Fronteira.

Holmberg, Allan R. 1948. The Sirionó. In *The tropical forest tribes*. Vol. 3 of *Handbook of South American Indians,* edited by Julian H. Steward, pp. 455–463. Bureau of American Ethnology, Bulletin 143. Washington, DC: US Government Printing Office.

———. 1969 (orig. 1950). *Nomads of the long bow*. Garden City, NJ: The Natural History Press.

———. 1985 (orig. 1950). *Nomads of the long bow*. Prospect Heights, IL: Waveland.

Holt, H. B. 1983. A cultural resource management dilemma: Anasazi ruins and the Navajos. *American Antiquity* 48(3):594–599.

Hooper, Elaine R. 2008. Factors affecting the species richness and composition of neotropical secondary succession: A case study of abandoned agricultural land in Panama. In *Post-Agricultural succession in the Neotropics,* edited by R. W. Myster, pp. 141–164. New York: Springer.

Horn Fitz Gibbon, F. von. 1955. *Breves notas sobre la lengua de los Indios Pausernas*. Publicaciones de la Sociedad de Estudios Geográficos y Históricos. Santa Cruz, Bolivia: Imprenta "Emília."

Hornborg, Alf. 2005. Ethnogenesis, regional integration, and ecology in prehistoric Amazonia. *Current Anthropology* 46(4):589–620.

Huber, Jacques. 1900. *Arboretum amazonicum*. Belém, Brazil: Museu Paraense de História Natural e Etnographia.

———. 1904. Materiaes para a Flora Amazônica: Notas sobre a patria e distribuição geographica das arvores fructiferas do Pará. *Boletim do Museu Goeldi (Museu Paraense)* 4:392–395.

———. 1909. Mattas e madeiras amazônicas. *Boletim do Museu Goeldi (Museu Paraense) de História Natural e Ethnographia* 6:91–225.

Huber, O., Julian A. Steyermark, Ghillean T. Prance, and C. Alés. 1984. The vegetation of the Sierra Parima, Venezuela-Brazil: Some results of recent exploration. *Brittonia* 36(2):104–139.

Hughes, J. D. 2001. *An environmental history of the world: Humankind's changing role in the community of life.* London: Routledge.

Hurst, W. J. 2001. *The identification of cacao residue in samples of archaeological interest.* Paper read at 101st Annual Meeting of the American Anthropological Association, 24 November, New Orleans.

Huston, Michael A. 1994. *Biological diversity: The coexistence of species on changing landscapes.* Cambridge: Cambridge University Press.

Huxley, Francis. 1956. *Affable savages: An anthropologist among the Urubu Indians of Brazil.* New York: Viking.

Im Thurn, E. F. 1967 (orig. 1883). *Among the Indians of Guiana, Being sketches chiefly anthropologic from the interior of British Guiana.* New York: Dover.

Irvine, Dominique. 1989. Succession management and resource distribution in an Amazonian rain forest. In *Resource management in Amazonia: Indigenous and folk strategies,* edited by Darrell A. Posey and William Balée, pp. 223–237. Bronx: New York Botanical Garden.

Isaac, B. 1977. The Siriono of eastern Bolivia: A reexamination. *Human Ecology* 5(2):137–154.

Isbell, William H. 2008. Conclusion. In *Handbook of South American archaeology,* edited by H. Silverman and W. H. Isbell, pp. 1137–1158. New York: Springer.

Jackson, Jean. 1983. *The fish people: Linguistic exogamy and Tukanoan identity in northwest Amazonia.* Cambridge: Cambridge University Press.

Jennings, D. L. 1976. Cassava. In *Evolution of crop plants,* edited by Norman W. Simmonds, pp. 81–84. London: Longman.

Jensen, Cheryl. 1990. *O desenvolvimento histórico da língua Wayampí.* Campinas, Brazil: Editora UNICAMP.

———. 1999. Tupí-Guaraní. In *The Amazonian languages,* edited by R. M. W. Dixon and A. Y. Aikhenvald, pp. 125–163. Cambridge: Cambridge University Press.

Johnson, Allen. 1989. How the Machiguengua manage resources: Conservation or exploitation of nature? In *Resource management in Amazonia: Indigenous and folk strategies,* edited by Darrell A. Posey and William Balée, pp. 213–222. Bronx: New York Botanical Garden.

Jones, W. O. 1959. *Manioc in Africa.* Stanford, CA: Stanford University Press.

Junqueira, André B., Glenn H. Shepard Jr., and Charles R. Clement. 2010. Secondary forests on anthropogenic soils in Brazilian Amazonia conserve agrobiodiversity. *Biodiversity and Conservation* 19(7):1933–1961.

Justeson, J. S., W. N. Norman, L. Campbell, and T. Kaufman. 1985. *The foreign impact on*

Lowland Mayan languages and script. Middle American Research Institute, Publication 53. New Orleans: Tulane University.

Kaplan, Hillary, and Kim Hill. 1985. Food sharing among Aché foragers: Tests of explanatory hypotheses. *Current Anthropology* 26(2):223–246.

Kates, R. W., B. L. Turner II, and W. C. Clark. 1990. The great transformation. In *The earth as transformed by human action: Global and regional changes in the biosphere over the past 300 years,* edited by B. L. Turner II, pp. 1–17. New York: Cambridge University Press, with Clark University.

Kaufman, Terrence. 1990. Language history in Amazonia: What we know and how to know more. In *Amazonian linguistics: Studies in lowland South American languages,* edited by Doris L. Payne, pp. 13–67. Austin: University of Texas Press.

Kawa, Nicholas C., and Augusto Oyuela-Caycedo. 2008. Amazon dark earth: A model of sustainable agriculture of the past and future? *The International Journal of Environmental, Cultural, Economic, and Social Sustainability* 4(3):9–16.

Kensinger, Kenneth M. 1995. *How real people ought to live: The Cashinahua of eastern Peru.* Prospect Heights, IL: Waveland.

Kidder, Tristram R., and Gayle J. Fritz. 1993. Subsistence and social change in the Lower Mississippi Valley: The Reno Brake and Osceola sites, Louisiana. *Journal of Field Archaeology* 20(3):281–297.

Kiltie, R. A. 1980. More on Amazon cultural ecology. *Current Anthropology* 21(4):541–544.

Kipnis, Renato. 1990. *Comparative ethnoecology in eastern Amazonia.* Supplementary report to the Jessie Smith Noyes Foundation, William Balée, P.I. On file, Comparative Ethnoecology Project, Museu Goeldi, Belém, Brazil.

Kitamura, P. C., and C. H. Müller. 1984. Castanhais nativas de Marabá-PA: Fatores de depredação e bases para a sua preservação. *Documentos* 30:5–32. EMBRAPA/CPATU, Belém, Brazil.

Kloos, Peter. 1977. The Akuriyo way of death. In *Carib-speaking Indians: Culture, society, and language,* edited by Ellen B. Basso, pp. 114–122. Anthropological Papers of the University of Arizona, 28. Tucson: University of Arizona Press.

Knight, Chris. 1991. *Blood relations: Menstruation and the origins of culture.* New Haven, CT: Yale University Press.

Kohn, Edward. 2002. Natural engagements and ecological aesthetic among the Ávila Runa of Amazonian Ecuador. PhD diss., University of Wisconsin-Madison.

Kolata, Alan. 1987. Tiwanaku and its hinterland. *Archaeology* 40(1):36–41.

———. 1993. *The Tiwanaku: Portrait of an Andean civilization.* Cambridge, MA: Blackwell.

Kozák, V., D. Baxter, L. Williamson, and R. L. Carneiro. 1979. *The Héta Indians: Fish in a dry pond.* Anthropological Papers of the American Museum of Natural History, vol. 55, pt. 6. New York: American Museum of Natural History.

Krause, F. 1911. *In den Wildnissen Brasilien: Bericht und Ergebrisse der Leipziger Araguaya-Expedition.* Leipzig, Germany.

Kreike, Emmanuel. 2003. Hidden fruits: A social ecology of fruit trees in Manibia and

Angola, 1880s–1990s. In *Social history and African environments,* edited by William Beinart and Joann McGregor, pp. 27–42. Athens: Ohio University Press.

Kubitzki, Klaus. 1985. The dispersal of forest plants. In *Key environments: Amazonia,* edited by Ghillean T. Prance and Thomas Lovejoy, pp. 192–206. New York: Pergamon.

Langstroth, Roberto P. 1996. Forest islands in an Amazonian savanna of northeastern Bolivia. PhD diss., University of Wisconsin, Madison.

Lathrap, Donald. 1968. The "hunting" economies of the tropical forest zone of South America: An attempt at historical perspective. In *Man the hunter,* edited by Richard B. Lee and Irven DeVore, pp. 23–29. Chicago: Aldine.

———. 1970. *The Upper Amazon.* London: Thames and Hudson.

———. 1977. Our father the cayman, our mother the gourd: Spinden revisited, or a unitary model for the emergence of agriculture in the New World. In *The origins of agriculture,* edited by C. A. Reed, pp. 713–752. The Hague: Mouton.

Latinis, D. Kyle. 2000. The development of subsistence models for Island Southeast Asia and Near Oceania: The nature and role of arboriculture and arboreal-based economies. *World Archaeology* 32(1):41–67.

Lee, Kenneth. 1979. 7,000 años de historia del hombre de Mojos: Agricultura en pampas estériles: Informe preliminar. Revista "Panorama Universitario" no. 1, *Universidad Beni,* pp. 23–26.

Lee, Richard B. 1993. *The Dobe Ju/'hoansi.* Fort Worth, TX: Harcourt Brace College.

Lee, Richard B., and Irven DeVore. 1968. Problems in the study of hunters and gatherers. In *Man the hunter,* edited by Richard B. Lee and Irven DeVore, pp. 3–12. Chicago: Aldine.

Lehmann, Johannes, Dirse C. Kern, Bruno Glaser, and William I. Woods, eds. 2003. *Amazonian dark earths: Origins, properties, management.* Dordrecht: Kluwer Academic.

Lenko, K., and Nelson Papavero. 1979. *Insetos no folclore.* São Paulo, Brazil: Conselho Estadual de Artes e Ciências Humanas.

Lemle, Miriam. 1971. Internal classification of the Tupi-Guarani linguistic family. In *Tupi studies I, Summer Institute of Linguistics publications in linguistics and related fields 29,* edited by D. Bendor Samuel, pp. 107–129. Norman, OK: Summer Institute of Linguistics.

Lentz, David L., and C. R. Ramírez-Sosa. 2002. Cerén plant resources: Abundance and diversity. In *Before the volcano erupted: The ancient Cerén village in Central America,* edited by Payson Sheets, pp. 33–44. Austin: University of Texas Press.

Leonardi, V. 1999. Historiadores e os Rios: Natureza e ruína na Amazônia Brasileira. Brasília: Fundação Universidade de Brasília.

Lepofsky, Dana. 1992. Arboriculture in the Mussau Islands, Bismarck Archipelago. *Economic Botany* 46(2):191–211.

Léry, Jacques de. 1960 (orig. Fr. 1586). *Viagem à terra do Brasil.* Trans. S. Milliet. São Paulo, Brazil: Livraria Martins Editora.

Lévi-Strauss, Claude. 1950. The use of wild plants in tropical South America. In *Physical anthropology, linguistics and cultural geography of South American Indians.* Vol. 6 of

Handbook of South American Indians, edited by Julian H. Steward, pp. 465–486. Bureau of American Ethnology, Bulletin 143. Washington, DC: US Government Printing Office.

———. 1978. *The origin of table manners: Introduction to a science of mythology,* vol. 3. Trans. J. and D. Weightman. New York: Harper and Row.

Levins, R., and R. Lewontin. 1985. *The dialectical biologist.* Cambridge, MA: Harvard University Press.

Linares, Olga. 1976. "Garden hunting" in the American tropics. *Human Ecology* 4(4): 331–349.

Lisboa, Christovão de. 1904 (orig. 1626–1627). Tres cartas de Frei Christóvão de Lisboa (2 de outubro de 1626, 2 e 20 de janeiro de 1627). *Annaes da Bibliotheca Nacional do Rio de Janeiro* 26:395–411.

———. 1967. *História dos animais e árvores do Maranhão.* Lisbon: Publicações do Arquivo Histórico Ultramarino e Centro de Estudos Históricos Ultramarinos.

Lisboa, Pedro L. B., U. N. Maciel, and Ghillean T. Prance. 1987. Perdendo Rondônia. *Ciência Hoje* 6(36):48–56.

Lizarralde, Manuel. 2001. Biodiversity and loss of indigenous languages and knowledge in South America. In *On biocultural diversity: Linking language, knowledge, and the environment,* edited by Luisa Maffi, pp. 265–281. Washington, DC: Smithsonian Institution Press.

Lombardo, U., and H. Prümers. 2010. Pre-Columbian human occupations in the eastern plains of the Llanos de Moxos, Bolivian Amazonia. *Journal of Archaeological Sciences* 3: 1875–1885.

Loukotka, Čestmír. 1929. Le Setá: Un nouveau dialecte Tupi. *Journal de la Société des Americanistes* 21(2):373–398.

———. 1968. *Classification of South American Indian languages.* Los Angeles: Regents of the University of California.

Loureiro Fernandes, L. 1959. The Xetá: A dying people of Brazil. *Bulletin of the International Committee on Urgent Anthropological and Ethnological Research* 2:22–26.

———. 1964. Les Xetá et les palmiers de la forêt de Dourados: Contribution à l'ethnobotanique du Paraná. *VI Congrès International des Sciences Anthropologiques et Ethnologiques,* Tome II, pp. 39–43. Paris: Musée de l'Homme.

Lovejoy, Arthur O. 1936. *The great chain of being: A study of the history of an idea.* New York: Harper and Brothers.

Lukesch, A. 1976. Bearded Indians of the tropical forest: The Asurini of the Ipiaçaba. Graz, Austria: Akademische Druck.

Lunardi, F. 1938. I Siriòno. *Archivio per l'antropologia e l'etnologia,* Firenze 6:178–223.

Lyon, Patricia J. 1985. Editor's note. In *Native South Americans: Ethnology of the least known continent,* edited by Patricia J. Lyon, p. 3. Prospect Heights, IL: Waveland.

Maack, R. 1968. *Geografia física do Estado do Paraná.* Curitiba, Brazil: Banco de Desenvolvimento do Paraná, Universidade Federal do Paraná e Instituto de Biologia e Pesquisas Tecnológicas.

Maffi, Luisa. 2001. Introduction: On the interdependence of biological and cultural diversity. In *On biocultural diversity: Linking language, knowledge, and the environment*, edited by Luisa Maffi, pp. 1–50. Washington, DC: Smithsonian Institution Press.

Magalhães, Antônio Carlos. 1991. Os Parakanã e os Akwawa em Paranatin. *Boletim do Museu Paraense Emílio Goeldi, Antropologia* 7(2):181–207.

Mann, Charles. 2005. *1491: New revelations of the Americas before Columbus.* New York: Alfred A. Knopf.

Markley, K. S. 1956. Mbocayá or Paraguay cocopalm: An important source of oil. *Economic Botany* 10(1):3–32.

Marquardt, William H., ed. 1992. *Culture and environment in the domain of the Calusa.* Monograph 1, Institute of Archaeology and Paleoenvironmental Studies. Gainesville: University of Florida.

Martin, M. K. 1969. South American foragers: A case study in cultural devolution. *American Anthropologist* 71(2):243–260.

Martin, Paul S. 1984. Prehistoric overkill: The global model. In *Quaternary extinctions: A prehistoric revolution,* edited by Paul S. Martin and Richard G. Klein, pp. 354–403. Tucson: University of Arizona Press.

Maxwell, Kenneth R. 1973. *Conflicts and conspiracies: Brazil and Portugal, 1750–1808.* Cambridge: Cambridge University Press.

May, Peter H., Anthony B. Anderson, Michael J. Balick, and J. M. F. Frazão. 1985. Subsistence benefits from the babassu palm (*Orbignya martiana*). *Economic Botany* 39(2): 113–129.

Maybury-Lewis, David. 1967. *Akwẽ-Shavante society.* Oxford, UK: Clarendon.

McClatchey, Will, Mynkee Qusa Bankidan Sirikolo Jr., Lazarus Kaleveke, and Carefree Pitanapi. 2006. Differential conservation of two species of *Canarium* (Burseraceae) among the Babatana and Ririo of Laur (Choiseul), Solomon Islands. *Economic Botany* 60(3):212–226.

McDonald, D. 1977. Food taboos: A primitive environmental protection agency (South America). *Anthropos* 72:735–748.

McEwan, C., C. Barreto, and E. G. Neves, eds. 2001. *Unknown Amazon: Culture and nature in ancient Brazil.* London: The British Museum Press.

McMichael, C. H., D. R. Piperno, M. B. Bush, M. R. Silman, A. R. Zimmerman, M. F. Raczka, and L. C. Lobato. 2012. Sparse pre-Columbian human habitation in western Amazonia. *Science* 336:1429–1431.

McNeil, Cameron L., ed. 2006. *Chocolate in Mesoamerica: A cultural history of cacao.* Gainesville: University Press of Florida.

Meggers, Betty J. 1954. Environmental limitation on the development of culture. *American Anthropologist* 56(5, pt. 1):801–824.

———. 1971. *Amazonia: Man and culture in a counterfeit paradise.* Arlington Heights, IL: AHM.

———. 1996. *Amazonia: Man and culture in a counterfeit paradise.* 2nd ed. Washington, DC: Smithsonian Institution Press.

Meggers, Betty J., O. F. Dias, E. T. Miller, and C. Perota. 1988. Implications of archaeological distributions in Amazonia. In *Proceedings of a workshop on neotropical distri-*

bution patterns, edited by W. R. Heyer and P. E. Vanzolini, pp. 275–294. Rio de Janeiro: Academia Brasileira de Ciências.

Meggers, Betty J., and Clifford Evans. 1957. *Archaeological investigations at the mouth of the Amazon.* Washington, DC: Smithsonian Institution.

———. 1973. *A reconstituição da pré-histórica amazônica: Algumas considerações teóricas.* Belém, Brazil: Museu Paraense Emílio Goeldi, Instituto Nacional de Pesquisas da Amazônia, and Conselho Nacional de Pesquisas.

Melancon, T. 1982. Marriage and reproduction among the Yanomamo Indians of Venezuela. PhD diss., Pennsylvania State University, University Park.

Mello, A. A. S. 1996. Genetic affiliation of the language of the Indians Aurê and Aurá. *Opción* 12(19):67–81.

———. 2002. Evidências fonológicas e lexicais para o sub-agrupamento interno Tupi-Guarani. In *Línguas indígenas Brasileiras: Fonologia, gramática e história,* edited by A. S. A. C. Cabral and Aryon D. Rodrigues, pp. 338–342. Atas do I Encontro Internacional do Grupo de Trabalho sobre Línguas Indígenas ANPOLL. Belém, Brazil: EDUFPA.

Mertz, E. T., L. S. Bates, and O. E. Nelson. 1964. Mutant gene that changes protein composition and increases lysine content of maize endosperm. *Science* 145:279–280.

Métraux, Alfred. 1928. *La civilisation matérielle des tribus Tupi-Guarani.* Paris: Librairie Orientaliste Paul Geuthner.

———. 1948. The Guarani. In *The tropical forest tribes.* Vol. 3 of *Handbook of South American Indians,* edited by Julian H. Steward, pp. 69–94. Bureau of American Ethnology, Bulletin 143. Washington, DC: US Government Printing Office.

———. 1979 (orig. 1950). *A religião dos tupinambás e suas relações com as demais tribus tupí-guaranís.* 2nd ed. Trans. E. Pinto. Coleção Brasiliana 267. São Paulo, Brazil: Companhia Editora Nacional.

Métraux, Alfred, and Herbert Baldus. 1946. The Guayaki. In *The marginal tribes.* Vol. 1 of *Handbook of South American Indians,* edited by Julian H. Steward, pp. 435–444. Bureau of American Ethnology, Bulletin 143. Washington, DC: US Government Printing Office.

Milton, Katharine. 1991. Comparative aspects of diet in Amazonian forest dwellers. *Philosophical Transactions of the Royal Society of London, B* 334:253–263.

Mitchell, John D., and Scott A. Mori. 1987. *The cashew and its relatives* (Anacardium: *Anacardiaceae).* Memoirs of the New York Botanical Garden 42. Bronx: New York Botanical Garden.

Moore, Denny A., and I. L. Maciel. 1987. Relatório da consulta lingüística à Frente de Atração Rio Tapirapé. Brasília: Fundação Nacional do Índio.

Moran, Emilio F. 1989. Models of native and folk adaptation in the Amazon. In *Resource management in Amazonia: Indigenous and folk strategies,* edited by Darrell A. Posey and William Balée, pp. 22–29. Bronx: New York Botanical Garden.

———. 1990a. Ecosystem ecology in biology and anthropology: A critical assessment. *The ecosystem approach in anthropology: From concept to practice,* edited by Emilio F. Moran, pp. 3–40. Ann Arbor: University of Michigan Press.

———. 1990b. Levels of analysis and analytical level shifting: Examples from Amazonian ecosystem research. *The ecosystem approach in anthropology: From concept to practice,* edited by Emilio F. Moran, 279–308. Ann Arbor: University of Michigan Press.

Morcote-Rios, G., and R. Bernal. 2001. Remains of palms (Palmae) at archaeological sites in the New World: A review. *Botanical Review* 67:309–350.

Moreira Neto, Carlos de A. 1988. *Índios da Amazônia: De Maioria a Minoria (1750–1850).* Petrópolis, Brazil: Vozes.

Mori, Scott A., Brian M. Boom, André M. de Carvalho, and Talmón S. Dos Santos. 1983. Southern Bahian moist forests. *Botanical Review* 49(2):155–232.

Motamayor, J. C., A. M. Risterucci, V. Laurent, A. Moreno, and C. Lanaud. 2000. The genetic diversity of *criollo* cacao and its consequence in quality breeding. *Memorias del Primero Congreso Venezolano del Cacao y su Industria,* cacao.sian.info.ve/memorias/html/03.html (accessed 2 June 2002).

Mulder, Monica B. 1987. Adaptation and evolutionary approaches to anthropology. *Man* 22(1):25–41.

Müller, Regina P. 1985. Asurini do Xingu. *Revista de Antropologia* 27/28:91–114.

———. 1993. Os Asurini do Xingu: História e arte. Campinas, Brazil: Editora UNICAMP.

Murphy, Robert F. 1960. *Headhunter's heritage: Social and economic change among the Mundurucú Indians.* Berkeley: University of California Press.

———. 1979. Lineage and lineality in lowland South America. In *Brazil: Anthropological perspectives. Essays in honor of Charles Wagley,* edited by Maxine Margolis and William Carter, pp. 217–224. New York: Columbia University Press.

Myazaki, N. 1979. Pesquisa preliminar arqueo-etnohistórica na região do alto Rio Xingu. *XLII Congrès International des Americanistes,* 1X-A, pp. 181–183. Paris: Société des Americanistes.

Myers, Norman. 1988. Threatened biotas: "Hot spots" in tropical forests. *The Environmentalist* 8(3):187–208.

Nassif, Ricardo C. 1993. *Interim report to the Biodiversity Support Program/World Wildlife Fund.* Prepared for Museu Paraense Emílio Goeldi, Belém, Brazil.

Nations, James, and Ronald Nigh. 1980. Tropical rainforests. *Bulletin of the Atomic Scientists* 36(3):12–19.

Navarrete, Acácio A., Fabiana S. Cannavan, Rodrigo G. Taketani, and Siu M. Tsai. 2010. A molecular survey of the diversity of microbial communities in different Amazonian agricultural model systems. *Diversity* 2(5):787–809.

Neel, James V. 1977. Health and disease in unacculturated Amerindian populations. In *Health and disease in tribal societies,* edited by Katherine Elliott and Julie Whelan, pp. 155–177. Ciba Foundation Symposium 49. New York: Elsevier.

Neitschmann, Bernard. 1973. *Between land and water: The subsistence ecology of the Miskito Indians, eastern Nicaragua.* New York: Seminar.

Neves, Eduardo Góes, and James B. Petersen. 2006. Political economy and pre-Columbian landscape transformations in central Amazonia. In *Time and complexity in historical ecology: Studies in the Neotropical lowlands,* edited by William Balée and Clark L. Erickson, pp. 279–309. New York: Columbia University Press.

Neves, Walter Alves. 2006. Origens do homem nas Américas: Fósseis vs. moléculas? In *Nossa origem. O povoamento das Américas: Visões multidisciplinares,* edited by Hilton Pereira da Silva and C. Rodrigues-Carvalho, pp. 45–76. Rio de Janeiro: Vieira e Lent.

Nicholaides, J. J., III, P. A. Sanchez, D. E. Bandy, J. H. Villachica, A. J. Coutu, and C. S. Valverde. 1983. Crop production systems in the Amazon Basin. In *The dilemma of Amazonian development,* edited by Emilio F. Moran, pp. 101–153. Boulder, CO: Westview.

Nichols, Johanna. 1992. *Linguistic diversity in space and time.* Chicago: University of Chicago Press.

Nimuendaju, Curt. 1948. The Turiwara and Aruã. In *The tropical forest tribes.* Vol. 3 of *Handbook of South American Indians,* edited by Julian H. Steward, pp. 193–198. Bureau of American Ethnology, Bulletin 143. Washington, DC: US Government Printing Office.

Noelli, Francisco S. 1996. As hipóteses sobre o centro de origem e rotas de expansão dos Tupi. *Revista de Antropologia* 39(2):7–53.

Nolan, Justin M. 2001. Pursuing the fruits of knowledge: Cognitive ethnobotany in Missouri's Little Dixie. *Journal of Ethnobiology* 21:9–51.

Nordenskiöld, Baron Erland von. 1924. The ethnography of South America seen from Mojos in Bolivia. *Comparative Ethnographical Studies* 3:1–254.

———. 1931. Origin of the Indian civilizations in South America. *Comparative Ethnographical Studies* 9:1–77.

Noronha, José M. de. 1856. Roteiro da viagem da cidade do Pará até as últimas colonias dos domínios portugueses em os rios Amazonas e Negro. In *Notícias para a História e Geographia das Nações Ultramarinas,* vol. 6. Lisbon.

Notes, M. 1999. An alternative hypothesis for the origin of Amazonian bird diversity. *Journal of Biogeography* 26:476–485.

NSB (National Science Board). 1989. *Loss of biological diversity: A global crisis requiring international solutions.* Washington, DC: National Science Foundation.

Nuckholls, Janis B. 1996. *Sounds like life: Sound-symbolic grammar, performance, and cognition in Pastaza Quechua.* Oxford: Oxford University Press.

———. 1999. The case for sound symbolism. *Annual Review of Anthropology* 28:225–252.

———. 2010. The sound-symbolic expression of animacy in Amazonian Ecuador. *Diversity* 2(3):353–369.

Nugent, Stephen. 2009. Utopias and dystopias in the Amazonian social landscape. In *Amazon peasant societies in a changing environment: Political ecology, invisibility and modernity in the rainforest,* edited by Cristina Adams, Rui Murrieta, Walter Neves, and Mark Harris, pp. 21–32. New York: Springer.

Oberg, Kalervo. 1953. *Indian tribes of northern Mato Grosso, Brazil.* Institution of Social Anthropology, Publication 15. Washington, DC: US Government Printing Office.

Ohnuki-Tierney, E. 1984 (orig. 1974). *The Ainu of the northwest coast of southern Sakhalin.* Prospect Heights, IL: Waveland.

Oliveira, J. C. de. 1951. *Folclore amazônica,* vol. 1. Belém, Brazil: São José.

Pace, Richard. 1998. *The struggle for Amazon town: Gurupá revisited.* Boulder, CO: Lynne Rienner.

Palmatary, H. 1960. The archaeology of the Lower Tapajós Valley, Brazil. *Transactions of the American Philosophical Society,* n.s., 50(3). Philadelphia: American Philosophical Society.

Parker, Eugene. 1992. Forest islands and Kayapó resource management in Amazonia: A reappraisal of the *apêtê. American Anthropologist* 94(2):406–428.

Pärssinen, Martti, Denise P. Schaan, and Alceu Ranzi. 2009. Pre-Columbian geometric earthworks in the upper Purus: A complex society in western Amazonia. *Antiquity* 83(322):1084–1095.

Pennington, T. D. 1981. *Meliaceae.* Flora Neotropica Monograph 28. Bronx: New York Botanical Garden.

Peres, Carlos A., Toby A. Gardner, Jos Barlow, Jansen Zuanon, Fernanda Michalski, Alexander C. Lees, Ima C. Vieira, Fatima M. S. Moreira, and Kenneth J. Feeley. 2010. Biodiversity conservation in human-modified forest landscapes. *Biological Conservation* 143:2314–2327.

Pesce, C. 1985. *Oil palms and other oilseeds of the Amazon.* Trans. and ed. D. V. Johnson. Algonac, MI: Reference Publications.

Peters, Charles M. 2000. Pre-Columbian silviculture and indigenous management of neotropical forests. In *Imperfect balance: Landscape transformations in the pre-Columbian Americas,* edited by David L. Lentz, pp. 203–223. New York: Columbia University Press.

Pickersgill, Barbara. 1976. Pineapple. In *Evolution of crop plants,* edited by Norman W. Simmonds, pp. 15–18. London: Longman.

Pinheiro, C. V. B., and Michael J. Balick. 1987. *Brazilian palms.* Contributions from the New York Botanical Garden, vol. 17. Bronx: New York Botanical Garden.

Pinkley, Homer. 1973. The ethno-ecology of the Kofán Indians. PhD diss., Harvard University, Cambridge, MA.

Piperno, Dolores R., A. J. Ranere, I. Holst, and P. Hansell. 2000. Starch grains reveal early root crop horticulture in the Panamian tropical forest. *Nature* 407:894–897.

Pires, João Murça. 1973. Tipos de vegetação da Amazônia. *Publicações Avulsas* 20:179–202. Belém, Brazil: Museu Paraense Emílio Goeldi, Instituto Nacional de Pesquisas da Amazônia, and Conselho Nacional de Pesquisas.

Pires, João Murça, and Ghillean T. Prance. 1977. The Amazon forest: A natural heritage to be preserved. In *Extinction is forever,* edited by Ghillean T. Prance, pp. 158–194. Bronx: New York Botanical Garden.

———. 1985. The vegetation types of the Brazilian Amazon. In *Key environments: Amazonia,* edited by Ghillean T. Prance and Thomas Lovejoy, pp. 109–145. New York: Pergamon.

Plucknett, D. L. 1976. Edible aroids. In *Evolution of crop plants,* edited by Norman W. Simmonds, pp. 10–12. London: Longman.

Politis, Gustavo. 2001. Foragers of the Amazon: The last survivors or the first to succeed? In *Unknown Amazon: Culture in nature in ancient Brazil,* edited by Colin McEwan, Cristiana Barreto, and Eduardo Goes Neves, pp. 26–49. London: British Museum Press.

———. 2007. *Nukak: Ethnoarchaeology of an Amazonian people.* Walnut Creek, CA: Left Coast.

Posey, Darrell A. 1979a. Ethnoentomology of the Gorotire Kayapó of Central Brazil. PhD diss., University of Georgia, Athens.

———. 1979b. Kayapó mostra aldeia de origem. *Revista de Atualidade Indígena* 3(16): 52–58.

———. 1983a. Indigenous ecological knowledge and development of the Amazon. In *The dilemma of Amazonian development,* edited by Emilio F. Moran, pp. 135–144. Boulder, CO: Westview.

———. 1983b. Keeping of stingless bees by the Kayapó Indians of Brazil. *Journal of Ethnobiology* 3(1):63–73.

———. 1984a. Indigenous knowledge and development: An ideological bridge to the future. *Ciência e Cultura* 35(7):877–894.

———. 1984b. Keepers of the campo. *Garden* 8(6):8–12, 32.

———. 1985a. Indigenous management of tropical forest ecosystems: The case of the Kayapó Indians of the Brazilian Amazon. *Agroforestry Systems* 3:139–158.

———. 1985b. Ethnoecology as applied anthropology in Amazonian development. *Human Organization* 43(2):95–105.

———. 1986. Topics and issues in ethnoentomology with some suggestions for the development of hypothesis-generation and testing in ethnobiology. *Journal of Ethnobiology* 6:99–120.

———. 1994. Comments. In *Ethnobotany and the search for new drugs,* edited by D. J. Chadwick and J. Marsh, pp. 58–59. Ciba Foundation Symposium 185. New York: John Wiley and Sons.

———. 1998. Diachronic ecotones and anthropogenic landscapes in Amazonia: Contesting the consciousness of conservation. In *Advances in historical ecology,* edited by William Balée, pp. 104–118. New York: Columbia University Press.

———. 2002. *Kayapó ethnoecology and culture.* Ed. K. Plenderleith. London: Routledge.

Posey, Darrell A., and William Balée, eds. 1989. *Resource management in Amazonia: Indigenous and folk strategies.* Advances in Economic Botany 7. Bronx: New York Botanical Garden.

Posey, Darrell A., and J. M. F. de Camargo. 1985. Additional notes on the classification and knowledge of stingless bees (Meliponinae, Apidae, Hymenoptera) by the Kayapó Indians of Gorotire, Pará, Brazil. *Annals of the Carnegie Museum* 54:247–274.

Powis, T. G., F. Valdez Jr., T. R. Hester, W. J. Hurst, and S. M. Tarka Jr. 2002. Spouted vessels and cacao use among the Preclassic Maya. *Latin American Antiquity* 13(1): 85–106.

Prance, Ghillean T. 1975. The history of the INPA capoeira based on ecological studies of Lecythidaceae. *Acta Amazônica* 5(3):261–263.

Prance, Ghillean T., William Balée, Brian M. Boom, and Robert L. Carneiro. 1987. Quantitative ethnobotany and the case for conservation in Amazonia. *Conservation Biology* 1(4):296–310.

Prance, Ghillean T., and H. O. R. Schubart. 1977. Notes on the vegetation of Amazonia

I: A preliminary note on the origin of the open white sand *campinas* of the lower Rio Negro. *Brittonia* 30(1):60–63.

Priest, Perry N. 1987. A contribution to comparative studies in the Guaraní linguistic family. *Language Sciences* 9(1):17–20.

Priest, Perry N., and Ann Priest. 1980. *Textos siriono*. Riberalta, Bolivia: Instituto Lingüístico de Verano, Ministerio de Educación y Cultura.

Projeto GURUPI. 1975. *Relatório final de Etapa, 1*. Belém, Brazil: Companhia de Pesquisa de Recursos Minerais, Superintendência Regional de Belém.

Projeto RADAM. 1973. *Levantamento de recursos naturais, 3*. Rio de Janeiro: Ministério das Minas e Energia Departamento Nacional de Produção Mineral.

Pulliam, H. 1974. On the theory of optimal diets. *American Naturalist* 108:59–74.

Purseglove, J. W. 1969. *Tropical crops: Dicotyledons*. New York: J. Wiley and Sons.

Queiroz, Helder Lima de. 1989. *Relatório anual 1989: Projeto de manejo de animais de caça em área indígena*. Belém, Brazil: Departamento de Ecologia.

———. 1992. A new species of Capuchin monkey, genus *Cebus* Erxleben, 1777 (Cebidae: Primates) from eastern Brazilian Amazonia. *Goeldiana (Zoologia)* 15:1–13.

Quinlan, M. 2005. Considerations for collecting freelists in the field: Examples from ethnobotany. *Field Methods* 17:219–234.

Raffaelli, David G., and Christopher L. J. Frid. 2008. The evolution of ecosystem ecology. In *Ecosystem ecology: A new synthesis,* edited by D. G. Raffaelli and C. L. J. Frid, pp. 1–18. Cambridge: Cambridge University Press.

Raffles, Hugh. 2002. *In Amazonia*. Princeton, NJ: Princeton University Press.

Raffles, Hugh, and Antoinette M. G. A. WinklerPrins. 2003. Further reflections on Amazonian environmental history: Transformation of rivers and streams. *Latin American Research Review* 38:165–187.

Rainbird, Paul. 2004. *The archaeology of Micronesia*. Cambridge: Cambridge University Press.

Rambo, A. Terry. 1985. *Primitive polluters: Semang impact on the Malaysian tropical rain forest ecosystem*. Anthropological Papers, 76, Museum of Anthropology, University of Michigan. Ann Arbor: University of Michigan.

Redford, Kent H. 1991. The ecologically noble savage. *Cultural Survival Quarterly* 15(1):46–48.

Redford, Kent H., and John G. Robinson. 1987. The game of choice: Indian and colonist hunting in the Neotropics. *American Anthropologist* 89:650–667.

Reichel-Dolmatoff, G. 1996. *Yuruparí: Studies of an Amazonian foundation myth*. Cambridge, MA: Harvard University Press.

René-Moreno, G. 1888. *Biblioteca Boliviana: Catálogo del Archivo de Mojos y Chiquitos*. Santiago, Chile: Imprenta Gutenberg.

Reynolds, P. K. 1927. *The banana*. Boston: Houghton Mifflin.

Ribeiro, Darcy. 1956. Convívio e contaminação: Efeitos dissociativos da depopulação provocada por epidemias em grupos indígenas. *Sociologia* 18(1):3–50.

———. 1970. *Os Índios e a civilização*. Rio de Janeiro: Editora Civilização Brasileira.

———. 1976. Os índios Urubus: Ciclo annual das atividades de subsistência de uma tribo da floresta tropical. In *Uiráa sai à procura de Deus,* by D. Ribeiro, pp. 31–59. Rio de Janeiro: Paz e Terra.

———. 1996. *Diários Índios: Os Urubus-Kaapor.* São Paulo, Brazil: Companhia das Letras.

Riester, Jürgen. 1977. *Los Guarasug'we: Crónica de sus ultimos días.* La Paz, Bolivia: Amigos del Libro.

Rindos, David. 1984. *The origins of agriculture: An evolutionary perspective.* New York: Academic.

Rival, Laura. 2002. *Trekking through history: The Huaorani of Amazonian Ecuador.* New York: Columbia University Press.

Rivers, William Halse R. 1926 (orig. 1912). The disappearance of useful arts. In *Psychology and ethnology,* by W. H. R. Rivers, edited by G. Elliot Smith, pp. 190–210. London: Kegan Paul, Trench, Trubner.

Rivet, Paul. 1924. Les indiens Canoeiros. *Journal de la Société des Americanistes,* n.s., 16: 169–181.

Rivière, Peter. 1987. Of women, men, and manioc. In *Natives and neighbors in South America: Anthropological essays,* edited by H. O. Skar and F. Salomon, pp. 178–201. Etnologiska Studier 38. Göteborg: Göteborgs Etnografiska Museum.

Rizzini, Carlos T. 1963. Nota prévia sôbre a divisão fitogeográfica do Brasil. *Revista Brasileira de Geografia* 25(1):3–64.

Rodrigues, Aryon D. 1978. A língua dos índios Xetá como dialeto Guarani. *Cadernos de Estudos Lingüísticos* 1:7–11.

———. 1984/1985. Relações internas na família linguística Tupi-Guarani. *Revista de Antropologia* 27–28:33–53.

———. 1986. *Línguas brasileiras: Para o conhecimento das línguas brasileiras.* São Paulo, Brazil: Edições Loyola.

———. 1999. Tupí. In *The Amazonian languages,* edited by R. M. W. Dixon and A. Y. Aikhenvald, pp. 107–124. Cambridge: Cambridge University Press.

Rodrigues, Aryon D., and A. S. A. C. Cabral. 2002. Revendo a classificação interna da família Tupí-Guaraní. In *Línguas indígenas Brasileiras: Fonologia, gramática e história,* edited by A. S. A. C. Cabral and Aryon D. Rodrigues, pp. 327–337. Atas do I Encontro Internacional do Grupo de Trabalho sobre Línguas Indígenas ANPOLL. Belém, Brazil: EDUFPA.

Rogers, D. J., and S. G. Appan. 1973. *Manihot, Manihotoides (Euphorbiaceae).* Flora Neotropica Monograph 13. New York: Hafner.

Roosevelt, Anna C. 1980. *Parmana: Prehistoric maize and manioc subsistence along the Amazon and Orinoco.* New York: Academic.

———. 1987. Chiefdoms in the Amazon and Orinoco. *Chiefdoms in the Americas,* edited by R. D. Drennan and C. A. Uribe, pp. 153–185. Lanham, MD: University Presses of America.

———. 1989. Resource management in Amazonia before the conquest: Beyond ethnographic projection. In *Resource management in Amazonia: Indigenous and folk strategies,* edited by Darrell A. Posey and William Balée, pp. 30–62. Bronx: New York Botanical Garden.

———. 1992. Secrets of the forest. *The Sciences* 32:22–28.

———. 1998. Ancient and modern hunter-gatherers of lowland South America: An evolutionary problem. In *Advances in historical ecology*, edited by William Balée, pp. 190–212. New York: Columbia University Press.

Roosevelt, Anna C., R. A. Housley, M. Imazio da Silveira, S. Maranca, and R. Johnson. 1991. Eighth millennium pottery from a prehistoric shell midden in the Brazilian Amazon. *Science* 254:1621–1624.

Roosevelt, Anna C., M. Lima da Costa, C. Lopes Machado, M. Michab, N. Mercier, H. Valladas, J. Feathers, W. Barnett, M. Imázio da Silveira, A. Henderson, J. Silva, B. Chernoff, D. Reese, J. Holman, N. Toth, and K. Schick. 1996. Paleoindian cave dwellers in the Amazon: The peopling of the Americas. *Science* 272:373–384.

Rosch, E. 1978. Principles of categorization. In *Cognition and categorization*, edited by E. Rosch and B. B. Lloyd, pp. 29–48. Hillsdale, NJ: Lawrence Erlbaum Associates.

Roskoski, J. P. 1982. Nitrogen fixation in a Mexican coffee plantation. *Plant Soil* 67:283–291.

Ross, Eric. 1978. Food taboos, diet, and hunting strategy: The adaptation to animals in Amazon cultural ecology. *Current Anthropology* 19(1):1–36.

Rostain, S. 2008. The archaeology of the Guianas: An overview. In *Handbook of South American archaeology*, edited by H. Silverman and W. H. Isbell, pp. 279–302. New York: Springer.

Roth, Walter E. 1915. An inquiry into the animism and folk-lore of the Guiana Indians. *Thirtieth Annual Report of the Bureau of American Ethnology to the Secretary of the Smithsonian Institution, 1908–1909*, pp. 103–386. Washington, DC: US Government Printing Office.

———. 1924. An introductory study of the arts, crafts, and customs of the Guiana Indians. *Thirty-eighth Annual Report of the Bureau of American Ethnology to the Secretary of the Smithsonian Institution, 1916–1917*, pp. 25–720. Washington, DC: US Government Printing Office.

Ruddle, Kenneth. 1974. *The Yukpa cultivation system: A study of shifting cultivation in Colombia and Venezuela*. Ibero-Americana, 52. Berkeley: University of California Press.

Ruhlen, Merritt. 1991. *Classification*. Vol. 1 of *A guide to the world's languages*. Stanford, CA: Stanford University Press.

Rydén, Stieg. 1941. *A study of the Siriono Indians*. Göteborg: Elanders Boktryckeri Aktiebolag.

Saldarriaga, Juan G., and D. C. West. 1986. Holocene fires in the northern Amazon Basin. *Quaternary Research* 26:358–366.

Saldarriaga, Juan G., D. C. West, M. L. Tharp. 1986. *Forest succession in the Upper Rio Negro of Colombia and Venezuela*. ORNL-TM 9712. Oak Ridge, TN: Oak Ridge National Laboratory.

Salomão, R. P., M. F. F. Silva, and Nelson A. Rosa. 1988. Inventário ecológico em floresta pluvial tropical de terra firme, Serra Norte, Carajás, Pará. *Boletim do Museu Paraense Emílio Goeldi, sér. Bot.* 4(1):1–46.

Sapir, Edward. 1949. Time perspective in aboriginal American culture: A study in method. In *Selected writings of Edward Sapir*, edited by David Mandelbaum, pp. 389–462. Berkeley: University of California Press.

Sauer, Carl. 1969. *Land and life: A selection from the writings of Carl Ortwin Sauer.* Ed. John Leighly. Los Angeles: University of California Press.

Schaan, Denise P. 2006. São tartarugas até lá embaixo! Cultura, simbolismo e espacialidade na Amazônia pré-Colombiana. *Revista de Arqueologia Americana* 24:99–124.

———. 2008. The nonagricultural chiefdoms of Marajó Island. In *Handbook of South American archaeology,* edited by H. Silverman and W. H. Isbell, pp. 339–357. New York: Springer.

———. 2010. Long-term human induced impacts on Marajó Island landscapes, Amazon Estuary. *Diversity* 2(2):182–206.

———. 2011. *Sacred geographies of ancient Amazonia: Historical ecology of social complexity.* Walnut Creek, CA: Left Coast.

Schermair, Anselmo. 1958. *Vocabulario sirionó-castellano.* Innsbruck, Austria: Innsbrucker Beiträge zur Kulturwissenschaft.

———. 1962. *Vocabulario castellano-sirionó.* Innsbruck, Austria: Innsbrucker Beiträge zur Kulturwissenschaft.

Schultes, Richard E. 1984. Amazonian cultigens and their northward and westward migration in pre-Columbian times. In *Pre-Columbian plant migration,* edited by D. Stone, pp. 19–37. Papers of the Peabody Museum of Archaeology and Ethnology, 76. Cambridge, MA: Harvard University Press.

Schultz, D. R., and P. I. Arnold. 1977. Venom of the ant *Pseudomyrmex* sp.: Further characterization of two factors that affect human complement proteins. *Journal of Immunology* 119:1690–1699.

———. 1978. Ant venom (*Pseudomyrmex* sp.) as an activator of C1 and an inactivator of the C36 inactivator: Its use in rheumatoid arthritis. In *Clinical aspects of the complement system,* edited by W. Opferkuch, K. Rother, and D. R. Schultz, pp. 172–186. Stuttgart, Germany: Georg Thieme.

Schulz, J. P. 1960. *Ecological studies on rain forest in northern Surinam.* Amsterdam: N. V. Noord-Hollandsche Uïtgevers Matschappij.

Scott, James C. 2009. *The art of not being governed: An anarchist history of upland Southeast Asia.* New Haven, CT: Yale University Press.

Selin, Helaine, ed. 1997. *Encyclopedia of the history of science, technology, and medicine in non-Western cultures.* Dordrecht: Kluwer.

Shepard, Glenn H., Jr., and Henri Ramirez. 2011. "Made in Brazil": Human dispersal of the Brazil nut (*Bertholletia excelsa,* Lecythidaceae) in ancient Amazonia. *Economic Botany* 65(1):44–65.

Shepard, Glenn H., Jr., Douglas W. Yu, Manuel Lizarralde, and Mateo Italiano. 2001. Rain forest habitat classification among the Matsigenka of the Peruvian Amazon. *Journal of Ethnobiology* 21(1):1–38.

Simmonds, Norman W. 1982. *Bananas.* 2nd ed. London: Longman.

Simões, Mário F., and Fernanda Araújo-Costa. 1987. Pesquisas arqueológicas no baixo Rio Tocantins (Pará). *Revista de Arqueologia* 4(1):11–28.

Simões, Mário F., C. G. Corrêa, and A. L. Machado. 1973. *Achados arqueológicos no baixo Rio Fresco (Pará).* O Museu Goeldi no Sequicentenário, Publicações Avulsas, 20. Belém,

Brazil: Museu Paraense Emílio Goeldi, Instituto Nacional de Pesquisas da Amazônia, and Conselho Nacional de Pesquisas.

Simões, Mário F., and D. F. Lopes. 1987. Pesquisas arqueológicas no baixo/médio Rio Madeiras (Amazonas). *Revista de Arqueologia* 4(1):117–134.

Sington, D. (Director). 2002. *The secret of El Dorado*. Videotape. London: BBC Horizon Series.

Slater, Candace. 1994. *Dance of the dolphin: Transformation and disenchantment in the Amazonian imagination*. Chicago: University of Chicago Press.

Smith, J. J. 1993. Using ANTRHOPAC 3.5 and a spreadsheet to compute a free-list salience index. *Cultural Anthropology Methods Journal* 5:1–3.

Smith, Nigel J. H. 1974. Agouti and babassu. *Oryx* 12(5):581–583.

———. 1980. Anthrosols and human carrying capacity in Amazonia. *Annals of the Association of American Geographers* 70(4):553–566.

———. 1982. *Rainforest corridors*. Berkeley: University of California Press.

Smith, Nigel J. H., J. T. Williams, and D. L. Plucknett. 1991. Conserving the tropical cornucopia. *Environment* 33(6):7–9, 30–32.

Smith, Nigel J. H., J. T. Williams, D. L. Plucknett, and J. P. Talbot. 1992. *Tropical forests and their crops*. Ithaca, NY: Comstock.

Smole, William. 1976. *The Yanoama Indians: A cultural geography*. Austin: University of Texas Press.

———. 1980. Musa cultivation in pre-Columbian South America. *Geosciences and Man* 21:47–50.

Soares de Sousa, Gabriel. 1974. *Notícia do Brasil*. São Paulo, Brazil: Departamento de Assuntos Culturais do MEC.

———. 1987. *Tratado descritivo do Brasil em 1587*. 5th ed. Coleção Brasiliana 117. São Paulo, Brazil: Companhia Editora Nacional.

Sombroek, Wim G. 1966. *Amazon soils: A reconnaissance of the soils of the Brazilian Amazon region*. Wageningen, The Netherlands: Centre for Agricultural Publications and Documentation.

———. 2000. Amazon landforms and soils in relation to biological diversity. *Acta Amazônica* 30:81–100.

Spaulding, A. C. 1960. The dimensions of archaeology. In *Essays in the science of culture in honor of Leslie A. White,* edited by Gertrude E. Dole and Robert L. Carneiro, pp. 437–456. New York: Crowell.

Spix, J. B., and Carl F. P. von Martius. 1938 (orig. 1831). Viagem pelo Brasil, 1817–1820, vol. 3. Trans. L. F. Lahmeyer. 2nd ed. São Paulo, Brazil: Imprensa Nacional.

Sponsel, Leslie E. 1986. Amazon ecology and adaptation. *Annual Review of Anthropology* 15:67–97.

———. 1992. The environmental history of Amazonia: Natural and human disturbances, and the ecological transition. In *Changing tropical forests,* edited by H. K. Steen and R. P. Tucker, pp. 233–251. Durham, NC: Forest History Society.

Staden, Hans. 1930 (orig. 1557). *Viagem ao Brasil*. Trans. A. Lofzren. Rio de Janeiro: Oficina Industrial Graphica.

———. 1974 (orig. 1557). *Duas viagens ao Brasil*. Trans. G. de Carvalho Franco. Belo Horizonte, Brazil: Livraria Itatiaia.

Stahl, Peter W. 2002. Paradigms in paradise: Revising the standard model of Amazonian prehistory. *The Review of Archaeology* 23(2):39–51.

———. 2008. Animal domestication in South America. In *Handbook of South American archaeology*, edited by H. Silverman and W. H. Isbell, pp. 121–130. New York: Springer.

———. 2009. Adventive vertebrates and historical ecology in the pre-Columbian neotropics. *Diversity* 1:151–165.

Stearman, Allyn M. 1984. The Yuquí connection: Another look at Sirionó deculturation. *American Anthropologist* 86(3):630–650.

———. 1989. *Yuquí: Forest nomads in a changing world*. New York: Holt, Rinehart and Winston.

Stevens, P. F. 2001–present. *Angiosperm phylogeny website*. http://www.mobot.org/MOBOT/research/APweb/. Version 9 (accessed June 2008).

Steward, Julian H. 1938. *Basin-plateau aboriginal sociopolitical groups*. Bureau of American Ethnology, Bulletin 120. Washington, DC: Smithsonian Institution.

———. 1947. American culture history in the light of South America. *Southwestern Journal of Anthropology* 3:85–107.

———. 1948. Culture areas of the tropical forests. In *The tropical forest tribes*. Vol. 3 of *Handbook of South American Indians*, edited by Julian H. Steward, pp. 883–899. Bureau of American Ethnology, Bulletin 143. Washington, DC: US Government Printing Office.

———. 1949. South American cultures: An interpretative summary. In *The comparative ethnology of South American Indians*. Vol. 5 of *Handbook of South American Indians*, edited by Julian H. Steward, pp. 669–772. Bureau of American Ethnology, Bulletin 143. Washington, DC: US Government Printing Office.

———. 1955. *Theory of culture change: The methodology of multilinear evolution*. Urbana: University of Illinois Press.

———. 1977 (orig. 1970). Cultural evolution in South America. In *Evolution and ecology: Essays on social transformation*, edited by Jane C. Steward and Robert F. Murphy, pp. 128–150. Urbana: University of Illinois Press.

Steward, Julian H., ed. 1946–1950. *Handbook of South American Indians*, 7 vols. Bureau of American Ethnology, Bulletin 143. Washington, DC: US Government Printing Office.

Steward, Julian H., and Louis C. Faron. 1959. *Native peoples of South America*. New York: McGraw-Hill.

Stone, Doris. 1984. Pre-Columbian migration of *Theobroma cacao* Linnaeus and *Manihot esculenta* Crantz from northern South America into Mesoamerica: A partially hypothetical view. In *Pre-Columbian plant migration*, edited by D. Stone, pp. 67–83. Papers of the Peabody Museum of Archaeology and Ethnology, 76. Cambridge, MA: Harvard University Press.

Stradelli, Ermano. 1929. Vocabularios da Lingua Geral Portuguez-Nheêngatúe Nheêngatu-Portuguez. *Revista do Instituto Histórico e Geographico Brasileiro* 104(158):11–768.

Street, J. 1969. An evaluation of the concept of carrying capacity. *Professional Geographer* 21:104–107.

Strömer, C. (O. F. M.). 1932. Die Sprache der Mundurukú: Wörterbuch, Grammatik und Texte eines Indianeridioms am oberen Tapajoz, Amazonasgebiet. *Anthropos,* Tomo XI Band. St. Gabriel, Austria.

SUDAM (Superintendência do Desenvolvimento da Amazônia). 1976. *Polamazônia (Programa de Pólos Agropecuários e Agrominerais da Amazônia): Pré-Amazônia Maranhense.* Belém, Brazil: SUDAM.

Susnik, B. 1994. *Formación y dispersión étnica.* Vol. 1 of *Interpretación etnocultural de la complejidad Sudamericana Antigua.* Asunción, Paraguay: Museu Etnográfico Andrés Barbero.

Sweet, David G. 1974. A rich realm of nature destroyed: The Middle Amazon Valley, 1640–1750. PhD diss., University of Wisconsin, Madison.

Taylor, G. 1985. Apontamentos sobre o Nheengatú Falado no Rio Negro, Brasil. *Amerindia* 10:5–24.

Terrell, J. E., J. P. Hart, S. Barut, N. Cellinese, A. Curer, T. Denham, C. M. Kusimba, K. Latinis, R. Oka, J. Palka, M. E. D. Pohl, K. P. Pope, P. R. Williams, H. Haines, and J. E. Staller. 2003. Domesticated landscapes: The subsistence ecology of plant and animal domestication. *Journal of Archaeological Method and Theory* 10(4):323–368.

Thompson, J. Eric S. 1956. Notes on the use of cacao in Middle America. *Notes on Middle American Archaeology and Ethnology* 128:95–116.

Toniolo, Raquel (Director). 2011. *Unnatural histories: Amazon.* London: BBC Four (Documentary, first aired 23 June 2011).

Tonkinson, Robert. 1991. *The Mardu Aborigines: Living the dream in Australia's desert.* Fort Worth, TX: Holt, Rinehart and Winston.

Toral, André. 1986. Situação e perspectivas de sobrevivência dos Avá-Canoeiro. São Paulo, Brazil: Centro Ecumênico de Documentação e Informação.

Townsend, Wendy. 1995. Living on the edge: Sirionó hunting and fishing in lowland Bolivia. PhD diss., University of Florida, Gainesville.

Tsai, S. M., B. O'Neill, F. S. Cannavan, D. Saito, N. P. S. Falcão, D. G. Kern, J. Grossman, and J. Thies. 2009. The microbial world of *terra preta.* In *Amazonian dark earths: Wim Sombroek's vision,* edited by W. I. Woods, W. G. Teixeira, J. Lehmann, C. Steiner, A. M. G. A. WinklerPrins, and L. Rebellato, pp. 299–308. New York: Springer.

Uhl, Christopher. 1982. Recovery following disturbances of different intensities in Amazon rain forest of Venezuela. *Interciencia* 7:19–24.

Uhl, Christopher, and Robert Buschbacher. 1985. A disturbing synergism between cattle ranch burning practices and selective tree harvesting in the eastern Amazon. *Biotropica* 17(4):265–268.

UNEP (United Nations Environment Programme). 1997. *Global environment outlook.* New York: Oxford University Press.

Urban, Greg. 1996. On the geographical origins and dispersion of Tupian languages. *Revista de Antropologia* 39(2):61–104.

Urton, Gary. 1997. *The social life of numbers: A Quechua ontology of numbers and philosophy of arithmetic.* Austin: University of Texas Press.

Uzendoski, Michael. 2005. *The Napo Runa of Amazonian Ecuador.* Urbana: University of Illinois Press.

Van Steenis, C. G. G. J. 1958. Rejuvenation as a factor for judging the status of vegetation types: The biological nomad theory. *Proceedings of a Symposium on Humid Tropics Vegetation,* pp. 212–215. Paris: UNESCO.

Vansina, Jan. 1990. *Paths in the rainforests: Toward a history of political tradition in equatorial Africa.* Madison: University of Wisconsin Press.

Vanstone, J. W. 1972. The first Peary collection of Polar Eskimo material culture. *Fieldiana, Anthropology* 36(1):31–80.

Vasconcellos, Simão de. 1865. *Chronica da companhia de Jesus no Estado do Brasil.* Lisbon: A. J. Fernando Lopes.

Vellard, J. 1934. Les indiens Guayaki. *Journal de la Société des Americanistes* 26:223–292.

———. 1939. *Une Civilisation du Miel.* Paris: Gallimard.

Vickers, William T. 1980. An analysis of Amazonian hunting yields as a function of settlement age. In *Working Papers on South American Indians,* no. 2, edited by Raymond B. Hames, pp. 7–29. Bennington, VT: Bennington College.

Vidal, Lux. 1977. *Morte e vida de uma sociedade indígena brasileira: Os Kayapó-Xikrin do Rio Catete.* São Paulo, Brazil: Hucitec.

———. 1984–1985. O I encontro tupi: Uma apresentação. *Revista de Antropologia* 28:1–4.

Vidal, S. M. 2000. Kuwe Duwa Kalumi: The Arawak sacred routes of migration, trade, and resistance. *Ethnohistory* 47(3–4):635–667.

Viveiros de Castro, Eduardo B. 1984–1985. Proposta para um II encontro tupi. *Revista de Antropologia* 27–28:403–407.

———. 1986. *Araweté: Os deuses canibais.* Rio de Janeiro: Jorge Zahar.

———. 1992. *From the enemy's point of view: Humanity and divinity in an Amazonian society.* Chicago: University of Chicago Press.

———. 1996. Images of nature and society in Amazonian ethnology. *Annual Review of Anthropology* 25:179–200.

———. 1998a. Cosmological deixis and Amerindian perspectivism. *Journal of the Royal Anthropological Institute,* n.s., 4(3):469–488.

———. 1998b. Dravidian and related kinship systems. In *Transformations of kinship,* edited by Maurice Godelier, T. R. Trautmann, and F. E. Tjon Sie Fat, pp. 332–385. Washington, DC: Smithsonian Institution Press.

———. 2004. Perspectival anthropology and the method of controlled equivocation. *Tipiti: Journal of the Society for the Anthropology of Lowland South America 2:1–22.*

Wagley, Charles. 1971. *An introduction to Brazil.* 2nd ed. New York: Columbia University Press.

———. 1976 (orig. 1953). *Amazon town: A study of man in the tropics.* New York: Oxford University Press.

———. 1977. *Welcome of tears: The Tapirapé Indians of Central Brazil.* New York: Oxford University Press.

Watanabe, H. 1972. *The Ainu ecosystem: Environment and group structure.* Seattle: University of Washington Press.

Werner, Dennis. 1984. *Amazon journey: An anthropologist's year among Brazil's Mekranoti Indians.* New York: Simon and Schuster.

Wessels Boer, J. G. 1965. *Palmae. Flora of Suriname,* vol. 5, pt. 1, edited by J. Lanjouw. Leiden, The Netherlands: E. J. Brill.

Whitten, Norman E., Jr. 1976. *Sacha Runa: Ethnicity and adaptation of Ecuadorian jungle Quichua.* Urbana: University of Illinois Press.

Whitten, R. G. 1979. Comments on the theory of Holocene refugia in the culture history of Amazonia. *American Antiquity* 44(2):238–251.

Wichmann, Søren. 1999. A conservative look at diffusion involving Mixe-Zoquean languages. In *Archaeology and language II: Correlating archaeological and linguistic hypotheses,* edited by R. Blench and M. Spriggs, pp. 297–323. London: Routledge.

Wierzbicka, A. 1997. *Understanding cultures through their key words: English, Russian, Polish, German, and Japanese.* New York: Oxford University Press.

Wilbert, Johannes. 1969. *Textos folklorico de los indios Waraos.* Latin American Studies, 12. Los Angeles: University of California.

Willey, Gordon R. 1971. *South America.* Vol. 2 of *An introduction to American archaeology.* Englewood Cliffs, NJ: Prentice-Hall.

Willey, Gordon R., and Jeremy A. Sabloff. 1980. *A history of American archaeology.* 2nd ed. San Francisco: W. H. Freeman.

Wiseman, F. M. 1985. Agriculture and vegetation dynamics of the Maya collapse in central Petén, Guatemala. In *Prehistoric Lowland Maya environment and subsistence economy,* edited by M. Pohl, pp. 63–71. Cambridge, MA: Peabody Museum of Archaeology and Ethnology, Harvard University Press.

Witkowski, Stanley R., and Cecil H. Brown. 1983. Marking-reversals and cultural importance. *Language* 59(3):569–582.

Wolf, Eric. 1982. *Europe and the people without history.* Los Angeles: University of California Press.

Woods, William, and Joseph M. McCann. 1999. The anthropogenic origin and persistence of Amazonian dark earths. *Yearbook of the Conference of Latin Americanist Geographers* 25:7–14.

Woods, William I., Wenceslau G. Teixeira, Johannes Lehmann, Christoph Steiner, Antoinette M. G. A. WinklerPrins, and Lilian Rebellato, eds. 2009. *Amazonian dark earths: Wim Sombroek's vision.* New York: Springer.

Young, A. M. 1994. *The chocolate tree: A natural history of cacao.* Washington, DC: Smithsonian Institution Press.

Zent, Eglée L., and Stanford Zent. 2004. Amazonian Indians as ecological disturbance agents: The Hotï of the Sierra de Maigulaida Venezuelan Guayana. In *Ethnobotany and conservation of biocultural diversity,* edited by L. Maffi and T. J. S. Carlson, pp. 79–111. Bronx: New York Botanical Garden.

Zimmerer, Karl S. 1996. *Changing fortunes: Biodiversity and peasant livelihood in the Peruvian Andes.* Berkeley: University of California Press.

Zipf, G. 1949. *Human behavior and the principle of least effort.* Cambridge, MA: Addison Wesley.

PERMISSIONS

Grateful acknowledgement is made to the following for kind permission to reprint in whole or in part certain chapters herein, as follows, from:

Chapter 2. Darrell A. Posey and William Balée (eds.), *Resource Management in Amazonia: Indigenous and Folk Strategies. Advances in Economic Botany, vol. 7.* ©1989, The New York Botanical Garden, Bronx, New York.

Chapter 3. *L'Homme* 126–128, avr.-déc. 1993, XXXIII (2–4), pp. 231–54.

Chapter 4. *The Conservation of Neotropical Forests* (Kent H. Redford and Christine Padoch, eds.). Copyright ©1992 Columbia University Press.

Chapter 5. *Ethnoecology: Knowledge, Resources, and Rights* (Ted L. Gragson and Ben G. Blount, eds). Copyright © 1999 The University of Georgia Press.

Chapter 6. *On Biocultural Diversity: Linking Language, Knowledge, and the Environment* (Luisa Maffi, ed.). Copyright© 2001 Smithsonian Institution.

Chapter 7. *Nature Across Cultures: Views of Nature and the Environment in Non-Western Cultures* (Helaine Selin, ed.). Copyright ©2003. Kluwer Academic Publishers.

Chapter 8. *Ethnohistory* 47:2 (pp. 399–422). Copyright © 2000 The American Society for Ethnohistory.

Chapter 9. *Amazon Peasant Societies in a Changing Environment: Political Ecology, Invisibility and Modernity in the Rainforest* (Cristina Adams, Rui Murrieta, Walter Neves, and Mark Harris, eds.). Copyright © 2009 Springer Science and Business Media.

Chapter 10. *Diversity* 2010 2(2): 163–81 Copyright ©2010 MDPI AG (Basel, Switzerland).

INDEX

Acacia polyphylla, 47, 81
açaí palm, 9
Aché people: artifactual landscapes used by, 80–83; dependence on maize, 79; dependence on palm trees, 83–85; regression into nomadic lifestyle, 77, 78, 85–86; social units of, 76
Acre, earthworks in, 5, 183
Acrocomia aculeata, 39, 40, *114*
Acrocomia sclerocarpa, 80–81
Acrocomia sp. (mucajá palm), 83
Acuña, Cristoval de, 151
adaptationist theories, 32–37, 72
adult sex ratios, 87
Africa, cultural forests of, 129, 181, 182–83
African slaves, 147, 152; Afro-Brazilian refugee slaves, 59
agouti, 23, 25, 40, 42, 47, 48, 52, 68, 69, *185*
agricultural regression: in Amazonian societies, 89–92; development of foraging societies, 85–88; loss of horticultural knowledge, 101–2; process of, 78–79; reduced population density, 85–86; sociopolitical forces as factor in, 76; in Tupí-Guaraní societies, 76–78; use of term, 73–74, 208n2
agricultural societies: ancient knowledge of, 3–4; folk names for generic plant types, 100–101; language as evidence of lifestyle of, 92–96, 101–2; language families as defining characteristics of, 135; names for traditional domesticates, 93; recognition of diversity as traditional knowledge, 127; village societies, 72
agricultural technology, 132–34, 161
Akawaio people, 141
Akwẽ-Shavante people, 78, 79
Alcorn, Janice B., 52

Alexa imperatricis, 46, 47, 48, 49
alpha diversity, 159, 161–63, 164, 165, 178, 179, 180
Amazonia: European colonization of, 95, 181–82; geographical location of, 123; origins of place names in, 26; Pleistocene colonization of, 181
Amazonian cobra ant. *See* cobra ant
Amazonian culture: adaptationist theories, 32–37, 72; agricultural technology as traditional knowledge, 161; species recognition as traditional knowledge, 162–63, 165–67; theories of origins, 32. *See also* prehistoric societies
Amazonian Dark Earth. *See* terra preta
Amazonian ethnology, standard model of, 11–12, 72
Amazonian forests: extent of anthropogenic forests in, 3–5; impact of indigenous peoples on, 1–3, 66–67; and landscape transformation, 5–6; map, *118*; need for protection, 184; threats to, viii–ix, xiii, 174–75. *See also* anthropogenic forests; cultural forests; *terra firme* forests
Amazonian landscapes: distribution of species in, 177; impact of contact on, 103; impact of hunter-gatherers on, 161–62; impact of migration on, 103; indigenous languages as expression of biodiversity in, 103, 126–27; traditional knowledge of, 127–30, 144–45, 164. *See also* historical ecology
Amazonian languages: as evidence of environmental knowledge, 119, 123–25, 131; impact of cacao trade on, 153; impact of Luso-Brazilian society on, 158; importance of planting words, 130; plant nomenclature and classification,

125–26; retention of traditional knowledge, 134–36. *See also names of specific languages*

Amazonian peoples: dietary restrictions imposed by Franciscan missionaries, 138–39; environmental understanding of, 123, 131; impact of early agriculture on, 127–30; impact of European conquest on population density, 128–29; landscape transformation by, 119–20; loss of genealogical knowledge, 133–34; recognition of biodiversity by, 124–25, 162–63; taboos of, 34, 126, 136–40. *See also* indigenous peoples; *names of specific peoples and societies*

ambaibo (*Cecropia concolor*), 109

Anacardium occidentale (cashew). *See* cashew (*Anacardium occidentale*)

Anacardium spp. (wild cashew), 67, 96–97, *172*, *185*, *203*. *See also* cashew (*Anacardium occidentale*)

Ananas comosus (pineapple), 77, *93*, 96–97, 128, *202*

ancient knowledge. *See* traditional knowledge (TK)

Anderson, Anthony B., 2, 19, 66, 69

Andrade, E. B., 40

animals: agouti, 23, 25, 40, 42, 47, 48, 52, 68, 69, *185*; capybara, 34, 126; deer, 34, 52, 68–69, 72, 131, 210n2 (chap. 7); management of game species, 51–52, 182; monkeys, 23, 47, 48, 68, 125, 131; peccaries, 23, 30, 34, 51–52, 86; snakes, 10, 14, 30, 125–26; tapir, 34, 40, 52. *See also* tortoises

annatto-false (*Bixa arborea*), 94, *186*

annatto tree (*Bixa orellana*): Araweté cultivation of, 86; as early Amazonian domesticate, 128; Guajá names for, *93*, *94*, *186*, *202*; Tupí-Guaraní cognates for, 77, 109, *136*

Annona sp. (soursop), 68, 96, 128, 182, *186*

anthropogenic forests: bacuri forests as, 49, *50*; bamboo forests as, 41–42, *50*; Brazil nut forests as, 5, *50*; cacao forests as, 49, *50*; evidence of past human occupation in, 1–3, 19–20, 38–39, 72–74, 80–85, 173; extent of, 3–5, *50*, 53; Ka'apor recognition of species diversity in, 169–73, *172*, *203–5*; liana forests as, 43–49, *50*; low *caatinga* forests as, 43, *50*; mechanisms in development of, 5–6; palm forests as, 39–41, *50*, 82–85, *114*; pequi forests as, *50*, 50; seed dispersal mechanisms in, 171; sustainability of, 119; tree species in, *118*. *See also* cultural forests; fallow forests; forest islands

anthropogenic soils, 35, 129, 180. See also *terra preta*

ants: ritual and medicinal use of, 108, 120, *137*, 140–42, 209n5. *See also* cobra ant; fever ants

ant trees (*Tachigali* and *Triplaris* spp.), 108, *191*, *204*

apêtê. *See* forest islands

Apinayé people, 84

arable land, foraging societies' use of, 91–92, 107

Araújo Brusque, F. C. de, 85

Araweté forests: domesticated tree crops of, 24, 25–26; ecologically important species in, *47*; geographic location, *118*; study plots in, 43, 46–49; *terra preta* soils of, 44–45, 207n5

Araweté language: cacao words in, 153, *154*, *155*, 156; names for animals and insects, *137*; names for traditional domesticates, *136*; plant nomenclature and classification, 24, 97–98, 108–9; terms for forest and soil types, 28, 66

Araweté people: Day of the Bloody Leaves, 22–23; dependence on babaçu palm, 84; dependence on Brazil nut trees, 25; food habits of, 31; impact of epidemic diseases on, 36, 86; Kapayó raid on isolated family of, 86–87; language family as defining characteristic of, 135; maize cultivation of, 28, 45–46, 78, 79, 80; physical appearance of, 26, 27, *112*; prehistoric sites inhabited by, 44, 79–80; religious beliefs of, 28–29; ritual food taboos of, 139; site of relocated village, 27; as trekking people, 106

arboriculture, use of term, 182
Areca sp. (betel nut), 181
Arenga pinnata (sugar palm), 181
arrow cane (*Gynerium sagittatum*), 110, 128, *136*
arrowroot, 9, 30–31
artifacts, defined, 79–80
artifactual landscapes, defined, 80–85
artifactual resources, defined, 80–85
Artocarpus sp. (breadfruit), 181, 182
Aspelin, Paulo, 44
Assurini forests: as cultural forests, 18–20; domesticated tree crops of, 23–24; ecologically important species in, *48*; geographic location, *118*; study plots in, 43, 46–49
Assurini language, xi–xii, 14; cacao words in, 153, *154*, 154, *155*, 156; corn words in, 20–21; names for domesticated tree crops, 24; names for traditional domesticates, 21, *136*; plant nomenclature and classification, 97, 108–9, 210n7; terms for forest types, 66
Assurini people. xvii; Aurê and Aurá expelled by, 17–18; Day of the Bloody Leaves, 22–23; diet during time of hardship, 23; food crops cultivated by, 20–21, 207n6; impact of epidemic diseases on, 36; prehistoric sites inhabited by, 44; religious beliefs of, 28–29; relocation of, 18, 21–23; site of old village, 26, 27–28
Astrocaryum gynacanthum, 65, *83*, 170, *171*, *201*
Astrocaryum paramaca, 107, 140
Astrocaryum vulgare (tucumã palm). *See* tucumã palm (*Astrocaryum vulgare*)
Attalea maripa (inajá palm). *See* inajá palm (*Attalea maripa*)
Attalea speciosa (babaçu palm). *See* babaçu palm (*Attalea speciosa*)
Attraction Post for the Wandering Indians of the Tapirapé River, 16
Aurê and Aurá (indigenous wanderers), xii, 14–18, 31, *113*, 155, 212n2
Avá-Canoeiro people, 76, 77
axeheads, stone: as evidence of past human occupation, 29, 46, 183–84; found at sites of ancient villages, 44, 183–84; indigenous peoples' beliefs about, 66, 79, 183, 207n8
axeheads, steel: 16

babaçu flour. *See* flour: babaçu flour
babaçu forests, *50*
babaçu palm (*Attalea speciosa*): in Arawetê forests, *47*; Arawetê uses of, *31*, 86; in Assurini forests, *48*; Assurini uses of, 23, 49; Aurê and Aurá's words for, 16; as disturbance indicator, 46, 60, 83–85; ecological importance of, 64, 170, *171*; foraging societies' uses of, 23, 48, 73, 87; Guajá names for, *83*, *192*; Guajá uses of, 30, 31, 66, *83*, 83–85; as indicator of past occupation, 18, 39, 40, 41, 64; in Ka'apor forests, 170, *171*; photo, *113*, *114*; in pre-Amazonian forests, 55; seed dispersed by animals, 42, 68; in study plots, 57, 58, 64, *65*; in vine forests, 19
bacaba (*Oenocarpus distichus*), *65*, 68–69, *83*, *171*, *190*, *205*
Bactris gasipaes (peach palm), 4, 39, 128
Bactris spp., *83*, *189*, *190*, *191*, *201*
bacuri (*Platonia insignis*): as disturbance indicator, 49, 64; Guajá names for, *190*; in Ka'apor forests, 170, *171*, *205*; seed dispersed by animals, 68; in study plots, 59, *65*
bacuri forests, *50*
Bagassa guianensis, 24–25, *65*, *171*, *172*, *191*, *203*
Baldus, Herbert, x
Balée, William, 4–5, 92, 107
bamboo (*Guadua glomerata*), 41–42, 81, 96, 109, 182, *201*
bamboo (*Guadua* sp.), 109, 182
bamboo forests, 41–42, *50*, 51, 182
banana (*Musa* sp.): cultivation by aboriginal Tupí-Guaraní, 77, 78–79; Guajá names for, *93*, 94–95, *96*, *202*; Ka'apor names for, 101; Sirionó names for, 109; Tupí-Guaraní names for, 109. *See also* wild banana (*Phenakospermum guyannense*)

Index 251

Barbosa Rodrigues, João, 99
Barlow, Jos, 179
Basso, Ellen, 135
beans, 20
Beckerman, Stephen, vii, 37, 72
beetle larvae, 31
belief systems. *See* religious beliefs
Bergson, Henri, 166
Berlengua, Tomás de, 94
Berlin, Brent, 100, 101, 208n8
Bertholletia excelsa (Brazil nut tree). *See* Brazil nut tree (*Bertholletia excelsa*)
beta diversity, described, 162
betel nut (*Areca* sp.), 181
biodiversity: in African cultural forests, 182–83; and decrease in human management, 166, 173; and earthworks, 183; in fallow forests, 59; impact of European conquest on domesticates, 128–29; impact of human intervention on, 2–3, 12, 53–55, 101–2, 178; Ka'apor tree recognition study, 167–73, *171, 172, 203–5*; and landscape transformation, 163–64, 174, 184; language as expression of, 103; and past human occupation, 159–60; recognition of as traditional knowledge, 124–25, 127–31, 164, 165–67, 173; in Southeast Asian cultural forests, 181–82; in study plots, 59
biomass, 162
Bismarck Archipelago, 181
bitter manioc, 31, 79, 100, 106
Bixa arborea (annatto-false), *94, 186*
Bixa orellana (annatto tree). *See* annatto tree (*Bixa orellana*)
Black, G. A., 55
black soils. See *terra preta*
Bletter, Nathaniel, 120
Block, David, 152
blue earth. See *terra preta*
Boorman, John, 22
Borassus sp.(fan palm), 182
borrowing of words: for cacao, 150, 153, 154, 155–56, 158; and language diversification, 135; and loss of traditional knowledge, 142–43

botanical vocabulary. *See* plant nomenclature and classification
bottle gourd (*Lagenaria siceraria*), 95, 108, 209n6
bows, 16–17, 18, 31, 82, 84, 86–87, 91, *113*, 170
Brazilian Amazon. *See* Amazonia
Brazilian National Indian Foundation (FUNAI). *See* FUNAI (National Indian Foundation)
Brazil nut forests, 42–43, *50*
Brazil nut tree (*Bertholletia excelsa*): in anthropogenic forests, 4, 5; Arawaeté uses of, 25, 86; associated with *terra preta*, 28; cultivation by early Amazonians, 128; as disturbance indicator, 42, 46, 60; ecological importance of, *47*; foraging societies' uses of, 86, 87; geographic distribution, 23–26; as indicator of past occupation, 18, 25; Ka'apor uses of, 25; as long-lived species, 24–25; seed dispersed by animals, 69; Tupí-Guaraní names for, 25; in vine forests, 19
breadfruit (*Artocarpus* sp.), 181, 182
breu manga (*Tetragastris altissima*), 59, *65*, 69, *171, 192*
Brochado, José P., 79
brocket-deer-manioc-stem (*Manihot* sp.), 94
bromeliads, 98. *See also* pineapple (*Ananas comosus*)
Brosimum acutifolium, 96, 109, *190, 192, 193*
Brown, Cecil H., 93, 100, 101, 208n8
Brown, S., 63, 64
Buchenavia parvifolia (manioc-stem), *94, 189*
burial customs of Ka'apor people, 7
burning, effects on forests, 38–39, 40–41. *See also* slash-and-burn agriculture; swiddens
Bush, Mark, 3–4

caboclos (peasant societies): emergence as separate society, 146, 147, 158; food taboos of, 139; landscape knowledge of, 124, 144; mythology and folklore of, 211n7
cacao (*Theobroma cacao*): as disturbance indicator, 182; early cultivation of,

252 Index

149–51; as evidence of past occupation, 49; Guajá names for, *185*; Ka'apor names for, 98; in orchardlike cultural forests, 182; subspecies of, 148; Tupí-Guaraní names for, 109, 153–56, *154*, 212n1. *See also* wild cacao (*Theobroma speciosum*)
cacao beans, 145, 150, 152
cacao forests, *50*, 157–58
cacao trade: and *drogas do sertão*, 145, 148, 152; European demand for chocolate, 120; impact on Ka'apor language and culture, 148, 157–58; role of Jesuit missionaries in, 148, 151–53
cacao words: investigations into origins of, 148–49; Ka'apor terms, 98; origins of, 120, 145, 150; Parintintin terms, *117*; Tupí-Guaraní terms, 109, 153–56, *154*
caiauê (*Elaeis oleifera*), 40, 207n3
calabash tree (*Crescentia cujete*), *93*, 108, 209n1
Camponesia brevipetiolata, 182
Canarium sp. (kenari nut), 181, 182, 183
Canela people, 30
capuchin monkeys, 125, 131
capybara, 34, 126
caraipé (*Licania* spp.), 44, *192*
carambola, 182
Carapa guianensis (crabwood), *65*, 101, 124, *171*
Carneiro, Robert L., 35
Caru River, 92
Carvalho, Genésio, 13–14
Carvalho, João, 14
Caryocar spp. (piquiá tree), 24–25, *190*
Caryocar villosum (pequi tree), *50*, *205*
Casearia sp., 69, *191*, *194*
cashew (*Anacardium occidentale*): as disturbance indicator, 182; domestication of, 128; Guajá names for, 96–97, *202*; Ka'apor origin myth for, 67; Tupí-Guaraní cognates for, 77. *See also* wild cashew (*Anacardium* spp.)
cassava. *See* manioc (*Manihot esculenta*)
Cassia multijuga, 81
catappa nut, 109
Cecropia concolor (ambaibo), 109

Cecropia obtusa, *65*, *171*
Cecropia palmata, 69
Cecropia sciadophylla, 60, *185*
Cecropia spp., 23, 60, 96, *116*, *204*
Ceiba sp., 183, *187*
cemeteries of Ka'apor people, 7
Cenostigma macrophyllum, 47, 48, *48–49*, *188*
ceramic forests, 129, 180
cerrados, ix, 2, 42, 163. *See also* savanna landscapes
Chagnon, Napoleon W., 34
Chamaecrista xinguensis, 48
charcoal, in soils, 49, 59, 129
cherimoya tree, wild, 10, 11
Chernela, Janet M., 52
chili peppers, 77, 128, *202*
Chimarrhis turbinata, *65*, *171*, *205*
Chiriguano language, 108–9
chocolate, 149–50, 151. *See also* cacao (*Theobroma cacao*)
Chrysophyllum caimito (star apple tree), 127
Citrus sinensis (feral orange), 82, 85
Citrus spp., 182, *188*, *202*
classification systems. *See* plant nomenclature and classification
Clastres, Pierre, 91
Clement, Charles R., 132, 150, 166, 176
Clements, Frederic E., 175–76, 178
climate change, 166
climax forests. *See* high forests
cobra ants, 120, *137*
Coccoloba paniculata, 69
coconut palm (*Cocos nucifera*), 95, 181, 182
Cocos nucifera (coconut palm), 95, 181, 182
cocoyam, 128
Coelho dos Santos, Sílvio, 44
coffee, 109, 151
Cola sp., 183
colonialism, 134, 145, 151–52. *See also* diseases and epidemics; Jesuit missionaries
colonial warfare, 80
Congo pea, 20
conquest, 134
conservation of traditional knowledge. *See* retention of traditional knowledge
contact: and development of cacao words,

Index 253

157–58; impact on Amazonian language and environment, 103, 142–43; impact on Ka'apor culture and language, 147, 148; impact on Sirionó language, 110–11; influence on plant nomenclature and classification, 107; and loss of traditional knowledge, 119, 120–21. *See also* diseases and epidemics

contact language. *See* LGA (Língua Geral Amazônica)

contingent diversity, 163, 165

Cooper, John, 91, 102

copaiba oil tree (*Copaifera* sp.), 109, *189*, *204*

Copaifera sp. (copaiba oil tree), 109, *189*, *204*

copal tree (*Hymenaea parvifolia*), 65, *171*, *172*, 173, *195*, *203*

copal trees (*Hymenaea* spp.), 64, 68, *203*

Cormier, Loretta, 73

corn. *See* maize (*Zea mays*)

corn words, 15, 20–21

Corrêa da Silva, Beatriz, 145–46

cotton (*Gossypium barbadense*): Araweté cultivation of, 86; Assurini uses of, 20; domestication of, 73; evidence of early cultivation, 21; Tupí-Guaraní cognates for, 21, *110*, 136

Couepia guianensis, 65, *171*, *192*

coumarin tree (*Dipteryx odorata*), 24–25, *172*, *203*

Couratari guianensis, 65, *171*, *190*, *205*

crabwood (*Carapa guianensis*), 65, 101, 124, *171*

Crescentia cujete (calabash tree), 93, 108, *202*

criollo (cacao ssp.), 148, 151

Crofts, M., 156

crop failure, 86

Cuatrecasas, José, 149

Cucurbita moschata (squash), 15, 20, 21, 93, *202*

cultural activity, impact on species diversity, 161–63

cultural arts, loss of, 106, 119, 166

cultural devolution, defined, 73

cultural diversity, 2–3, 90, 159

cultural ecology, 32–37, 72–73, 75–76, 87–90

cultural evolution, defined, 177

cultural forests: African, 129, 181, 182–83; defined, 181; disturbance indicators in, 19, 40, 182–83; need for protection of, 184; Southeast Asian, 181–82; species diversity in, 175; threats to, 174–75; use of term, xiv, 11. *See also* anthropogenic forests; fallow forests; forest islands

cultural taboos, 26, 34, 136–40

cupuaçu (*Theobroma grandiflorum*), 156–57, *189*, *204*

Daly, Douglas, 120

Darwin, Charles, 33, 173

Day of the Bloody Leaves, 22–23

Dean, Rebecca, 180

declinist paradigm, use of term, 183

deer, 34, 52, 68–69, 72, 131, 210n2 (chap. 7); deer bone, 16

deer manioc, 94, 128, 131, 208n4

deforestation, 175, 179

Denevan, William, 120

Dialium guianense (*jutaipororoca*), 65, 68, 69, *171*, *196*, *204*

Dietrich, Wolf, 209n1, 209n2

Dinizia excelsia, 58

Dioscorea sp. (wild yam), 127, *201*

Dioscorea trifida (yam). *See* yam (*Dioscorea trifida*)

Dipteryx odorata (coumarin tree), 24–25, *172*, *203*

diseases and epidemics: and agricultural regression, 78, 85–86; and decrease in population densities, 36, 128–29, 166; as factor in choosing village sites, 22; impact on Araweté people, 86; measles epidemic (1949), 10

disturbance. *See* human-mediated disturbance; landscape transformation

disturbance indicators, 46–49, 60–61, 63–64, 80–85, 182–83

ditches, 5, 38, 183

diversity: alpha, 161–63, 178; beta, 162; contingent, 163, 165; cultural, 2–3, 90, 159; ecological, 159–60; genetic, 162, 166; global, 166; phenotypic, 173; in study plots, 56–60. *See also* biodiversity

Dodecastigma integrifolium, 65, *171,* 190
dolphins, 125
domesticates: as disturbance indicator, 53, 166, 177; as evidence of agricultural regression, 110–11; Guajá names for, 99–101, *100, 202,* 208n3; Ka'apor names for, 99–101, *100;* in Ka'apor swiddens, 67–68; names for as traditional knowledge, 92–96, *93;* naming and classification systems for, 93–96, 127, 132; Tupí-Guaraní names for, 24, 92, *110, 136*
domestication, plant, 128–29, 130, 162, 166, 176
dooryard gardens, 68–69, 98, 123, 157
drogas do sertão, 145, 148, 152, 154
Ducke, Adolfo, 5, 24, 55
Duguetia spp., *172, 189, 190, 203*

earthworks, 5, 128, 163–64, 177–80, 183, 184. *See also* mounds and mound-builders
ecological diversity, 159–60
ecological importance of plant species: fallow forest species, 63–64, *65,* 84; high forest species, 63–64, *65;* methods of calculation, 56–57, 207n7; in study plots, 82, 170–73, *171*
economic forces, impact on traditional knowledge, 120, 132–34. *See also* cacao trade
economic landscape, 157–58
Egler, Walter, 27
Eisenberg, J., 34
Elaeis oleifera (caiauê), 40, 207n3
The Emerald Forest (film), 21–22, 166
enclosures, 5, 183, 210n1 (chap. 7)
endemic species, 179
endemic warfare, 22, 36
environmental conditions, language change and, 103, 105, 107–8, 111
environmental determinism, 89–90
environmental intervention of early societies, 53–55, 72–73
environmental knowledge: early agricultural practices as, 127–30; and human contribution to landscape diversity, 130–31; plant nomenclature and classification as, 125–27; recognition of diversity as, 124–25, 127–30; religious beliefs as, 126; role of language in, 123, 131
environmental limitation, 71–72, 75–76. *See also* limiting factor theory
epidemics and diseases. *See* diseases and epidemics
Erickson, Clark L., 128, 176
Eschweilera coriacea (matamatá). *See* matamatá (*Eschweilera coriacea*)
ethnobiological knowledge, 132–36
ethnology, Amazonian, 11–12, 72
European colonialism, 73–74, 95, 166. *See also* diseases and epidemics; Jesuit missionaries
European conquest, 128–29
European contact, 80. *See also* diseases and epidemics
European economic systems, 145
Euterpe edulis, 81
Euterpe oleracea, 47, 48, 65, 83, *190*
Evans, Clifford, 39, 42, 49
evolution, domestication of species as, 176–77
evolutionary ecology, 75–76, 87–88
exchange and borrowing of words, 142–43

facial ornamentation of Ka'apor people, 8
Fairhead, James, 182, 183
Falesi, Ítalo, 43
fallow forests: absence of domesticated cacao in, 157; compared to high forests, 60–64; described, 55–56; foraging societies' use of, 75–79, 87; indigenous cultures' perceptions of, 66, 128; as landscape, 123; and landscape transformation, 145; as orchards, 64, 131; recognition of biodiversity in, 160; species diversity study in, *57,* 57–60, 164; species similarity in study plots, 62. *See also* anthropogenic forests; cultural forests; forest islands
fan palm (*Borassus* sp.), 182
Faron, Louis G., 33
Fausto, Carlos, x
feral orange (*Citrus sinensis*), 82, 85

Fernandes, Florestan, x
fever ants, 108, *137*, 141, 142, 209n5
fibers, sources of, 84. *See also* cotton (*Gossypium barbadense*)
field corn. *See* maize (*Zea mays*)
fire-making technology, 106, 119, 133
fishing, 52, 128
flour, 81, 82, 84; babaçu flour, 31, 84, 208n4; manioc flour, 9, 10, 25, 84, 137; palm heart flour, 81
folklore and mythology, 67, 125–26, 166, 167, 211n7
folk plant names: as evidence of past occupation, 127; frequency in freelisting studies, 168; Guajá domesticate names, *202*; Guajá herb names, *199–201*; Guajá nondomesticate names, *201*; Guajá tree names, *185–96*; Guajá vine names, *197–201*; in Guajá vs. Ka'apor languages, 99–101, *100*; Ka'apor tree names, *203–5*; as traditional knowledge, 124–26
food preparation methods, 9, 10, 81, 84
food taboos and avoidances, 126, 136–40, 210n3, 211n4, 211n7
foraging societies: and agricultural regression, 77–78, 85–88, 90, 91–92, 133; arable lands used by, 91, 107; artifactual landscapes used by, 80–85; avoidance of sociopolitical restraints, 74; folk names for generic plant types, 100–101; historical agricultural lifestyles of, 77; and landscape transformation, 4; language as evidence of lifestyle of, 101–2; language families as defining characteristics of, 135; limited population densities in, 72, 86–88; preference for anthropogenic forests, 72–73; recognition of diversity as traditional knowledge, 127. *See also* hunter-gatherers; trekking societies
forastero (cacao ssp.), 148
forest fragmentation, 175
forest islands: African, 183; as anthropogenic forests, 43, 129; contingent diversity in, 163–64; as earthworks, 160; evidence of past management of, 66–67, 69; extent of in *terre firme*, 50; of Ka'apor people, 129–30; of Kayapó people, 69; as modified savannas, 2
forest peoples. *See* Amazonian peoples
1491 (Mann), 3
fragmentation, forest, 175
Franciscan missionaries, 138–39
freelisting exercises, 167–69, *171*, *172*, *203–5*
Friedrich, Paul, 89
fruit trees: in African cultural forests, 183; cultural uses of endemic species, 179; in Ka'apor forests, 170–71; in prehistoric village sites, 182; unintentional planting of, 130, 131, 151. *See also* orchards, fallow forests as
FUNAI (National Indian Foundation): Araweté relocated by, 27, 86; Assurini relocated by, 20; attempts to find and settle isolated Indians, 13–15, 17–18, 87; and Aurê and Aurá, 14–18; Guajá villages located near posts of, 92, 98; relations with indigenous peoples, 41, 44, *113*

Gaia hypothesis, 176
Galvão, Eduardo, x
game animals, 51–52, 182. *See also names of specific animals*
genealogical knowledge, 133–34
generic plant names, 99–101, *100*, 109
genetic diversity, 162, 166
Gentry, Alwyn, 24
geoglyphs, 128. *See also* earthworks
Geonoma baculifera, 83, *189*
Glanz, W., 34–35
global diversity, 166
globalization, viii, xiii, 175
global warming, 166
gold seekers, 78
Gomes, Mércio, x
Gómez-Pompa, Arturo G., 149
Gossypium barbadense (cotton). *See* cotton (*Gossypium barbadense*)
gourds, 15, 21, 95, 108, 209n6
gray dolphin, 125
Greenberg, Joseph, 134

Grenand, Françoise, x, xii, 107, 108, 154, 155
Grenand, Pierre, x, xii
Gross, Daniel R., 34
Guadua glomerata (bamboo), 41–42, 96, 109, 182, *201*
Guadua sp.(bamboo), 109
Guajajara people, 55, 57, 63, 84
Guajá language, xi, xii, 21; cacao words in, 153, 154, *154, 155,* 157; as evidence of horticultural past, 74, 91–93, 94, 96, 99–102; as evidence of recent nonhorticultural lifestyle, 96–99; names for nondomesticates, *94, 201–1,* 208n6; names for traditional domesticates, *93, 202,* 208n3, 208n4; names for trees, 16, *185–96;* plant nomenclature and classification, xii, 20, 99–101, *100, 115;* Tupí-Guaraní origin of, 91, 145
Guajá people, xi, xvii, 11; and agricultural regression, 73–74, 77–78, 85–86, 101–2; *A. vulgare* use of, 40; bamboo forests associated with, 41; dependence on babaçu palm, 23, 31, 41; dependence on nondomesticates, 66; dependence on palm trees, *83,* 83–85; fallow forests used by, 57, 67; food habits of, 23, 30, 31; at FUNAI posts, 98; as inhabitants of study plots, 55, 56–58; Ka'apor epithets for, 30; language family as defining characteristic of, 135; loss of firemaking technology by, 133; maize cultivation of, 79; nomadic lifestyle of, 41; puberty rites of, 140–41; recognition of diversity as traditional knowledge, 127; region inhabited by, 42, 55, 92, *118;* ritual consumption of tortoise meat, 76; social units of, 76
Guaraçug'wé language, 104, 209n1
Guaraní people: decreasing population densities, 85–86; language family as defining characteristic of, 135; migrations of, 105–6, 209n2; names for cultivated plants, 21; population density at time of contact, 80; social units of, 76
Guarayo language, 104, 209n1
guava (*Psidium guayava*), *93,* 109, 128, 182, *202*

guava-false-stem (*Myrciaria floribunda*), *94, 193*
Gurupi River, 56, 59
Gurupi River basin, *57*
Gurupiuna (Gurupi River basin), *57,* 61
Gustavia augusta, 58, 65, 191
Gynerium sagittatum (arrow cane), 110, 128, *136*

Haas, Mary, 210n1 (chap. 8)
habitat classification, 99
Hames, Raymond B., 32, 34, 37, 38
hammocks, 8, 20, 40, 84, 86–87
Handbook of South American Indians (Steward), 11–12
Handroanthus impetiginosus, 65, 169, 170, *171, 172,* 188, *203*
Handroanthus serratifolius, 69, *172, 203*
Headland, Thomas N., 208n8
head shaving rituals, 140–42
Helicostylis tomentosa, 172
Henderson, Andrew, x
herbs and grasses, folk names for, 99, *100,* 109, *199–201*
Hetá people, 40, 76, 77, 78, 80–83, 85–86, 87
high forests: absence of domestic cacao in, 157; compared to fallow forests, 60–64; described, 55–56; ecological importance of trees, *172,* 172–73; photos, *115;* recognition of by Ka'apor people, 128; species diversity study in, *57,* 57–60, 164; species similarity in study plots, *62*
Hilbert, P. P., 51
Himatanthus sucuuba, 69, *194*
historical ecology: and agricultural regression, 89–90; and foraging societies' use of fallow forests, 75–76, 87–88; impact of technology and commerce on, 120–21; influence of landscape on language, 103, 111; and Ka'apor linguistic and cultural background, 145–48; and landscape transformation, 144–45, 163–64; use of term, xv, 3. *See also* cacao words
hog plum (*Spondias mombin*): as distur-

bance indicator, 64; domesticated by early Amazonian peoples, 128; Guajá names for, *191*; in Ka'apor forests, *171, 172, 203*; present in anthropogenic forests, 170–71, 181; present in forest islands, 69; present in prehistoric sites, 40; retention of Tupí-Guaraní names for, 109; seed dispersed by animals, 68–69; in study areas, *65*; vitamin C content of fruits, 10

Holanda Ferreira, A. B. de, 208n5

horticultural societies. *See* agricultural societies

horticultural technology, 132–34

horticultural village societies, 72. *See also* agricultural societies

Hoti people, 78, 79, 184

Huber, Jacques, 25, 33

human-mediated disturbance: effect on biodiversity, 53–55, 166; and primary succession, 175–76; and primary vs. secondary land transformation, 163–64; recognition of as traditional knowledge, 165–67; and species diversity, 175–76; use of term, 160

hunter-gatherers: dependence on past agriculture, 85, 88; impact on Amazonian landscapes, 4, 161–62; loss of cultural arts, 133; plant nomenclature and classification, 101, 126–27; theories of origins, 74, 75. *See also* foraging societies

hunting customs and practices, 34, 51–52, 138, 141, 211n4

Huston, Michael, 175

Hymenaea courbaril, *172, 188*

Hymenaea parvifolia (copal tree), *65, 171, 172,* 173, *195, 203*

Hymenaea spp. (copal trees), 64, 68, *203*

Ibibate Mound Complex, 177–80, 184

Igarapé Gurupiuna, 59

inajá palm (*Attalea maripa*): as disturbance indicator, 39, 64; ecological importance in study areas, *65*; Guajá dependence on, *83,* 83–85; Guajá names for, 68, *83*; in Ka'apor forests, 170, *171,* *172, 203*; seed dispersed by animals, 68; Tupí-Guaraní cognates for, 96

inaya. See inajá palm (*Attalea maripa*)

Indian black earth. *See terra preta*

Indians, wandering, 13–15, 17–18. *See also* Aurê and Aurá (indigenous wanderers)

indicator species, 4, 46–49, 60–61, 63–64, 80–85, 182–83

indigenous peoples: impact on Amazonian forests, 1–3; language as traditional knowledge, 119; limited population densities in, 33–37, 38, 71–74; mechanics of changes in language of, 74; need for protection, 184; recognition of biodiversity as traditional knowledge, 53–55, 184; resource use patterns of foragers, 75–76. *See also* Amazonian peoples; *names of specific peoples*

infanticide, 36

ingá (*Inga* spp.): in Araweté forests, *47*; Guajá names for, *187, 194, 196*; in Ka'apor forests, *204, 205*; seed dispersal by animals, 68–69; and soil fertility, 35; Tupí-Guaraní cognates for, 107, 109

initiation rites, 120

insects, 124; as food, 31, 81; as pests, 21–22, 86; repellent for, 48, 49; ritual and medicinal use of ants, 120, *137,* 140–42, 209n5. *See also* ants; cobra ant; fever ants

intermediate disturbance hypothesis, 179–80

intervention, environmental, 53–55, 72–73

invasive species, 162, 163, 166

Ipomoea batata (sweet potato). *See* sweet potato (*Ipomoea batata*)

Ipomoea sp. (morning glory), 67, *198*

Iriarte, José, 5

Ischnosiphon sp. (*warumã*), 30, *200*

Ixora spp., 182

Jacaranda spp., *172, 195, 203*

Jacaratia spinosa (wild papaya). *See* wild papaya (*Jacaratia spinosa*)

Jaccard coefficients, use in species diversity study, 164

Jacundá (Assurini woman), 23
jerivá palm (*Syagrus romanzoffiana*), 81–82, 84
Jesuit missionaries: and development of LGA, 146–47; feral orange trees planted by, 82; impact on Amazonian peoples and languages, 104–5, 152–53, 158; role in cacao trade, 151–52, 157–58
Junqueira, André B., 176
jutaipororoca (*Dialium guianense*), 65, 68, 69, *171*, *196*, *204*

Ka'apor forests, 7, 39, 46, 49, 55, 57–59, 63, 82, 85, *118*; author's study area in, 1–3, 30, 56, 71, *115*; ecologically important species in, 82; evidence of human occupation in, 9–11; geographic location, *118*; mechanisms in development of, 5–6; origins of domesticates in, 67–70; species diversity in, 169, *171*, *172*, *203*–5. See also *ka'a-te* forests; *taper* forests
Ka'apor language, xi, xii, xiii, 14, 27, 93, 99, 125; accent, 8; cacao words in, 120, 150, 153–58, *154*, *155*; corn words in, 20–21; cultural influences on, 145–48; epithets for other peoples, 29–30; as evidence of past agricultural lifestyle, 67–68, 101–2; impact of LGA on, 148; and language of Aurê and Aurá, 14–16, 17, 31; names for domesticates, 67–68, 92, 93, 94, 95, 98–99, 101, 127, *136*; names for forest trees, 2, 10, 25, 40, *118*, 124, 128, *203*–5; names for generic plants, 92, 93, 99–101, *100*; names for nondomesticates, 101; names for wild plants, 67–68, 95, 127; planting words in, *130*, 131; plant nomenclature and classification, 107–9, *115*, 124–26; recognition of tree species diversity, 167–69, *171*, *172*; terms for forest types, 66; Tupí-Guaraní origin of, 91, 92, 96, 106, 116, 209n1
Ka'apor people, xvii, 1–2, 7–8, 63, 85, 92; and agroforestry, 63, 64, 67–70, 85, 98; association with babaçu and bamboo forests, 23, 41–42, 58, 84; burial customs of, 7; and cacao trade, 158; concept of "other", 29–30; decreasing population densities, 85–86; dependence on bacuri, 49, 59; dependence on Brazil nut trees, 25; and epidemic diseases, 10; evidence of past agricultural lifestyle, 64–70, 84–85; facial ornamentation of, 8, *112*; fallow forests used by, 5–6, 31, 58, 66, 69, 82, 84, 85; food habits of, 10, 20, 23, 28, 30–31, 34, 40, 49, 66, 69; food preparation methods of, 9, 10, 58, 84; gardening practices of, 69–70, 86, 92, 99, 101, *130*; hunting practices of, 34, 40, 51–52; as inhabitants of study area, 55, 56–60; life cycle of, 8; linguistic and cultural background of, 145–48; maize cultivation of, 28; marriage customs of, 8; massacre of, 9; measles epidemic among, 10, 42; mythology and folklore of, 67, 125–26, 211n7; photo, *117*; puberty rites of, 120, 140, 141, 142; recognition of biodiversity as traditional knowledge, 127, 159–60, 169–73, *172*, *203*–5; religious beliefs of, 7–8, 183; ritual and medical use of ants, 120, *137*, 140, 141–42, 209n5; ritual consumption of tortoise meat, 34, 76, *117*, 137–40, 210n3, 211n4; ritual taboos of, 34; social units of, 76; steel tools used by, 183; warriors, 78, 98
ka'a-te forests, 58, 59, 66, 128, 165, 172, 173. *See also* high forests
Kakuri (Ka'apor man), 30
Kamaracĩ (Araweté man), 26, 28, 29
Karajá people, 51
Kayapó forests, 2, 66–67, 69, 129, 130, 131
Kayapó people, 69, 78, 79, 86–87, 141; and ants, 141; diseases of, 36; enemies of, 22, 24, 86; hunting, 51; modification of landscape, 2, 129; plant cultivation by, 42, 64, 66–67, 69, 78, 80, 129, 130, 131; species identification by, 2, 125; study of, 2; and tortoises, 211n6
kenari nut (*Canarium* sp.), 181, 182
Kensinger, Kenneth, 211n7

kirawa fiber (*Neoglaziovia variegata*), 86, 140, 141
Kuikuru people, 35, 80, 135

labor, indigenous, 147, 148, 157–58
lacandonica (cacao ssp.), 149, 151
Lacmellea sp., 154
Lagenaria siceraria (bottle gourd), 95, 108, 209n6
landscape, use of term, 103–4, 123, 144–45
landscape domestication, 176
landscape engineering, 177–78
landscape transformation: and anthropogenic forest development, 5–6, 173, 183–84; author's study of, 1–3; extent of Amazonian cultural forests, 3–5; and historical ecology, 180–81; impact on species diversity, 174; and Kayapó people, 2, 129; and primary succession, 159, 175–77; use of term, 159. *See also* human-mediated disturbance; primary landscape transformation; secondary landscape transformation
language: development of families of, 210n1 (chap. 8); as expression of landscape, 103–4, 111; impact of contact and migration on, 103; as reflection of environmental conditions, 105–8; source of traditional knowledge, 119, 134–36. *See also* borrowing of words; plant nomenclature and classification
late successional plants, 80–83
Lathrap, Donald, 35, 74, 91, 208n2
Leach, Melissa, 182, 183
Lecythis chartacea, 65, 192
Lecythis idatimon, 172, 190, 203
Lecythis pisonis, 65, 171, 195, 203
lesser giant hunting ant, 137
Lévi-Strauss, Claude, 90, 100
LGA (Língua Geral Amazônica). *See* Língua Geral Amazônica (LGA)
liana forests, 43–49, 50, 51, 81
Licania spp (*caraipé*), 44, 192
lima bean (*Phaseolus* sp.), 93, 202
lime tree, 27–28, 85
limiting factor theory, 33–37, 38

Linares, Olga, 72
Língua Geral, 138
Língua Geral Amazônica (LGA): and Ka'apor language, 145–46, 148; and origins of cacao words, 153, 154, *154*, 155, 156, 157; source of vocabulary items, 152–53
Lisboa, Christovão de, 138–39, 211n5
Llanos de Mojos, 104, 119–20, 132–33, 209n2. *See also* Ibibate Mound Complex
localized species. *See* endemic species
logging, effect on species diversity, 178
Lombardo, U., 179
Lopes, D. F., 207n2
Lovejoy, Arthur, 169
low *caatinga* forests, 43, *50*
Lugo, A. E., 63, 64
lumping of plant names, 97–99
Lusiã (Ka'apor youth), 9–10
Luso-Brazilian society, 147, 148, 151, 158

Maack, R., 80, 81
Mabea sp., 65, 154, *171*
Maffi, Luisa, viii
Magalhães, Antônio Carlos, 17
Mahle, Francis, 5
mahogany, 13, 124
Ma'ir (Ka'apor culture hero), 67
maize (*Zea mays*): Amazonian varieties of, 20, 44; Assurini names for, 20–21; Aurê and Aurá's words for, 15; cultivation of, 20, 45–46, 78–79, 86, 87; dependence of semisedentary peoples on, 79; evidence of early cultivation of, 21, 77; Guajá names for, *93*, *202*; origin myths for, 29; as protein source, 72; *terra preta* soils preferred for, 28, 45, 207n6; Tupí-Guaraní cognates for, 20–21, 77, *110*, *136*
Major (SPI/FUNAI employee), 58
Malesian cultural forests, 181–82
mango, 109, 182
Manihot esculenta (manioc). *See* manioc (*Manihot esculenta*)
Manihot spp., *94*, 199, 200

260 Index

Manilkara huberi (sapote tree), 10, 11, 24–25, 96, *186, 203*
manioc (*Manihot esculenta*), 95, 96, 124; Assurini names for, 20; Aurê and Aurá's words for, 15; bitter, 31, 79, 100, 106, 133, 135, 137; cultivation of, 20, 67, 77, 78–79, 86, 95, 98, 135; deer manioc, 128, 131, *199,* 208n4, 210n2 (chap. 7); domestication of, 28, 73, 128; evidence of early cultivation of, 21, 77; fields, 9, 14; flour, 9, 10, 25, 84, 137; Guajá names for, *94,* 100, *202,* 208n4; as introduced crop species, 95; Ka'apor names for, 20, 100; preparation of, 9, 10, 30, 41, 58, 84; soils for cultivation of, 28; as staple of indigenous peoples, 31, 78–79, 106, 124; sweet, 20, 78–79, 98, 100, *202,* 208n3; tree, *189;* Tupí-Guaraní cognates for, 15, 21, 77, 208n3, 208n4; wild, 68, 210n2 (chap. 7)
manioc flour. *See* flour: manioc flour; manioc (*Manihot esculenta*): flour
manioc-stem (*Buchenavia parvifolia*), *94, 189,* 208n4
Mann, Charles, 3
Marajó Island, 39–40, 49, 162, 177–78. *See also* Ibibate Mound Complex
Maranhão (state), 55–56, 145, 146
marginal tribes, 76, 91
marking of terms: as characteristics of Tupí-Guaraní languages, 111; domesticates and nondomesticates distinguished by, 97, 108, 109; as evidence of species diversity, 165; indicators of word origins, 93–94, 95
marking reversals, 93, 95, 97
Markley, K. S., 81
marriage customs of Ka'apor people, 8
Mascagnia sp., 69, *198*
matamatá (*Eschweilera coriacea*): in Assurini forests, *48;* as food for game animals, 48; Guajá names for, *194;* indigenous peoples' uses for, 49; in Ka'apor forests, *171, 172,* 173, *203;* in study plots, 58, 59, 63–64, *65*
Matisia spp., 48

Matsigenka people, 126
Maués people, 140–41
Mauritia flexuosa (moriche palm), 39, 85
Mawé people, 80
Maxwell, Kenneth R., 151
May, Peter H., 84
Maya culture, 149
Maytenus sp., 69, *194*
McMichael, C. H., 3
measles epidemic (1949), 10
medicine, *137,* 140–42, *209n5*
Meggers, Betty J., 12, 35, 39, 42, 49, 75–76
menstruation and menarche, 34, 76, 135, 136–40, 141, 142, 210n3, 211n7
Metroxylon spp. (sago palm), 181
migration, linguistic impact of, 103, 107, 110–11
missionaries, 138–39. *See also* Jesuit missionaries
missionization, 78, 145
Mixe-Zoquean language, 149, 150–51, 153
modeling, in Tupí-Guaraní languages, 93–96, 110–11
monkeys, 23, 47, 48, 68, 125, 131
monocropping and monocultures, 132–34, 178
Moore, Denny A., 13–14, 15, 17, 107
Morgan, L. H., 54
moriche palm (*Mauritia flexuosa*), 39, 85
morning glory (*Ipomoea* sp.), 67
mounds and moundbuilders: as evidence for prehistoric land management, 38, 129, 163–64, 183, 184; Ibibate Mound Complex, 177–80, 184; of Marajó Island, 39–40, 49, 162, 177–78. *See also* earthworks
mucajá palm (*Acrocomia* sp.), 83
Müller, Regina P., 44
Munduruku people, 80, 156
Murphy, Robert F., viii
Musa sp (banana). *See* banana (*Musa* sp.)
Myazaki, N., 50
Myrciaria floribunda (guava-false-stem), *94, 193*
Myrciaria obscura, 65, *187*

mythology and folklore, 67, 125–26, 166, 167, 211n7

Nassif, Ricardo, 92
National Indian Foundation (FUNAI). *See* FUNAI (National Indian Foundation)
native forests. *See* high forests
native warfare, 22, 36
Nazca lines, 128, 210n1 (chap. 7)
Neea oppositifolia, 48, *194*
Neea sp., 65, 69, *171*, *194*
Neoglaziovia variegata (*kïrawa* fiber), 86, 140, 141
Newtonia psilostachya, 65, *171*, *187*
Nichols, Johanna, 210n1 (chap. 8)
Nimuendajú, Curt, x
nomadic peoples. *See* foraging societies; hunter-gatherers; trekking societies
nondisturbance, indicator species of, 63–64
nondomesticates: folk generic names for, 201–2; Guajá dependence on, 66–67; Guajá names for, 93–94, *94*, 96–101, *201*; Ka'apor names for, 124; naming and classification of, 109, 127, 154, 155–56, 157; planting mechanisms for, 130, 131, 151; presence in cultural forests, 177, 182; retention of names for, 132; Sirionó names for, 109, 111; in swiddens and fallow forests, 67–68, 69; Tupí-Guaraní names for, 93–96
Nordenskiöld, Baron Erland von, 103, 105
Nukak people, 184

occupation mounds. *See* mounds and moundbuilders
Oenocarpus distichus (bacaba), 65, 68–69, *83*, *171*, *190*, *205*
old fallow. *See* fallow forests
Old World crops, 95
Old World diseases. *See* diseases and epidemics
Olmec civilization, 149
optimal foraging theory, 37–38
orchards, fallow forests as, 64, 74, 129, 131, 182

ordeals (ritual puberty rites), 140–41
Ori-ru (Ka'apor headman), 41
Oryza sativa (rice), 15, 96, *202*
"other," concept of, 29–30
overdifferentiation, 124–25

Pacific almond, 182
palm forests, 39–41, *50*, 82–85
palm fruits, 82
palm heart flour. *See* flour: palm heart flour
palm trees, 81, 82–85, *83*, 170–71, 182. *See also* names of specific palm species
papaya, 86, 97, 128, 182. *See also* wild papaya (*Jacaratia spinosa*)
Parakanã people, 17, 133, 139–40
Parinari sp., 183, *186*, *203*
Parintintin language, 153, *154*, 156, 157
Parker, Eugene, 69
pau d'arco tree, 9
peach palm (*Bactris gasipaes*), 4, 39, 128
peanut, 31, 77, 79, 128
peasant societies. *See caboclos* (peasant societies)
peccaries, 23, 30, 34, 51–52, 86
pequi forests, *50*
pequi tree (*Caryocar villosum*), 50, *205*
Peres, Carlos, 4
perspectivism, in indigenous belief systems, 166
Phaseolus sp. (lima bean), *93*, *202*
Phenakospermum guyannense (wild banana), 94, 95, 96, *201*
phenotypic diversity, 173
P.I. Awá, *57*, 61, 63
P.I. Canindé, *57*, 115
pigeon pea (*Cajanus cajan*), 20
P.I. Guajá, *57*, 61
Pindaré River, 56, *57*, 61, 63, 92
pineapple (*Ananas comosus*), 77, *93*, 96–97, 128, *202*
pioneer species, 60–61
piquiá tree (*Caryocar* sp.), 24–25
Pires, João Murça, 24
Pisonia sp., 65, *171*, *194*
plantain, 109, *116*
plant diversity. *See* biodiversity

planting words, *130*, 131
plant nomenclature and classification: and agricultural regression, 91–92, 93; as evidence of language origins, 106–9, *110*, 207n7; folk plant names, 99–101, *100*, 124–26, 127, 168; generic plant names, 99–101, *100*, 109, 132; Ka'apor tree recognition study, 167–69, *171*, *172*; as landscape history, 165; and societal changes, 102, 120, 121; species recognition as traditional knowledge, 166; and Tupí-Guaraní societies, 97–99
plant vocabulary. *See* plant nomenclature and classification
Plato, vii, 165–66, 169, 173
Platonia insignis (bacuri). *See* bacuri (*Platonia insignis*)
Platypodium elegans, 65, *171*, 195
political and social institutions, 74, 90, 91
population densities: and agricultural regression, 85–86; of Amazonian peoples, 71–74; and domestic species diversity, 128–29; of foraging societies, 90; limiting factor theory, 33–37, 38; and Old World diseases, 166
Portuguese language, 130, 146, 152–53, 155
Posey, Darrell A., 2, 66, 69
potsherds, 18, 44, 46, 49, 59, 129
pottery, 18, 29, 44, 207n4
Pourouma guianensis, 65, *171*, 185
Pourouma minor, 65, *171*
Pouteria bilocularis, 65
Pouteria caimito (wild star apple tree), 127, *192*
Pouteria macrophylla, 65, *171*, *172*, 185, *203*
Pouteria spp., *192*, *204*, *205*
pre-Amazonian forests, 55–60
pre-Clovis cultures, 161
prehistoric societies: and agricultural regression, 208n1 (chap. 5); primary landscape transformation by, 38, 129, 163–64, 183, 184; recognition of diversity as traditional knowledge, 165; South American origins of, 161
prestige principle, 150, 152–53, 154
Priest, Perry N., 209n2

primary forests, 55–56. *See also* high forests
primary landscape transformation: in African cultural forests, 182–83; defined, 2, 179; earthworks as, 177–80; forest islands as, 163–64; use of term, 160, 176. *See also* human-mediated disturbance
primary succession, 160, 175–77
pristine forests, 53, 55, 56, 76, 208n1 (chap. 3). *See also* high forests
protein limitation, 34–35, 71–72
Protium decandrum, 65, *171*, 196
Protium giganteum, 65
Protium pallidum, 65, *171*, 196
Protium polybotryum, 65, 196
Protium spp., 154, *189*, *192*, *196*, *204*
Protium trifoliolatum, 65, *171*, 196
Proto-Mixe-Zoquean language, 153. *See also* Mixe-Zoquean language
Proto-Tupí-Guaraní: influence on modern languages, 92, 105, 106, 107, 210n1 (chap. 8); as mother language, 21, 24, 25, 92; names for domesticates, 94, 100, 101; and origin of cacao words, 156, 157
Prumers, H., 179
Prunus spp., 182
Psidium guayava (guava), 93, *94*, 128, 182, *202*
psychological importance of species names, 168, *172*
puberty rites, 136–41. *See also* food taboos and avoidances
pumpkins, 77
pyrophytic plant species, 19, 40, 182–83

Queiroz, Helder Lima de, *117*

rainforests. *See* Amazonian forests
Ramirez, Henri, 5
Ranzi, Alceu, 5
rare species. *See* endemic species
red-footed tortoise, *137*, 137–40, 210n3
Redford, Kent, 53, 54
religious beliefs: of Ka'apor people, 7–8; origins of stone axeheads, 66, 79, 207n8; ritual consumption of tortoise meat, 76; shape-shifting, 125–26, 166; as traditional knowledge, 126, 167; of

Tupian people, 7. *See also* mythology and folklore; taboos
The Republic (Plato), 166
Reserva Indígena Alto Turiaçu, 11, 58
Reserva Indígena Caru, 57
resource management strategies, 32–37, 51–52, 72, 75–76
retention of traditional knowledge: languages as methods of measuring, 120, 134–36; ritual and medicinal use of ants, 140–42; ritual consumption of tortoise meat, 136–40; word borrowing and exchange, 142–43
Reynolds, P. K., 94
Ribeiro, Darcy, x, 10, 14, 133
rice (*Oryza sativa*), 14, 96, 178; words for, 15, 96, 103
Rindos, David, 64, 66
ringed villages, 183
Rinorea spp., *172, 195, 203*
Rivers, William H. R., 89, 91
Rizzini, Carlos T., 55
Rodrigues, Aryon D., x, 209n1
Rollinia exsucca, 69, *190, 204*
Roosevelt, Anna C., 35
rope plant, 128
Roskoski, J. P., 207n1
Ross, Eric, 34
rubber tapping, 146

Saccharum officinarum (sugar cane), 15, 96, *202*
Sacoglottis spp., 69, *187, 204*
sago palm (*Metroxylon* spp.), 181
Santos, Salomão, 13
sapote tree (*Manilkara huberi*), 10, 11, 24–25, 96, *186, 203*
Sauer, Carl, 181
savanna landscapes, 2, 29, 55, 80, 129, 163, 178, 183; savanna-like forests, 43
Schaan, Denise, 5, 177–78
Schaden, Egon, x
Schefflera morototoni, 96, *190, 204*
Schefflera sp., 69
Schultes, Richard Evans, 150
Scott, James C., 73
The Search for El Dorado (film), 3

secondary forests, 175. *See also* fallow forests
secondary landscape transformation: in Africa, 181, 182–83; and forest island biodiversity, 164; in Ka'apor forests, 170; in Southeast Asia, 181–82; use of term, 2, 160, 176. *See also* human-mediated disturbance; swiddens
secondary succession, 3–4, 160, 175
seed dispersal mechanisms: animals, 25, 38–39, 40, 68–69, 131; intentional or unintentional human actions, 130, 131, 151, 171
semidomesticates, 67–68, 96–97, 177
Senna silvestris, 172, 186, 203
Sera dos Dourados, 80–82
Serō (Ka'apor headman), 8–10
Setá people, 77
shape-shifting, in indigenous belief systems, 125–26, 166
Sheffler, M., 156
shelters: of Aurê and Aurá, *113*; use of palms for, 23, 83
Shepard, Glenn H., Jr., 4, 5
Simaba cedron, 65, 171, 185
Simarouba amara, 65, 69, *187, 204*
similarity, species, 60–64, *62,* 164
Simões, Mário F., 207n3
Sington, D., 3
Siona-Secoya people, 37–38
Sirionó language: mechanics of change in, 74; names for traditional domesticates, *136*; plant nomenclature and classification, 108–11, *110,* 169; Tupí-Guaraní origin of, 103–6, 209n1, 209n2, 210n7
Sirionó people: ceremonial uses of endemic trees, 179; forest island use by, 129–30; language family as defining characteristic of, 135; maize cultivation of, 77; puberty rites of, 140–41; ritual and medical use of animals, *137,* 142, 209n5; ritual food taboos of, 139–40; trekking lifestyle of, 104–7; useful arts lost by, 102, 133
slash-and-burn agriculture, 19, 18, 123, 127. *See also* swiddens
slave raids, 77, 78

slaves. *See* African slaves; slave raids
Smith, J. J., 168–69
Smith, Nigel J. H., 44–45, 68
Smith's s, 168–69, 170, 203
snakes, 10, 14, 30, 125–26
social and political institutions, 74, 90, 91
soil fertility, 35, 45–46, 71–72, 75–76
soil limitation, theories of, 34–35, 37–38
soils, 35, 129, 180. See also *terra preta*
Sombroek, Wim G., 33, 42
Sorocea guilleminiana (turumbúri tree), 179, 184
soursop (*Annona* sp.), 68, 96, 128, 182
South America, origins of human presence in, 161
Southeast Asia, cultural forests of, 181–82
Spanish language, cacao words and, 150, 153
Spaulding, A. C., 80
species recognition, as traditional knowledge, 165–67
species similarity, 60–64, *62*, 164
species turnover, 176, 178, 179–80
spiritual beliefs. *See* religious beliefs
Spondias mombin (hog plum). *See* hog plum (*Spondias mombin*)
squash (*Cucurbita moschata*), 15, 20, 21, 77, *93*, *202*
standard model of Amazonian ethnology, 11–12, 72
star apple tree (*Chrysophyllum caimito*), 127. *See also* wild star apple tree (*Pouteria caimito*)
Stearman, Allyn M., 119, 209n2
Sterculia pruriens, *47*, 49, *65*, *171*, *191*
Steward, Julian Haynes, 11–12, 33–34, 75, 76, 91, 208n1 (chap. 3)
Street, J., 38
study plots: high vs. fallow forests, 60–64; inventory methods used, 56–57; locations of, 55–56, *57*, *115*; results, 57–60; species similarity in pairs of, *62*
succession: and Amazonian land transformation, 3; primary, 160, 175–76; secondary, 3–4, 160, 175
sugar cane (*Saccharum officinarum*), 15, 96, *202*
sugar palm (*Arenga pinnata*), 181

superorganisms, use of term, 176
Sweet, David, vii, 2, 53
sweet manioc, 20, 78–79, 100
sweet potato (*Ipomoea batata*), 14, 98; Aché raids for, 85; Assurini cultivation of, 20; Aurê and Aurá's words for, 15; cultivation of, 77, 98, 101; domesticated, 109, *110*, *136*, *202*, 208n3; dependence of semisedentary peoples on, 79; evidence of early cultivation of, 21; Guajá names for, *202*, 208n3; Guajá raids for, 85; Ka'apor names for, 67, 101; nondomesticated, 109; Tupí-Guaraní cognates for, 21, 109, *110*, *136*, 208n3
swiddens, 34, 46, 47, 58, 68, 69, 75, 81, 87, 123; and agoutis, 42, 69; Araweté, 45–46; burning of, 69; cultivation, 51, 69, 183; game management in, 51–52, 137–38; hunting, 51–52; Ka'apor cultivation practices in, 64, 66, 67–68, 69, 86, 98, 101–2, *130*, 131, 183; as landscape, 123; and landscape transformation, 68, 69, 70, 127–28, 144–45, 170, 179, 184; and optimal foraging theory, 37; as orchards, 131; prehistoric, 39, 96, 183; raids on, 41, 85–86, 91, 96, 97, 98; relocation of, 75; soil fertility in, 35, 207n5; species diversity in, 101–2, 164; steel tools used in, 183–84; *terra preta*, 44, 46, 50–51, 207n5; traditional domesticates in, 67–68

taboos, 34, 136–40; food/eating, 23, 34, 126, 137–39, 210n3, 211n6–7; game, 34, 211n4; incest, 31
Tachigali spp. (ant trees), 108, *191*, *204*
takamã (*Astrocaryum vulgare*). *See* tucumã palm (*Astrocaryum vulgare*)
Takomãi (Assurini woman), 23
takwar. *See* bamboo (*Guadua glomerata*)
tamarind, 109, 182
tapa cloth, 90
taper forests, 2–3, 128, 169, *203*–5. *See also* anthropogenic forests; cultural forests
taper fruit tree. *See* hog plum (*Spondias mombin*)

tapir, 34, 40, 52
Tapirapé Indians, 20, 36
Tapirira guianensis, 69, *188*
Taralea oppositifolia, 65, *171*, *188*, *204*
Tatuavī (Araweté man), 26, 28
technology, 132–34, 159, 161. *See also* traditional knowledge (TK)
Teixeira, Wenceslau, 5
TEK (traditional ethnobiological knowledge), 132–36. *See also* traditional knowledge (TK)
Tembé language, *136*, 153, *154*, *155*, 156, 157
Tembé people: as inhabitants of study area, 55, 56, 57; language family as defining characteristic of, 135; location of forests of, *118*; plant nomenclature and classification, 97–98, 107–9; village of, *115*
terra firme forests: extent of anthropogenic forests in, 32, *50*, 50–51, 53, 207n3, 207n10; as landscape, 123; types of, 55–56. *See also* fallow forests; high forests
Terra Indígena Alto Guamá, *115*
Terra Indígena Alto Turiaçu, 11, 58
terra preta: as anthropogenic soil, 4, 163, 164; as artifactual resource, 80; Brazil nut groves associated with, 28; chemical composition of, 45, 207n5; as evidence of past occupation, 12, 44–49; as modified soils, 72; palms and cultivated species grown in, *114*; as preferred soil for maize cultivation, 28, 207n6
Tetragastris altissima (breu manga), 59, *65*, 69, *171*, *192*
Tetragastris panamensis, 65, *190*
Theobroma angustifolium, 149
Theobroma bicolor, 149, 150
Theobroma cacao (cacao). *See* cacao (*Theobroma cacao*)
Theobroma grandiflorum (*cupuaçu*), 156–57, *189*, *204*
Theobroma speciosum (wild cacao). *See* wild cacao (*Theobroma speciosum*)
Theobroma subincanum, 150, 156
Thevet, André, 138

Thompson, J. Eric S., 148
threatened species. *See* endemic species
Timaeus (Plato), 166
TK (traditional knowledge). *See* traditional knowledge (TK)
tobacco, 20, 128, 130
tools and utensils, 82, 183
tortoises: as food, 16, 23, 29, 210n3; food for, *48*; in ka'apo mythology, 211n7; medical uses of, 210n3; ritual consumption of, 34, 76, 136–40, *137*, 211n6–7; ritual hunting practices, 34, 142, 211n4; shells, 16, 29, 45, *117*; taboos, 211n4, 211n6
tortoise tree, *195*
tortoise vine, *198*
traditional ethnobiological knowledge (TEK), 132–36. *See also* traditional knowledge (TK)
traditional knowledge (TK): agricultural technology as, 159; and Amazonian environments, 119–20; and biodiversity, 159; and landscape transformation, 160, 181; recognition of diversity as, 162, 173; recognition of landscapes as, 127–30, 164; recognition of past disturbance as, 165; recognition of species names as, 165–67, 168; religious beliefs as, 167; retention of, 120; theories of Amazonian origin, 161; use of term, 119–21, 132. *See also* environmental knowledge; historical ecology; retention of traditional knowledge
tree recognition study, 167–69, *171*, *172*, *203–5*
trees: ecological importance of, 170–71, *171*, *172*; generic names for, 99, 109; Guajá names for, *100*, *185–96*; Ka'apor names for, *203–5*
trekking societies: and agricultural regression, 133; and landscape transformation, 4; maize cultivation of, 78–79; as marginal tribes, 76; population densities of, 86–88; Sirionó as, 104–7, 129–30. *See also* foraging societies
Trema micrantha, 35, *194*
Trichilia lecointei, *48*, *189*

Trichilia quadrijuga, 65, *171,* 189
Trichilia spp., 46, *47,* 49, *189, 191, 194*
Triplaris americana (ant trees), 108
tropical forest products, 145, 148, 152, 154
tropical forests. *See* Amazonian forests
tucumã palm (*Astrocaryum vulgare*), 10–11; as disturbance indicator, 18, 39–40, 64, 83–85; foraging societies' uses of, 73, 87, 140–41; Guajá dependence on, 83–85, 140–41; Guajá names for, *191*; in Ka'apor forests, 170, *171, 172, 203*; Ka'apor uses for, 10–11; in study plots, 58, 64, *65*; Tupí-Guaraní cognates for, 96, 107
Tupí-Guaraní languages: cacao words in, *154, 155*; conservation of traditional knowledge, 120; as evidence of past occupation, 31; evolution of, 25; and language of Aurê and Aurá, 17–18, 212n2; linguistic evidence of horticultural past of foraging societies, 64–68, 92, *93, 94, 94,* 96; names for animals and insects, *137*; names for domesticates, 24, 92, 94–97, *136*; names for semidomesticates, 96–97; names for wild plants, 93–96, 208n5; plant nomenclature and classification, 97–98, *110, 115,* 210n7; subgroups of, 1, 103–6, 145–46, 209n1, 209n2
Tupí-Guaraní peoples: and agricultural regression, 76–78; diversity of modern societies, 76–77; evidence of plant management by, 64–68; food taboos of, 139–40; general identifying characteristics of, 136; pottery of, 44, 207n4; puberty rites of, 120, 142; religious beliefs of, 7–8; traditional knowledge as defining characteristic of, 135–36
Tupi language family, 145, 156
Tupinambá language, 95, 138, 146, 152–53
Tupinambá people, 76, 77–78, 135, 138, 140–41
Tupinology, ix
Turiaçu River, *57,* 58, 61, 92
turumbúri tree (*Sorocea guilleminiana*), 179, 184

Uayás people, 77–78
Uhl, Christopher, 39
underdifferentiation, 124–25
Unonopsis guatterioides, 48, 48
Urutawi, *57,* 58

Van Steenis, C. G. G. J., 60
Vellard, J., 82
Vickers, William T., 32, 37
village horticulture, 75–76, 101–2, 182
vine forests, 18–19, 25, 43–49, *50,* 51, 81
vines, folk names for, 99, *100,* 109, *197–201*
virility ordeals, 141
Vitex flavens, 69
Viveiros de Castro, Eduardo B., x, 11, 29, 44, 78, 135, 139
vocabulary, 103, 120, 165. *See also* plant nomenclature and classification

Wagley, Charles, x, 36
wandering Indians, 13–15, 17–18. *See also* Aurê and Aurá (indigenous wanderers)
Waorani people, 106
Warao people, 85, 141
warfare: colonial, 80; endemic, 22, 36; role in agricultural regression, 78, 85–86
warumã (*Ischnosiphon* sp.), 30, *200*
Wayana people, 140, 141
Wayãpi language: cacao words in, 153, *154,* 154–56, *155,* 157, 158; names for animals and insects, *137*; names for traditional domesticates, *136*; plant nomenclature and classification, 107–9, 210n7; Tupí-Guaraní origin of, 209n1
Wayãpi people, *118,* 120, 140–41, 145–46
weevil larvae, 81
Wessels Boer, J. G., 40
Western society, expansion into Amazonia, 54, 73–74, 95
white-lipped peccaries, 52. *See also* peccaries
wild banana (*Phenakospermum guyannense*), 94, 95, 96, *199,* 208n5. *See also* banana (*Musa* sp.)
wild cacao (*Theobroma speciosum*): in Araweté forests, *47*; in Assurini forests, *48*; as disturbance indicator, 46;

Guajá names for, 98, *185*; indigenous peoples' uses for, 48, *65*; in Ka'apor dooryard gardens, 68; in Ka'apor forests, 69, *171, 172, 203*; Ka'apor names for, 98; limited commercial value of, 151; in orchardlike cultural forests, 182; seed dispersed by animals, 68–69, 131; Tupí-Guaraní cognates for, 154–55, *155*, 212n1, 212n3. *See also* cacao (*Theobroma cacao*); cupuaçu (*Theobroma grandiflorum*); *Theobroma angustifolium*

wild cashew (*Anacardium* spp.), 67, 96–97, *172, 185, 203*. *See also* cashew (*Anacardium occidentale*)

wild fig trees, 90

wild papaya (*Jacaratia spinosa*): in Aché-Heta habitat, 82; Guajá names for, *186*; in Ka'apor forests, *171, 172*, 173, *203*; in study plots, 58, *65*; Tupí-Guaraní names for, 109, 208n6. *See also* papaya

wild plants. *See* nondomesticates

wild star apple tree (*Pouteria caimito*), 127, *192*. *See also* star apple tree (*Chrysophyllum caimito*)

wild yam (*Dioscorea* sp.), 127, *201*. *See also* yam (*Dioscorea trifida*)

Wolf, Eric, 72

women, food taboos and avoidances and, 34, 136–40

Woods, William, 5

Xikrin-Kayapó people, 36, 80

Xingu River, 13, 17, 18, 22, 23, 24, 26, 27, 29, 35, 36, 41, 43, 50, 55, 59, 86, 97, 112, 133, 135, 142, 147, 148, 158, 212n2

yam (*Dioscorea trifida*), 98; Araweté cultivation of, 86; Assurini cultivation of, 20; folk species of, 127; Guajá names for, *93, 202*; Ka'apor cultivation of, 85; Ka'apor names for, 127; Tupí-Guaraní cognates for, 21, 77. *See also* wild yam (*Dioscorea* sp.)

Yanomamö people, 36, 37–38, 79, 106, 142

Ye'kwana people, 37–38

yellow-footed tortoise, 34, *137*, 137–40, 211n7

Yupará (Ka'apor man), 35, 168

Yupaú language, *117*, 133, 212n3

Yuquí people, 102

Yuquí language, 135, 209n2

Zea mays (maize). *See* maize (*Zea mays*)

Zipf's law, 168